Lecture Notes in Computer Science   **12616**

More information about this subseries at http://www.springer.com/series/7409

Madhusudan Singh · Dae-Ki Kang ·
Jong-Ha Lee · Uma Shanker Tiwary ·
Dhananjay Singh · Wan-Young Chung (Eds.)

Intelligent Human Computer Interaction

12th International Conference, IHCI 2020
Daegu, South Korea, November 24–26, 2020
Proceedings, Part II

Springer

Editors
Madhusudan Singh ⓘ
Woosong University
Daejeon, Korea (Republic of)

Dae-Ki Kang ⓘ
Dongseo University
Busan, Korea (Republic of)

Jong-Ha Lee ⓘ
Keimyung University
Daegu, Korea (Republic of)

Uma Shanker Tiwary ⓘ
Indian Institute of Information Technology
Allahabad, India

Dhananjay Singh ⓘ
Hankuk University of Foreign Studies
Yongin, Korea (Republic of)

Wan-Young Chung ⓘ
Pukyong National University
Busan, Korea (Republic of)

ISSN 0302-9743   ISSN 1611-3349   (electronic)
Lecture Notes in Computer Science
ISBN 978-3-030-68451-8   ISBN 978-3-030-68452-5   (eBook)
https://doi.org/10.1007/978-3-030-68452-5

LNCS Sublibrary: SL3 – Information Systems and Applications, incl. Internet/Web, and HCI

This Springer imprint is published by the registered company Springer Nature Switzerland AG
The registered company address is: Gewerbestrasse 11, 6330 Cham, Switzerland

Preface

The science and technology of Human Computer Interaction (HCI) has taken a giant leap forward in the last few years. This has given impetus to two opposing trends. One divergent trend is to organize separate conferences on focused topics such as Interaction Design, User-Centered Design, etc., which earlier would have been covered under HCI. The other convergent trend is to assimilate new areas into HCI conferences, such as Computing with Words, Prosocial Agents Development, Attention-based Applications, etc. IHCI is one of the rare conferences focusing on those issues of Intelligence and Human Computer Interaction which exist at the crossroads of the abovementioned trends. IHCI is an annual international conference in the Human Computer Interaction field, where we explore research challenges emerging in the complex interaction between machine intelligence and human intelligence. It is a privilege to present the proceedings of the 12th International Conference on Intelligent Human Computer Interaction (IHCI 2020), organized on site and online by the Korea Institute of Convergence Signal Processing (KICSP) during November 24–26, 2020 at EXCO Daegu, South Korea. The twelfth instance of the conference was on the theme of "Intelligent Interaction for Smart Community Services", having 9 special sessions related to the main theme of the conference.

Out of 175 submitted papers, 93 papers were accepted for oral presentation and publication by the Program Committee, in each case based on the recommendations of at least 3 expert reviewers. The proceedings are organized in 9 sections corresponding to the 9 tracks of the conference. The 12th IHCI conference included keynote and invited talks with powerful expert session chairs who have worked in both industry and academia. It attracted more than 200 participants from more than 27 countries.

IHCI has emerged as the foremost worldwide gathering of the field's academic researchers, graduate students, top research think tanks, and industry technology developers. Therefore, we believe that the biggest benefit to the participant is the actualization of their goals in the field of HCI. That will ultimately lead to greater success in business, which is ultimately beneficial to society. Moreover, our warm gratitude should be given to all the authors who submitted their work to IHCI 2020. During the submission, review, and editing stages, the EasyChair conference system proved very helpful. We are grateful to the technical program committee (TPC) and local organizing committee for their immeasurable efforts to ensure the success of this conference. Finally we would like to thank our speakers, authors, and participants for their contribution to making IHCI 2020 a stimulating and productive conference. This IHCI conference series cannot achieve yearly milestones without their continued support in future.

November 2020

Wan-Young Chung
Dhananjay Singh

Organization

General Chairs

Wan-Young Chung   Pukyong National University (PKNU), Busan, Korea
Dhananjay Singh   Hankuk University of Foreign Studies (HUFS), Seoul, Korea

Technical Program Chairs

Uma Shanker Tiwary   IIIT-Allahabad, Allahabad, India
Dae-Ki Kang   Dongseo University, Busan, Korea
Jong-Ha Lee   Keimyung University, Daegu, Korea
Madhusudan Singh   Woosong University, Daejeon, Korea

Steering Committee

Uma Shanker Tiwary   IIIT-Allahabad, India
Santanu Chaudhury   IIT Jodhpur, India
Tom D. Gedeon   Australian National University, Australia
Debasis Samanta   IIT Kharagpur, India
Atanendu Sekhar Mandal   CSIR-CEERI, Pilani, India
Tanveer Siddiqui   University of Allahabad, India
Jaroslav Pokorný   Charles University, Czech Republic
Sukhendu Das   IIT Madras, India
Samit Bhattacharya   IIT Guwahati, India

Special Session Chairs

Uma Shanker Tiwary   IIIT-Allahabad, Allahabad, India
Suzana Brown   The State University of New York, Korea
Mark D. Whitaker   The State University of New York, Korea
Arvind W. Kiwelekar   Dr. Babasaheb Ambedkar Technological University, India
Kenneth A. Yates   University of Southern California, USA
Mohd Helmy Abd Wahab   Universiti Tun Hussein Onn Malaysia, Malaysia
Masoud Mohammadian   University of Canberra, Australia
Eui-Chul Lee   Sangmyung University, Korea
Hakimjon Zaynidinov   Tashkent University of Information Technologies, Uzbekistan
Jan-Willem van 't Klooster   University of Twente, The Netherlands
Thierry Oscar Edoh   University of Bonn, Germany
Zia Uddin   Woosong University, Korea

Muhammad Sohaib	Lahore Garrison University, Pakistan
Jong-Ha Lee	Keimyung University, South Korea
Shyam Perugu	National Institute of Technology Warangal, India
Nagamani M.	University of Hyderabad, India
Irish Singh	Ajou University, Korea

Publicity Chairs

Mario José Diván	National University of La Pampa, Argentina
Amine Chellali	University of Évry Val d'Essonne, France
Nirmalya Thakur	University of Cincinnati, USA

Industry Chairs

Antonio Jara	HOPU, Spain
Sangsu Jung	VESTELLA, Korea
Gyanendra Kumar	infoTrust, Singapore
Prem Singh	COIKOSITY, India

Local Organizing Committee

Daejin Park	Kyungpook National University, Korea
Jonghun Lee	DGIST, Korea
Do-Un Jeong	Dongseo University, Korea
Hoon-Jae Lee	Dongseo University, Korea
Sang-Gon Lee	Dongseo University, Korea
Yeon Ho Chung	Pukyong National University, Korea
Andrew Min-Gyu Han	Hansung University, Korea
Paul Moon Sub Choi	Ewha Womans University, Korea
Jae Hee Park	Keimyung University, Korea
Sukho Lee	Dongseo University, Korea
Sang-Joong Jung	Dongseo University, Korea
Pamul Yadav	GREW Creative Lab, Korea
Hyo-Jin Jung	Daegu Convention and Visitors Bureau, Korea

Technical Program Committee

Jong-Hoon Kim	Kent State University, USA
N. S. Rajput	Indian Institute of Technology (BHU) Varanasi, India
Ho Jiacang	Dongseo University, Korea
Ahmed Abdulhakim Al-Absi	Kyungdong University, Korea
Rodrigo da Rosa Righi	Unisinos, Brazil
Nagesh Yadav	IBM Research, Ireland
Jan Willem van 't Klooster	University of Twente, The Netherlands
Hasan Tinmaz	Woosong University, Korea

Zhong Liang Xiang	Shandong Technology and Business University, China
Hanumant Singh Shekhawat	Indian Institute of Technology Guwahati, India
Md. Iftekhar Salam	Xiamen University, Malaysia
Alvin Poernomo	University of New Brunswick, Canada
Surender Reddy Salkuti	Woosong University, Korea
Suzana Brown	State University of New York, Korea
Dileep Kumar	MR Research, Siemens HealthCare, India
Gaurav Trivedi	Indian Institute of Technology Guwahati, India
Prima Sanjaya	University of Helsinki, Finland
Thierry Oscar Edoh	University of Bonn, Germany
Garima Agrawal	Vulcan AI, Singapore
David (Bong Jun) Choi	Soongsil University, Korea
Gyanendra Verma	NIT Kurukshetra, India
Jia Uddin	Woosong University, Korea
Arvind W. Kiwelekar	Dr. Babasaheb Ambedkar Technological University, India
Alex Wong Ming Hui	Osaka University, Japan
Bharat Rawal	Gannon University, USA
Wesley De Neve	Ghent University Global Campus, Korea
Satish Kumar L. Varma	Pillai College of Engineering, India
Alex Kuhn	State University of New York, Korea
Mark Whitaker	State University of New York, Korea
Satish Srirama	University of Hyderabad, India
Nagamani M.	University of Hyderabad, India
Shyam Perugu	National Institute of Technology Warangal, India
Neeraj Parolia	Towson University, USA
Stella Tomasi	Towson University, USA
Marcelo Marciszack	National Technological University, Argentina
Andrés Navarro Newball	Pontificia Universidad Javeriana Cali, Colombia
Marcelo Marciszack	National Technological University, Argentina
Indranath Chatterjee	Tongmyong University, Korea
Gaurav Tripathi	BEL, India
Bernardo Nugroho Yahya	HUFS, Korea
Carlene Campbell	University of Wales Trinity Saint David, UK

Keynote Speakers

P. Nagabhushan	IIIT-Allahabad, India
Maode Ma	Nanyang Technological University. Singapore
Dugan Um	Texas A&M University, USA
Ajay Gupta	Western Michigan University, USA

Invited Speakers

Yeon-Ho Chung	Pukyong National University, Korea
James R. Reagan	IdeaXplorer, Korea

Mario J. Diván	Universidad Nacional de La Pampa, Argentina
Jae-Hee Park	Keimyung University, Korea
Antonio M. Alberti	Inatel, Brazil
Rodrigo Righi	Unisinos, Brazil
Antonio Jara	HOP Ubiquitous, Spain
Boon Giin Lee	University of Nottingham Ningbo China, China
Gaurav Trivedi	Indian Institute of Technology Guwahati, India
Madhusudan Singh	Woosong University, Korea
Mohd Helmy Abd Wahab	Universiti Tun Hussein Onn Malaysia, Malaysia
Masoud Mohammadian	University of Canberra, Australia
Jan Willem van' t Klooster	University of Twente, The Netherlands
Thierry Oscar Edoh	University of Bonn, Germany

Organizing Chair

Dhananjay Singh	Hankuk University of Foreign Studies (HUFS), Korea

Contents – Part II

Assistive Living

Image Processing and Deep Learning

Human-Centered AI Applications

Contents – Part I

Natural Language, Speech, Voice and Study

Intelligent Usability and Test System

IoT System for Monitoring a Large-Area Environment Sensors and Control Actuators Using Real-Time Firebase Database

Giang Truong Le[1], Nhat Minh Tran[2], and Thang Viet Tran[3,4(✉)]

[1] NTT Hi-Tech Institute, Nguyen Tat Thanh University, 298–300 A Nguyen Tat Thanh,
Ward 13, District 4, Ho Chi Minh City 700000, Vietnam
[2] Department of Telecommunication, Sai Gon University, Ward 3, District 5,
Ho Chi Minh City 700000, Vietnam
[3] Faculty of Information Systems, University of Economics and Law (UEL), VNU-HCM,
Linh Xuan Ward, Thu Duc District, Ho Chi Minh City 700000, Vietnam
tvthang74@gmail.com
[4] Branch of Vietnam Research Institute of Electronics, Informatics, and Automation,
169 Vo van Ngan Street, Linh Chieu Ward, Thu Duc District, Ho Chi Minh City 700000, Vietnam

Abstract. In this paper, we propose a real-time IoT (Internet of Things) system with a built-in hardware security function, allows the ambient environment monitoring and control actuators via a wireless network at the frequency of 433 MHz, a real-time database is used to store and share environmental sensors data and users with a proposed Mobile App can send commands to turn on/off actuators through the Internet. The proposed system was developed and tested using an actuator device (AD) and two type of sensors, CO_2 sensor and air-temperature and -humidity sensor. The main components of the proposed IoT system consist of: a proposed IoT gateway, three of sensor nodes and a real-time Firebase database. Detailed components of the proposed sensor node included: a low-power system on chip unit (SoC) with an built-in Wi-Fi module and a RF antenna, a MQ135 sensor and a SHT10 sensor. Based on Firebase database, a Mobile App have been designed and used to monitor measured recordings of sensors and as control the AD. A wireless module operates at frequency of 433 MHz with top secret built-in security algorithm (AES128) is selected to encrypt the result of ambient environmental sensor data and as transfer commands to the AD mentioned. In addition, some advantages compared with wired connection networks, and in which the structure of the online database of the proposed IoT system has some additional features and more flexibility. The experiment results with the proposed Mobile App are implemented that the proposed IoT system can be used to monitor ambient environments that live to report of measured sensors and make the information visible anywhere as well as control the AD. Our IoT system can be able use to monitor the ambient environment in wirelessly solutions in large-areas with high-level secure algorithm and control electrical machines in many sectors such smart buildings, smart homes, smart agriculture.

Keywords: IoT system · Environment monitoring · Real-time firebase database · Environment sensor

© Springer Nature Switzerland AG 2021
M. Singh et al. (Eds.): IHCI 2020, LNCS 12616, pp. 3–20, 2021.
https://doi.org/10.1007/978-3-030-68452-5_1

1 Introduction

In the 3rd Industrial Revolution (IR) information resulting are collected and shared among devices through wired and wireless communication standards, and then in the IR 4th, those communication standards have been using and processing for various purposes. IoT is one of various technologies that it used sensors to sense and share necessary sensor data from human, ambient environment and physical phenomenon in the natural world. According to a recent forecast reports by Gartner [1], the number of installed IoT devices will be exceeding 25 billion in 2021. There are many applications and use cases of the IoT device include: smart homes, smart offices, smart agriculture, smart factories and smart cities. People are staying online longer than before, which opens up tremendous possibilities for projects related to the Internet of Things (IoT) [2]. Around 47% of the world's population is already using the Internet [3] and by 2020 it is foreseen that the number of devices connected to the Internet will be over 50 billion [4–6]. Users and researchers now may have access to a large number of data from multiple sources that were unavailable before, such as open-source EEG data repositories [7], virtual skeleton database [8], physiological or biometric data gathered with a wristband device or a watch. IoT comes up possibilities that were until recently only in the science fiction movies.

Nowadays, we are starting to build smart home applications, cities and vehicles that automated management and control themselves without any interacts from human. There are two architectures in IoT applications i) IoT devices connect direction to cloud server via internet network as shows in Fig. 1; ii) another as shown in Fig. 2, in which by using an IoT gateway hardware modules which allow transfer IoT information between a wireless network and the internet network, and which one we focused on this research, and as in particular, on smart building applications.

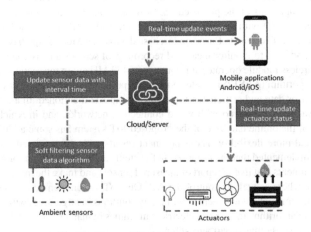

Fig. 1. The architechture of an IoT system

Figure 2 showed the architecture of the proposed IoT system, the main components of an IoT system usually consist of 1) an embedded sensor/actuator devices (they called

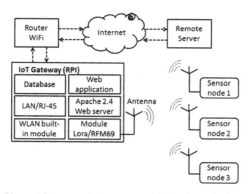

Fig. 2. The architecture of the proposed IoT using wireless network

"things" in IoT); 2) a gateway (also called hub), which is required when the embedded devices do not have a physical the internet network connectivity functions;3) an internet service/database running on an online Web server in which allow to manage and store the collected information of "things"; 4) one or more clients (either web/mobile app or browser-based), for the users to interact with the service via a mobile application which installed on mobile phones, tablets or personal computers (PCs).

Most of the IoT gateway has two main functions:1) send/receive the information between the IoT device by using a wireless protocols (e.g. Lora, Zigbee, Z-Wave and Bluetooth) and the TCP/IP protocol (Wi-Fi, Wired cable connection); 2) allow access to the internet network. In many cases the gateway can be co-located with the local internet router, sharing the utilization with conventional (non IoT) traffic such as Web browsing.

The gateway on the IoT architecture can be a fixed, dedicated device, such as the Samsung SmartThings Hub [7, 12]. In alternative, it can be a mobile device, shared device, like the functions of a smartphone. In this situation, data can be collected either from the mobile device's internal sensors, for example: finger print sensor, microphone, built-in cameras (front and back cameras), GPS (Global Positioning System), accelerometer, magnetometer, gyroscope, proximity sensor or light sensor) or through external sensor devices, which can be connected to the mobile device using wired (e.g., USB cable) or wireless technologies (e.g., Bluetooth, RFID, Infrared etc.). In some IoT healthcare devices, the sensors may be placed on the user's body (e.g., EGG, heart rate sensors, pressure sensors [8–11]) or in the environment (e.g., air temperature, air humidity, pH sensors). The authors mentioned in [11, 12], the mobile phone applications are an emerging area of research and development potentiated by the widespread proliferation of mobile devices, and are one of the important concerns of the IoT systems.

Another IoT architecture alternative consists of the direct connection of from an IoT device to the internet network, without the necessary of an IoT gateway [4, 7]. For examples are devices that possess connectivity to smart phone data cellular networks or an IEEE 802.11/Wi-Fi interface, such as the Libelium Sensor devices (Plus and Sense) [12–14].

Sensor/actuator devices which in modern vehicles are normally connected using a cabled network technology called CAN [10]. The increasing number of sensors on vehicles also increases the number of cables required, and as the weight and maintenance costs. The cables also impose constraints on the placement of sensors and we need a space for these wired cables.

By using wireless technologies may be is an idea to solve these problems [12, 13] and to increase the flexibility and mobility of the system, allowing also the installation of sensors in the machines and user for convenient reasons in the future and it helps reduce the maintenance costs, when and where desired, as well as the integration of wearable sensors in healthcare applications, is mentioned in [8, 15] to monitor the drivers' biomedical signals and even to provide alerts when the users are impossible to drive. In Fig. 2., an architecture of the proposed IoT system is shown that it can be applied building applications and factories designed to allow to monitor and control of actuators anywhere in the world through the Internet network.

A high-performance microcontroller unit ESP32, the main components of the system are including an ESP32 (Espressif Systems located in ShangHai, China) that low-cost SoC with built-in antenna Wi-Fi (WLAN network) [17], a mobile application with friendly user interfaces, an online database without the complexity scripting in the back-end side [16, 18].

This paper is organized as follows. In Sect. 2, we briefly introduce the similar existing system and two famous secure algorithm RSA and AES. In Subsect. 3.1 and 3.2 illustrate the hardware design and Mobile App of the proposed IoT system. A description of its main components includes sensors and AD and detailed characteristics of these components and Firebase database and client mobile application of the proposed IoT system in Subsect. 3.3 and 3.4. Next, Sect. 4 presents experimental results concerning the developed system and the detailed of the structure of database and IoT hardware components and security algorithm selection. Finally, Sect. 5 discuss about experiment results and conclusions.

2 Related Work

Recently, smart intrusion systems are also becoming one of the key research direction in the field of embedded and IoT systems due to their increasing smart functionality and connectivity. In an IoT system, most of all hardware devices and software applications have been developed to detect and prevent security attacks [22, 25]. The main concern related to reliable data transmission is security, the secure data transmission in which like personal information, smart locks, Credit Card, ATM etc. Those need complexity security algorithm, because most of the transaction is held through the internet network (TCP/IP) and wireless network.

Currently, various security techniques have been presented in the literature such as reference monitors, cryptography, maintain information flow track [10, 13, 21]. However, there are no one which is the best of security algorithm in the world, researchers have been based on two main features include hardware components and strong level of security algorithms.

In the follow section, a briefly describes about security algorithm that it integrated into one into the proposed IoT system and the follow introduction of famous existing algorithm will discuss in Sect. 2.1.

2.1 Existing Technologies

From the literature survey, recently there has been focused on security algorithm in IoT applications as well in last few years that presented and mentioned in some of commercial products, can provide more safety services and in particular, smart homes, smart building applications. There seem to be lack of each security algorithm such as operational time and data storage in the limit memory of microcontroller unit or the requirement of IoT application.

In [25], the authors carried out a model detection techniques for IP based sensor networks would require implementation of TCP/IP stack and signature storage which is not possible for power and resource constrained microcontroller based sensor nodes. A similar application, based intrusion detection techniques have been proposed in the literature, by inserting run-time checks in the application code in order to detect illegal memory accesses (IMAs). Existing anomaly detection solutions based on neural networks and compiler based techniques for the detection of IMAs have been explained in [26]. The other existing neural networks based intrusion detection solutions [27–29] are computationally intensive, rely on extensive profiling of the communication traffic, higher performance MCU and have been designed for such systems having ample resources where power consumption is not a design constraint. Therefore, such solution cannot be deployed for battery operated wireless sensor nodes and it became a great challenge in IoT system.

Through the models are shown in those studies, such smart algorithms require a network infrastructure for server uploads, database for storage and bi-directional data exchange the collected sensor data and commands to control actuators by users. Beside of the experimental efficient of the results from these systems, those systems have complexity layers from software to infrastructure of the hardware and it consumes a lot of power.

2.2 Security Technologies

In [30], the authors showed that if a mobile device is rooted (Android/iOS), it can be leaked and extracted the information of the secret keys of August Smart Lock from the configured file system of the devices. So, a security algorithm is needed for encryption information in IoT devices, especially in some situations such as bank applications, smart homes, companies, hospital, etc. In different situations of IoT applications have different requirements. Thus there are two main solutions for IoT security:i) hardware in which they use encryption/decryption chip (hardware logic) for encrypting data before send/receive data; ii) software, the algorithms are programming into flash memory of a micro-controller unit.

Such as the authors in [32, 33] showed the hardware-based security solutions, are not generic and require either dedicated hardware modules or specific modifications in the processor pipeline or cache architecture. Similarly, in [23–25, 34], hardware based on tagging data coming from untrusted sources and then tracking their usage as the application executes. These techniques require modifications in application's data, processor architecture and much memory layout. More the attackers can redesign the hardware components on the hardware boards. Therefore, they are not feasible for low-cost IoT systems and insecure.

For software security algorithms, there are two of the most widely used encryption/decryption algorithms today are RSA and AES. They are used in different situation, but both are highly effective and secure [22, 36].

RSA Algorithm: In this subsection, we will briefly explain the RSA algorithm as the following

Encryption: Side A sends the message in plaintext format M with the public key (e, n) to side B, side B receives the data and decrypts with the private key (d, n), with $0 <$ m<n. To encrypt the plaintext M, we have to calculate:

$$C = M^e \bmod n \tag{1}$$

So A will be used the public key to transfer the message in plaintext into cipher text.

Decryption: Side B received the message M in cipher text format, then B used private key to convert cipher text to plain text.

$$M = C^d \bmod n \tag{2}$$

The following is an example with the number to implement RSA algorithm above.

AES Algorithm: we will brief introduction about the AES security technique and discuss some features of AES algorithm.

Related to the introduction part, the Advanced Encryption Standard (AES) is a part of Rijndael block cipher and then developed by two Belgian cryptographers, the AED is using symmetric key encryption methods and now used in worldwide [22]. Various security technologies and mechanisms have been designed around cryptography algorithms in order to provide specific security services. Cryptography algorithms such as RSA, AES and SHA [22, 23] have been used in embedded systems to establish integrity of data within the system.

In [35], the National Security Agency (NSA) reviewed and suggested use of the either 128, 192 and 256 bit-keys length so that a processor process in 8 rounds for 128 bit-keys, 12 rounds for 192 bit-keys and 14 rounds 256 bit-keys. Recent reports [22, 27, 29, 30]. The authors have been focused on security mechanisms for the IoT devices have been discussed based on cryptography. With event a 128 bit-keys length is used, the crackers have to do checking 2^{128} possible keys values and by using a supercomputer today with high-performance processors, it takes on average more than 100 trillion years to check so that the AES has never been cracked, at least of years come. However, such techniques are not realistic for low-power IoT devices due to resource constraints, the limit of EEPROM memory and large power consumption overheads and in particular, real-time monitoring applications. Different IoT systems have different requirements. We take an Internet enabled environmental monitoring system as an example to demonstrate security requirements of such an IoT system, and believe other systems share similar attributes.

3 Hardware and Mobile App Design

In the proposed system, we focused on the architecture of the IoT in which using a IoT gateway to communicate with server and send/receive information to/from sensor nodes via an RF network.

The design of hardware of the proposed IoT node will detail in Sect. 3.1 as well as the characteristics of sensors and actuators in subsection. Next, we will introduce about the proposed IoT gateway function as such components in Sect. 3.2.

In addition, the operation protocol of the proposed IoT gateway with Google Firebase will introduce in Subsect. 3.3 and with Client Mobile App will describe in Sect. 3.4.

3.1 The Proposed IoT Node

The proposed IoT node was designed for two main functions that is consist of 1) collects environmental data by using sensors, 2) turn on/off actuators via physical relays on the AD board. The main components of the proposed IoT hardware moudule are shown in Fig. 3, an low-power MCU with built-in WiFi module (the GPIO pins operate at 3.3VDC) [17], a digital air–temperature and –humidity sensor (SHT10, Sensirion) [19], a gas sensor which allow measured carbone dioxide gas in the air (MQ135, Olimex) [20].

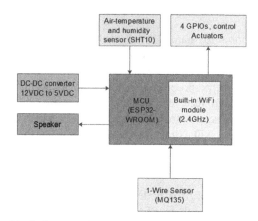

Fig. 3. Block diagram of the proposed sensors and actuators module

Beside of convenience features of the ESP32: ARM processor with 160 MHz oscillator, low-price MCU, a great develop KIT with 8bit/16bit parallel male header interfaces and it can use H-SPI serial interface with a RFM69HCW module without any TTL-CMOS logic converter (3.0VDC to/from 5.0VDC).

The RFM69HCW wireless module, operates at 3.3VDC and frequency of 433 MHz, will introduce in Subsect. 3.1.

Sensors: Sensors used in this paper include: 1) MQ135 sensor which allows measure CO concentration in the air, 2) SHT10 sensor which also allows measured humidity with the highly accurate result, especially, it is protected with a metal materials and can sense in higher humidity environment. Some characteristics of these sensors is shown in Table 1.

Table 1. Characteristics of the SHT10 [19] sensor and MQ135 sensor [20]

Description	SHT10	MQ135
Measuring range	−20 °C−100 °C, 0%–80%	$2\,k\Omega−20\,k\Omega$ (100 ppm NH$_3$)
Accuracy	±3%, ± 0.4 °C	Slope rate: ≤0.65, 5 ppm
Operating range	−40–123.8 °C, 0–100%	−55–125 °C; ≤ 95%
Supply voltage	2.4–5.5 VDC	5.0–24.0 VDC
Response time	RH ≤ 8 s, 5 s ≤ Temperature ≤ 30 s	NA
Sensor type	Digital	Analogue
Pin connection	4 pins	3 pins

Actuators: In the proposed IoT hardware module that supports four electromechanical relays is shown in Fig. 3, and these relays are designed to operate in on/off mode. To ensure complete safety, an opto-coupler was used to drive the state of each relay. The general characteristics of the proposed relays module are shown in Table 2. The ESP32 (MCU) [16] is programmed to receive from data from Firebase database through a built-in Wi-Fi module, the received data from the database is formatted in JSON (JavaScript Object Notation) and then it will extract the state of each relay. Finally, MCU will turn on/off each relay by sending a logic level (logic level "1": on, logic level "0": off) to the GPIOs accordingly.

Wireless Module RF69HCW: AIoT node cannot communicate with the IoT gateway without a wireless connection so that a RFM69HCW is chosen, it a transceiver module capable of operation over a wide frequency range, including the 315 MHz, 915 MHz, 868 MHz and 433 MHz license-free ISM (Industry Scientific and Medical) frequency bands. In particular, we choose a frequency band of 433 MHz which match with the radio frequency license for radio transmitters and transceivers used in Viet Nam.

The RFM69HCW transceiver can operate in five modes like state-machine mechanism: sleep, synthe-sizer, standby, receive, and transmit. While sleep mode is the radio's lowest power mode, drawing only 0.1µA as mentioned in the RFM69HCW's datasheet [21] and standby mode is the transceiver's idle mode, which the radio must pass through before entering both the receiving and transmitting modes. With the synthe-sizer mode occurs at wake-up from sleep mode or power on, to set up a frequency synthe-sizer and allow the phase-locked loop of the transceiver to lock. A useful feature of the RFM69HCW is a programmable bitrate in the air between 1.2 kbps to 300 kbps, and allows for setting of transmission output from 5 dBm to 20 dBm by updating values in

Table 2. Characteristics of the proposed actuators module

Characteristics	Value
Power supply	5–24 VDC
Actuator isolation	Opto-isolator (code: PC817)
Actuator power supply voltage	150–280VAC
Operating frequency	50 Hz
Maximum AC current	10 A
Number of actuators	04
Logic level	Low: off High: on

control registers that it mentioned in Sect. 3.3.6 in datasheet [21] (The hecxa address format of registers are $0 \times 5A$ and $0 \times 5B$).

The AES-128 encryption/decryption (ECB mode) process takes approximately 7.0 us per 16-byte blocks, each 8-byte block is encrypted independently ($128/8 = 16$ bytes) and approximate to 28 us. Thus for the size of data communication between IoT gateway and nodes is always fixed length of data. Therefore, there is no need a padding key is adding before encrypting with AES-128 and only the payload part of the packet is encrypted and preamble, sync word, length byte, address byte and CRC are not encrypted that they presented in datasheet (subsection 2.5.5.1).

In the Table 3 is shown the wireless parameters setting for RFM69HCW module in our experiment.

Table 3. The experimental parameters of the wireless module RFM69HCW

Description	Value
Power supply	2.4–3.6 VDC
Operating frequency band	433 MHz
Topology	Star
Number of nodes in RF network	1023
Number of address in RF network (Zone)	256
Encryption/decryption	AES128 (16 bit CRC)
RF bit rate	250 kpbs (max: 300 kbps)
Sleep mode current	0.1 μA
Interface with MCU	SPI
Max. transmission Power	20 dBm (100 mW)
Antenna length	164 mm

3.2 The Proposed IoT Gateway

The follow, we will show the design of the proposed IoT gateway in this project. In Fig. 4., the main block diagrams function of the proposed gateway is shown. A Raspberry Pi 3, a high-speed processor or a mini-computer was used as main micro-controller unit. In addition, a built-in LAN port that it allows a wired cable of internet connection and a built-in WiFi module, operates at the frequency of 2.4 GHz and a low-power wireless RFM69HCW to communicate with sensor nodes.

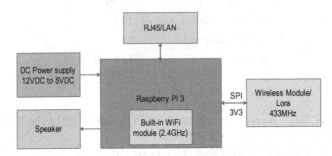

Fig. 4. The main block diagram of the proposed wireless IoT gateway

In the proposed gateway has to two main important functions: firstly, it allows send/receive the information between the IoT devices via low-power consumption wireless module. Accordingly the RFM69HCW's datasheet, shows some useful features as low-power consumption, compact size of module, long-range communication, and the most useful feature is integrated symmetric encryption (Advanced Encryption Standard) [21] and widely use in both practices and studies [22, 23] areas; secondly, the TCP/IP connection (the Raspberry Pi 3 supported both WiFi and LAN cable connection) to access to the internet network. This function, allows the proposed IoT gateway can send/receive information from server. The PCB of IoT gateway and the IoT node are only on board, but we configure different functionality and used differnece of processor as abovementioned.

Scalability is an important feature in most of the IoT service systems because of the huge number of the IoT devices connected to server (such as smart cities application, weather monitoring. For the proposed system, the Google firebase has been respected a tag (branch of structured database) to each sensor node or AD node, in which stored measurement sensor data in the fields as shown in Fig. 8. Moreover, currently Google firebase can support up to 200,000 simultaneous connections [16]. In addition, the main wireless communication standard between IoT Gateway and nodes and a good routing protocol was integrated. Scalability of routing protocol in IoT system is a critical issue. The proposed system, the star topology used and multi-path routing sensor node and AD nodes as presented in Fig. 5. Authors in [35], studied on long-range Lora standard for smart cities and suggested up-to 1,500 devices can exchange data in wireless network. In comparison, the RFM69HCW used in this paper can manage with 1023 IoT devices in single RF network and presented in Table 3. An interesting feature of

RFM69HCW module, by changing RF network indexing (network ID), the IoT Gateway can be communicate with other sensor node in difference network ID.

In next section, we will describe the protocol between the IoT gateway and real-time Google Firebase database as well as the protocol of the Mobile Application.

3.3 Internet Serverless and Database (Google Firebase)

Firebase is a real time database service which enables and supports users to develop multi-application such as web, mobile application (iOS/Android) and classification sensor data by integrating machine learning (ML) tool-kit functions with minimal server-side programming and complexity configurations to the back-end database. In this paper, we use the Firebase database stored the measured sensor data and implemented a mobile application. Scalability of the sensor database Google firebase database is limited of capacity based (up-to 10 GB for each account) and the proposed database has been provided one tag to one sensor node. In this paper, the proposed IoT system was set up for experimenting with limited of a 10 GB on database. However, a cloud storage configuration has been used to extend the number of capacity of sensor devices, the IoT measured data can be upload to cloud extension cloud services as well as the payment needed. The proposed IoT system claimed to real time to control (only On/Off AD devices) and used real time Google firebase database. The designed firebase database allows for updating top-level of database and as well as any subset of tags in database. Others, real time digital signal processing is defined in [36, 37] in which the samples data can be processed continuously in the time data from input and output the same set of samples independent of the processing delay.

Protocol of the proposed IoT node and Firebase database: In this subsection, we will show the operation protocol of the proposed IoT gateway, and how the mobile application can interact with Firebase database via the internet connection.

Firstly, as showing in Fig. 5, we numbered and separated the sequences with colors, sensor data collected from an IoT hardware node will process and convert into JSON format by MCU and then it will write these sensor values to the Firebase database via a Wi-Fi network. In case there is no internet connection available, only schedule data is cached and processed locally, a friendly schedule settings interface was designed as shown in Fig. 6. Once the new connection comes up again, data is uploaded and the cache is cleared. This is done to save up storage in the mobile device. The structure of the database is shown in Fig. 8. Data is designed and stored in JSON format in which easy encode and decode to extract the information and adopted with Firebase database. And in this paper, the collected data from the environment are stored under a tag called "*sensor*" that it is shown in the top of Fig. 8, and the parameters to control actuator called "*device*". Firebase allows registering listeners for data change in real-time that it can improve performance of MCU and reduce the time process JSON data from server response. In the client's mobile app, we subscribe to the sensor we want to listen to and we retrieve the changes in real-time. This approach is also used to update the change to GPIOs (relays) in AD as mentioned in the Sect. 3.4.

Next, we describe the flow of the operations of our mobile app and the IoT system. After the IoT node is powered on, it collects data from environment sensors such as

DHT22 and AM2302 sensors. Then they process and send the messages over the WiFi network to Google Cloud database (real-time database Firebase). The IoT node used machine-to-machine (M2M) protocol also listen to the messages carrying nodes' presence state from the Mobile Apps at the same time. In the following figures, we separate the flows of the proposed node into two of the sequences as shown in Fig. 5.

3.4 Client Mobile Application

The client mobile app was designed for both iOS and Android smartphone platforms and it is currently under development and improvement. It can be used to connect to the Firebase database and have access to all sensor data uploaded by the sensor module in real-time. It will also allow the configuration of system parameters and the control of relays in the hardware that are shown in Fig. 8, such as the user can remotely turn on/off the actuator devices through the internet network.

The proosed IoT node update to Firebase database

1. *update sensor data:* when the IoT node is powered on and established a connection to a router WiFi success, it read sensor values and then send those measured data to Firebase database.
2. *get sensor data:* because the proposed Mobile App used events handler C ++ library to detect any changeable on database, it can compose the presence message send by the proposed IoT nodes in step 1.
3. *update success:* like the acknowledgment message, if the proposed node updates sensor values success, the function in the flow of the operation of the IoT node will return true. Others, they will return false. According this value, the proposed node will go to next stages or rebroadcast the message, respectively. This steps certain the measured sensor data that has been updated to the database.
4. *get sensor data success:* almost the same functions in the number 3, but the sequence applied on the Mobile App side.

The user(the Mobile App) sends command control relays to the proposed IoT nodes

5. *user command to tag "/device":* an interface, allows interactions with the user was designed in this paper. To set the actuators via relays on the IoT node, the user can click the button on the screen of any smart phone. This action will process and upload to the tag *"/device"* on the Firebase database.
6. *detect events from tag "/device":* by using multipath events handlers library, the proposed IoT node hears the message which uploaded by Mobile App in the JSON format. Then it will decode the JSON message and update states (on/off) to physical devices on the IoT node.
7. *update data to tag "/status":* to certain the IoT node received the message, the acknowledgement message will upload back to the Firebase database with the different of the tag (*"/status"*).
8. *detect events from tag "/status":* this step confirmed to the Mobile App that the IoT node was received the message from the user (step 5).

9. *get confirm command "/status"*: this is the acknowledgement message between the IoT node and Firebase Cloud. It means the message has been uploaded to the tag named *"/status"* on the Firebase database.

As shown in Fig. 5, the diagram illustrates the interactivity between the mobile user and the Firebase backend system is presented. From the users' perspective, they are able to receive more than the sensor values and control relays on the proposed IoT node such as schedules, setting of the threshold values for each sensors.

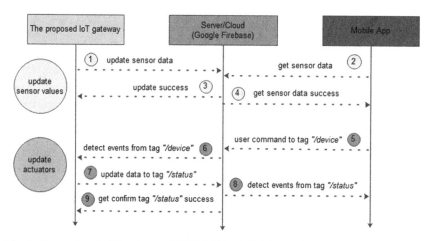

Fig. 5 The protocol of the proposed IoT Mobile application and the proposed IoT gateway

In addition, the user can easily establish a schedule with this mobile application via some simple interfaces on the smartphone. Since the user will not be always using the client's app we can use the Firebase Cloud Messaging (FCM) to send a push notification when there is an important event, for example, a critical change of sensor values. When this happens, the icon of this sensor will change in red color and an HTTP request to an FCM endpoint that will trigger a notification in the clients that subscribed to that event.

4   Experiment and Results

An IoT system for ambient monitoring and control actuators (connected to 4 relays) by using a real-time Firebase database, wireless module and a mobile application was presented (Fig. 7).

The measured sensor data will be stored in the Google Firebase database as shown in Fig. 8, the tag at the top-level in the screenshot named *"sensor"* refers to the direction to stored sensor values, next is the second-level key refer to the encrypted sensor ID for this module. In each encrypted sensor ID key, there are some values major and minor, such as *min, max, rssi, currTime, isConnected.*

Fig. 6. The main interface of the proposed mobile application and screenshot of the schedule settings in the proposed mobile application

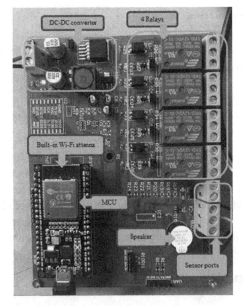

Fig. 7. The proposed sensor and actuator module in our system

The *currTime* key indicates the locally time that the proposed sensor module updated and the *isConnected* key means the sensor was connected (*isConnected: 1*) or disconnected (*isConnected: 0*) and *rssi* indicates the received signal strength indication of the Wi-Fi connection.

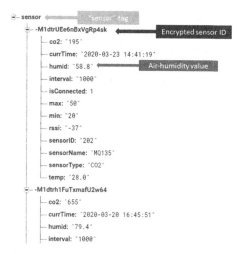

Fig. 8. A screenshot of the Firebase database showing the results of measurement of sensors

Table 4. The performance with different micro-controller units using symmetric algorithm

Description	ESP8266	ESP32	Arduino Pro Mini (Atmega128P)
Frequency clock of MCU (MHz)	80	80	8
Plaintext content (*)	Msg	Msg	Msg
Message length (bytes)	350	350	350
HASH algorithm	SHA256	SHA256	SHA256
Encryption with AES algorithm time (seconds)	0.002	0.001	0.02
Decryption with AES algorithm time (seconds)	0.003	0.001	0.03
Checksum (CRC) bits	16	16	16
Memory size (bytes)(Percent of MCU (%))	242065 bytes (23%)	229645 bytes (17%)	8052 bytes (28%)

(*) *The content of plaintext message is used to experiment, char Msg[] = "1234567890abcdefghijklmnopqrstuvwxyz\n1234567890abcdefghijklmnopqrstuvwxyz\n1234567 890abcdefghijklmnopqrstuvwxyz\n1234567890abcdefghijklmnopqrstuvwxyz\n1234567890abcdef ghijklmnopqrstuvwxyz\n1234567890abcdefghijklmnopqrstuvwxyz\n1234567890abcdefghijklmno pqrstuvwxyz\n1234567890abcdefghijklmnopqrstuvwxyz\n1234567890abcdefghijklmnopqrstuvwx yz\n1234567890abcd\n"*

The results of performance parameters as shown in Table 4 that combination of the SHA256 hash algorithm and symmetric algorithm (AES128) with different MCU. The ESP32-WROOM, operated with 80 MHz has better computation time for encryption/decryption 350 bytes sample data CRC in character format, 16 bits check sum (CRC) than others ESP8266 and Atmega1284P. Because of difference flash programming capacity, so the percentage of memory size has different values. As the results showed that the ESP32-WROOM has best performance time and AES algorithm with larger key length required more memory such AES192, AES256 [22]. With experiment results, it also means that in scenario when Atmega1284P is used, there are only AES128 algorithm can be applied.

5 Conclusion

In this paper, we have been depicted the proposed IoT system with strong security algorithm and by using real-time Firebase database for monitoring the result of environmental sensors, controlling AD over the internet network with a friendly proposed Mobile App. This system can be used to monitor ambient parameters in buildings, and with long-range wireless communication module. The proposed system is also used to agricultural fields. Through Mobile App, users can use the proposed mobile application to track recordings of sensor data and control actuators through the internet network. The Firebase database has been supported multiple clients, both local and remote applications can concurrently receive two tags of collective data (*"sensor"* and *"device"*) from the proposed IoT hardware through the proposed mobile app in real-time. This system is well designed and implemented as well as tested with the ESP32 processor, the proposed mobile application and ambient sensors. With the very high accuracy of the SHT10 sensor and MQ135 sensor (measured CO_2 in the air), this paper succeeds to provide the user with a solution by using a low-cost real-time database and experiment result about these sensors and actuators. In particular, most of buildings IoT systems is needed encrypted the private data so that a strong security algorithm has been chosen for this reason.

Based on experiment results, the real time Firebase database and the proposed network protocol to control actuators and collect sensor data abovementioned that it can be used to collect and monitoring sensors data and as control actuators such as pumps, AC fan in agricultural sectors and buildings, homes applications. The implemented wireless transceiver security module (RFM69HCW with AES128 algorithm) in this research was designed and tested with important specific setting parameters and different types of commercial sensors in real-world IoT applications. A mobile application with friendly user's interface was designed (supported Android/iOS platform). In this paper, we also provide useful comparison results and performance a symmetric algorithm with different MCUs and according the results of implementation and the limits payload of RFM69HCW module when the IoT devices in transmission mode in which the length of frame data is changed, the frame of data has small size.

Funding: This work was supported by IoT-2020 project, VIELINA-HCMC, Ho Chi Minh City, Viet Nam. The corresponding author is with the Branch Vietnam Research Institute of Electronics, Informatics, and Automation, 169 Vo Van Ngan

Street, Ward Linh Chieu, District Thu Duc, Ho Chi Minh City, Viet Nam (e-mail: tvthang74@gmail.com).

References

1. Middleton, P, Tsai, T, Yamaji, M, Gupta, A, Rueb, D.: Forecast: Internet of Things–Endpoints and Associated Services, Worldwide (2017)
2. Brahima, S.: ICT Facts and Figures-2016 (2017). http://www.itu.int/en/ITU-D/Statistics/Doc uments/facts/ICTFactsFigures2016.pdf
3. Evans, D.: The Internet of Things - How the Next Evolution of the Internet Is Changing Everything, CISCO, pp. 1–10 (2011)
4. Dibaba, H.: IoT Implementation with Cayenne Platform, Metropolia University of Applied Sciences (2018)
5. Samsung SmartThings Hub. https://www.smartthings.com/products/smartthings-hub
6. The Patient Repository for EEG Data + Computational Tools (PRED + CT). https://www.frontiersin.org/articles/10.3389/fninf.2017.00067/full
7. Open Source Brain-Computer Interface Database. https://openbci.com/
8. Di Natale, M., Zeng, H., Giusto, P., Ghosal, A.: Understanding and Using the Controller Area Network Communication Protocol: Theory and Practice, Springer, New York (2012)
9. Awesome Benefits of Internet of Things (IoT) for Web and Mobile App Development (2020). https://theiotmagazine.com/
10. Andrea, K., Andreas, P.: Mobile phone computing and the internet of things: a Survey. IEEE Internet Things J. **3**, 885–898 (2016)
11. Fan, T., Chen, Y.: A scheme of data management in the internet of things 2010. In: 2nd, IEEE International Conference on Network Infrastructure and Digital Content (2018)
12. Plus and Sense, Libelium company. http://www.libelium.com/libelium-integrates-sensors-with-ibm-bluemix-cloud-for-smart-cities/
13. Wang, Y., Jang, S.: An IoT-enabled smart elderly living environment. Int. J. Biosens.Bioelectron. **6**, 28–33 (2020)
14. Knupfer, S., Hensley, R., Hertzke, P., Schaufuss, P., Laverty, N., Kramer, N.: Electrifying Insights: How Automakers can Drive Electrified Vehicle Sales and Profitability. McKinsey&Company, pp. 1–28 (2017)
15. ESP32 A feature-rich MCU with integrated Wi-Fi andBluetooth connectivity for a wide-range of applications. https://www.espressif.com/en/products/socs/esp32/overview
16. Google Firebase. https://firebase.google.com
17. SHT10 sensor, Sensirion company. https://www.sensirion.com/en/environmental-sensors/humidity-sensors/digital-humidity-sensors-for-accurate-measurements/
18. MQ135 gas sensor (CO_2). https://www.olimex.com/Products/Components/Sensors/Gas/SNS-MQ135/resources/SNS-MQ135.pdf
19. Datasheet of transceiver module RFM69 ISM (2015). http://www.hoperf.cn/upload/rf/RFM69-V1.3.pdf
20. NIST, Advanced Encryption Standard:U.S. National Institute of Standards and Federal Information Processing Standards Publication (FIPS PUBS), vol. 197, (2001)
21. Security techniques. https://wiki.mikrotik.com/wiki/Manual:IP/IPsec#Diffie-Hellman_G roups
22. Amin, S.O., Siddiqui, M.S., Hong, C.S., Lee, S.: RIDES: Robust: intrusion detection system for IP-based ubiquitous sensor networks. Sensors **9**(5), 3447–3468 (2009)
23. Timotheou, S.: The random neural network: a survey. Comput. J. **53**, 251–267 (2010)

24. Kessler, G.C.: An Overview of Cryptography (2020). https://www.garykessler.net/library/crypto.html#rsamath
25. Viderberg, A.: Security evaluation of smart door locks, Master thesis Master in Com puter Science, School of Electrical Engineering and Computer Science (2019)
26. Ye, M. et al.: Security analysis of Internet-of-Things: a case study of August smart lock. In: 2017 IEEE Conference on Computer Communications Workshops (INFOCOM WKSHPS). 2017, pp. 499–504 (2017)
27. RSA, Public-Key Cryptography Standards (PKCS): RSA Cryptography Specifications Version 2.1 (2003)
28. Security techniques and mechanism. https://www.di-mgt.com.au/rsa_alg.html
29. Wu, S.X., Banzhaf, W.: The use of computational intelligence in intrusion detection systems: a review. Appl. Soft Comput. **10**, 1–35 (2010)
30. Schwartz, E.J., Avgerinos, T., Brumley, D.: In Security and Privacy (SP), 2010 IEEE Symposium on IEEE, pp. 317–331 (2010)
31. AES security. https://en.wikipedia.org/wiki/Advanced_Encryption_Standard
32. Casper Vietnam, Electric Equipment Co., Ltd. https://casper-electric.com/en
33. Temperature and humidity meter. https://www.thegioiic.com/products/dong-ho-nhiet-do-do-am-htc-2
34. Taherkordi, A., Eliassen, F..: Scalable modeling of cloud-based IoT services for smart cities. In: 2016 IEEE International Conference on Pervasive Computing and Communication Workshops (PerCom Workshops), Sydney, Australia, pp. 1–6 (2016)
35. Lavric, A.: LoRa (Long-Range) High- Density Sensors for Internet of Things. Hindawi Journal of Sensors, vol. 2019 (2019)
36. Real time in digital signal processing. https://en.wikipedia.org/wiki/Real-time_computing
37. Kuo, S., Lee, B.H., Tian, W.: Real-Time Digital Signal Processing: Implementations and Applications. Wiley, Hoboken (2006)

A Method for Localizing and Grasping Objects in a Picking Robot System Using Kinect Camera

Trong Hai Nguyen[1], Trung Trong Nguyen[2], and Thang Viet Tran[3(✉)]

[1] HUTECH Institute of Engineering, Ho Chi Minh City University of Technology (HUTECH), 475A Dien Bien Phu, Ward 25, Binh Thanh District, Ho Chi Minh City 700000, Vietnam
[2] Branch of Vietnam Research Institute of Electronics, Informatics, and Automation, 169 Vo Van Ngan Street, Linh Chieu Ward, Thu Duc District, Ho Chi Minh City 700000, Vietnam
[3] Faculty of Information Systems, University of Economics and Law (UEL), VNU-HCM, Linh Xuan Ward, Thu Duc District, Ho Chi Minh City 700000, Vietnam
tvthang74@gmail.com

Abstract. This paper proposes a method for localizing and grasping objects in a picking robot system using Kinect stereo camera sensor. To do this task, the followings are done. Firstly, an image processing system including Kinect camera sensor is described. Secondly, RGB color map and new depth map for image inpainting are obtained using Kinect SDK mapping function to align RGB image with depth image. Thirdly, a new rectangle representation algorithm based on the Kinect camera is proposed to localize the grasping point of the object. Finally, the effectiveness and the applicability of the proposed algorithms is verified by using experiment. The experimental results show that the proposed algorithm successfully detects an object and finds its grasping points with accuracy 100% for no transparent objects and with 95% for transparent objects. The proposed algorithm also has higher accuracy to successfully find the grasping points than the 2D planar method.

Keywords: Mobile robot · Object recognition · Object localization · Grasping object

1 Introduction

A robotic manipulation of objects typically involves object detection/recognition and grasping control. In the detection of the object, it is very similar with the basic process of pattern recognition: source data acquisition, preprocessing, feature extraction, classification training, and object detection. Thus, feature and classification have an important influence on the object detection. Good feature selection and accurate classification is important, which can significantly improve the object detection effects. According to the problem of feature selection, different applications will vary according to their respective emphasis. There are some popular features extraction: color feature [1], feature extraction based on histogram oriented gradients (HOG) [2], edge feature [3], optical flow features

© Springer Nature Switzerland AG 2021
M. Singh et al. (Eds.): IHCI 2020, LNCS 12616, pp. 21–26, 2021.
https://doi.org/10.1007/978-3-030-68452-5_2

[4], and texture features [5]. These features have their own advantages and limitations, so in reality they are often used in combination. There is no "generalpurpose-scheme" applicable in all situations. A review of object representation techniques using shape for object recognition can be found in [6]. This representation is often only suitable for "simple objects" (e.g., circles, crosses, rectangles, etc. in 2D or cylinders, cones in the 3D case). In order to recognize an object in a scene image, a matching step has to be performed at some point in the algorithm flow [7], i.e., the object model (or parts of it) has to be aligned with the scene image content such that either a similarity measure between model and scene image is maximized or a dissimilarity measure is minimized, respectively.

Robust robot grasping in novel and unstructured environments is an important research problem that has many practical applications. A key sub-problem is localization of the objects or object parts to be grasped. Localization is challenging because it can be difficult to localize graspable surfaces on unmodelled objects. Moreover, even small localization errors can cause a grasp failure. For RGBD image, several learning algorithm [8, 9] have shown promise in handling incomplete and noisy data and variations in the environment as well as grasping novel objects. The most dominant methods are based on single camera to receive color image and laser sensor for depth information. However, they are complex and expensive. In real-world grasping, the full 3D shape of the object is hard to perceive.

To solve this problem, RGB-D data from Kinect camera is used to localize grasping points of an object. The experimental results show that the proposed algorithm successfully detects an object and finds its grasping points with an acceptable small error.

2   Proposed Algorithm

2.1   System Description

Figure 1 shows the workspace of a picking robot system consisting of manipulator platform, stereo camera and a horizontal table with an object at the manipulator workspace. The Kinect sensor is placed on the table with 1 m height.

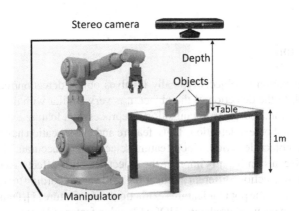

Fig. 1. Work space of a picking robot system

2.2 Mapping RGB Image into Depth Map

The transformation of the color frame has to be performed because an RGB color image has higher resolution than the depth map. This transformation is achieved based on a mapping function of the SDK, which enables to map a corresponding color to a corresponding pixel in depth space for RGB color image to complete the lost depth information from the depth map.

2.3 Proposed Object Grasp Detection Method

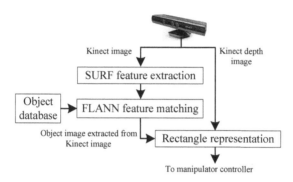

Fig. 2. Block diagram for grasp detection using Kinect camera sensor

Block diagram for a proposed method for object grasp detection using Kinect camera sensor is shown in Fig. 2. For object detection, to ensure robust and correct object identification, two algorithms to be effective for object matching are employed: Speeded Up Robust Features (SURF) algorithm first for matching object's features [3], and Fast Library for Approximate Nearest Neighbors (FLANN) algorithm for matching [4]. FLANN is a library available in OpenCv. A proposed rectangle representation algorithm for this paper consists of two phases based on depth map data from Kinect camera sensors. The first phase is for adaptive peak-depth detection to detect the peak-depth value. The second phase is for grasp detection to localize grasping points of an object.

Figure 3 shows an example of grasping rectangle G with length m_G, width n_G and it angle from the x-axis, θ

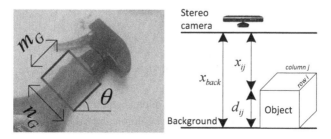

Fig. 3. Example of grasping rectangle.

Table 1 shows 7 step process in the proposed rectangle representation method. The reference distance at j^{th} column of rectangle G, d_j^{adt}, is an adaptive value given by Eq. (1)

Table 1. Proposed rectangle representation algorithm

Step	Contents
	First phase
1	Initializing the K, i, j, min, max values; new G;
2	If ($i > n_G$): $i = 0$; $j + +$; If ($j > m_G$): Goto step 1; If ($j > m$): Goto step 6;
	Else:
	- Input a new depth data; Update d_j^{adt};
3	Looking for the peak-depth:
	- Dist = max − current d_{ij} value;
	- If (Dist <= d_j^{adt}):
	Peak-depth = false ; $i + +$; Go to step 2;
	- Else:
	Peak-depth = true;
	Store the peak-depth (*pixelij*= max);
	$i + +$; Go to the step 4;
4	If ($i > n_G$): $i = 0$; $j + +$; If ($j > m_G$): Goto step 1; If ($j > m$): Goto step 6;
	Else:
	- Input a new depth data; Update d_j^{adt};
5	Looking for the valley-depth:
	- Dist = min + current d_{ij} value;
	- If (Dist >= d_j^{adt}):
	Valley-depth = false; $i + +$;Go to step 4;
	- Else:
	Valley-depth = true;
	Store the valley-depth (*pixelij*=min);
	$K = 0.8(\text{max}-\text{min})$; Go to the step 2;
	Second phase
6	Connect the peak and valley depth value (dimension of object image is $n \times m$)
	- For ($i = 1$ to m) and ($j = 1$ to n)
	pixelij and neighborhood 3x3 *pixel* have max value or min value: Connect;
	Storing the peak length or valley length
7	Looking for the rectangle representation
	- If (peak-depth length < width of physical size width of the gripper) and (valley length > width of physical size of the finger of gripper): Rectangle = true;
	- Else: Rectangle = false;

$$d_j^{adt} = K_j \frac{\sum_{i=1}^{n_G} d_{ij}}{n_G} \text{ and } d_{ij} = x_{ij} - x_{back} \tag{1}$$

where $j = 1 \sim mG$, K is a dynamic value that depends on the peak-valley amplitude, d_{ij} is the distance between i^{th} point depth x_{ij} at j^{th} column and back ground x_{back} in Fig. 3.

Figure 4 shows the results of the adaptive peak-depth detection method for rectangle representation algorithm. The d_j^{adt} value adapts itself to the changes of grasping rectangle G.

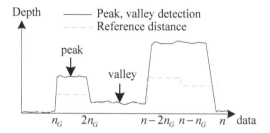

Fig. 4. Results of the adaptive peak-depth detection method.

3 Experimental Results

Physical size of the gripper is 40×10 mm, Fig. 5 show some grasping rectangle results of the proposed method. Table 2, 3 show the results of object detection and rectangle representation of the proposed method and compared with 2D planar method [2]. The experimental results show that the proposed algorithm successfully detects an object with accuracy 92.5% when the background distance is 0.6 m and finds the grasping points with accuracy 100% for no transparent object.

Fig. 5. Experiment results of grasping rectangle.

Table 2. Object detection results

Algorithm	Accuracy (%)	Background distance
Proposed	83.9	1 m
2D planar	62.6	1 m
Proposed	92.5	0.6 m
2D planar	70.3	0.6 m

Table 3. Rectangle representation results

Object	Proposed Accuracy (%)	2D planar Accuracy (%)
Spire bottles	95	50
Tennis bat	95	20
Biscuit box	100	95
Small ball	100	95
Shampoo bottle	100	95

4 Conclusions

This paper proposed a method for localizing and grasping objects in a picking robot system using Kinect camera. The experimental results showed that the proposed algorithm successfully detects an object with accuracy 92.5% when the background distance is 0.6 m and finds the grasping points with accuracy 100% for no transparent object. But, in this method, the depth data error is increased with the distance increased.

References

1. Chen, W.T., Liu, W.C., Chen, M.S.: Adaptive color feature extraction based on image color distributions. IEEE Trans. Image Process. **19**(8), 2005–2016 (2010)
2. Kadota, R., Sugano, H., Hiromoto, M., Ochi, H.: Hardware architecture for HOG feature extraction. In: International Conference on Intelligent Information Hiding and Multimedia Signal Processing, pp. 1330–1333 (2009)
3. Canny, J.F.: A computational approach to edge detection. IEEE Trans. Pattern Anal. Mach. Intell. **8**(6), 679–698 (1986)
4. Navid, N.V., Paulo, V.K.B., Jonathan, M.R.: A Study of feature extraction algorithms for optical flow tracking. In: Australasian Conference on Robotics and Automation (2012)
5. Verma, B., Kulkarni, S.: Texture feature extraction and classification. In: International Conference on Computer Analysis of Images and Patterns, pp 228–235 (2001)
6. Zhang, D., Lu, G.: Review of shape representation and description techniques. Pattern Recogn. **37**, 1–19 (2004)
7. Muja, M., Lowe, D.G.: FLANN - Fast library for approximate nearest neighbours (2009). http://www.cs.ubc.ca/~mariusm/uploads/FLANN/manual.pdf
8. Detry, R., Ek, C.H., Madry, M., Kragic, D.: Learning a dictionary of prototypical grasp-predicting parts from grasping experience. In: IEEE International Conference on Robotics and Automation, pp. 601–608 (2013)
9. Fischinger, D., Vincze, M.: Empty the basket-a shape based learning approach for grasping piles of unknown objects. In: IEEE/RSJ International Conference on Intelligent Robots and Systems, pp. 2051–2057 (2012)

A Comparative Analysis on the Impact of Face Tracker and Skin Segmentation onto Improving the Performance of Real-Time Remote Photoplethysmography

Kunyoung Lee[1] ⓘ, Kyungwon Jin[2] ⓘ, Youngwon Kim[3], Jee Hang Lee[4(✉)] ⓘ, and Eui Chul Lee[4(✉)] ⓘ

[1] Department of Computer Science, Graduate School, Sangmyung University, Seoul, South Korea
[2] Department of Artificial Intelligence and Informatics, Graduate School, Sangmyung University, Seoul, South Korea
[3] IT Convergence Components Research Center, Korea Electronics Technology Institute, Gwangju, South Korea
[4] Department of Human-centered Artificial Intelligence, Sangmyung University, Seoul, South Korea
{jeehang,eclee}@smu.ac.kr

Abstract. Remote photoplethysmography (rPPG) is well known means for measuring heart rate in remote, by analyzing physical changes such as skin color caused by cardiac activity. Previous rPPG studies usually focused on the method that how to accurately extract rPPG signals from video clips using the fixed cameras in the stable environment. In this paper, we conducted a comparative analysis using the deep learning-based object tracker and the skin segmentation method, which are image preprocessing techniques to extract rPPG signals from facial video. Experiment results showed that the noise problem caused by the unstable tracking trajectory and bounding box coordinate of tracker can be solved by skin segmentation. In addition, the skin segmentation that defines the skin color space in the YCbCr and HSV color models did not affect a real-time rPPG processing speed, and rPPG accuracy was significantly improved. In conclusion, we experimentally verified that faster rPPG measurements without loss of accuracy are possible using faster image preprocessing algorithms based on observations found through comparative analysis of skin segmentation and object trackers for real-time rPPG.

Keywords: Remote photoplethysmography · Biomedical monitoring · Remote sensing · Heart rate measurement

1 Introduction

Remote photoplethysmography (rPPG) is a camera-based, remote health monitoring that measures heartbeat by analyzing observable visual changes in the body caused by

© Springer Nature Switzerland AG 2021
M. Singh et al. (Eds.): IHCI 2020, LNCS 12616, pp. 27–37, 2021.
https://doi.org/10.1007/978-3-030-68452-5_3

cardiac activity [1]. Due to the coronavirus disease 2019 (Covid-19) pandemic, there is a growing demand on various basic health check solutions in remote, enabling both the diagnosis on the patient's symptoms and the early detection of a risk of subsequent infection. In this paper, we experimentally verified the efficient image preprocessing algorithms for the enhancement of rPPG performance. In general, the object tracking algorithms require less computation than that in object detecting algorithm, so it is suitable for fast face tracking achieving continuous tracking of changes in the facial skin color. In addition, skin segmentation is an image processing technique to spatially distinct the rPPG pulse signal and a noise without pulse information received from camera sensor. Thereby, we conducted a comparative analysis of face tracker algorithm and skin segmentation method to investigate to what extent they can play a substantial role in improving real-time rPPG.

Recently, the object tracking algorithms have made drastic progress due to Siamese architecture of deep learning [2–5]. However, to the best of our knowledge, there are a few attempts on deep learning-based tracking algorithms for improving real-time rPPG. we chose state-of-the-art trackers based on Siamese architecture and kernelized correlation filter (KCF) trackers [6] to that end. In addition, when extracting rPPG signal from the tracked Region-of-Interest (ROI) without skin segmentation, we confirmed the results corresponding to the previous study that rPPG performance is dependent on a kind of trackers [10]. The main reason is the unstable tracking trajectory and bounding box coordinates that differ for each kind of tracker causing noise that interferes with rPPG signal extraction. However, as the results of rPPG experiments using skin segmentation, there is no difference in rPPG performance except for specific trackers that were weak in facial rotation. In other words, by performing skin segmentation, it was possible to solve the problem that the rPPG performance was affected by the unstable trajectory of the tracker and frequent changes in the bounding box. In addition, the rPPG performance improvement obtained by performing image preprocessing with skin segmentation was remarkable.

Based on the comparative analysis of image preprocessing techniques on rPPG results, it would be worth considering that rPPG measurements could be performed without loss of accuracy even with a Siamese architecture-based tracker with a small number of model parameters and much faster inference times. Therefore, we finally conducted a performance comparison experiment between trackers with a backbone as a residual network with 50 convolutional layers (Resnet50) and trackers with a backbone network using less layers than Resnet50 and a faster inference time. As a result of the experiment, it was confirmed that in the case of the SiamRPN tracker [3], although the rPPG measurement speed increased about twice as fast from 86.9 frame per second(fps) to 168.1 fps on NVIDA GTX 1070 mobile, there was little loss of rPPG accuracy.

2 Methods

2.1 Comparative Analysis on Face Tracker and Skin Segmentation

For the performance comparison between image preprocessing methods, signal processing for rPPG signal extraction and face detection algorithm for initial ROI setting were set identically across all experiments. We selected SiamFC [2], SiamRPN [3], SiamMASK [4], SPM [5] and kernelized correlation filter (KCF) [6] trackers as a preprocessing means. We compared the contribution of each them to the enhancement on the performance of rPPG measurement. For the skin segmentation methods to conduct the performance comparison, we used two methods which can detect skin pixels with a little computation by defining a skin color space in a color model. The one is a method to define a skin color space using both the YCbCr and the HSV color models [8], and the other is a method to define a skin color space in the YCbCr color model in sole [9]

We conducted a comparative experiment between the method of measuring heart rate by extracting rPPG signal within the ROI using the tracking algorithm in sole, and the rPPG method using both tracking algorithm and skin segmentation. This comparative experiment confirms that skin segmentation is essential image processing for real-time rPPG. In addition, we conducted a performance comparison for the two skin segmentation methods. Also, through several observations obtained in the comparison analysis, we verify through experiments that faster rPPG measurements are possible without loss of rPPG accuracy.

2.2 RPPG Extraction for the Comparative Analysis

For comparative analysis of face tracking and segmentation methods in rPPG, the chrominance-based rPPG method was used to extract rPPG obtained from preprocessed frames [1]. Figure 1 shows the rPPG extraction process used for the comparative analysis, divided into 4 steps. In Fig. 1, the first step is face detection. The bounding box for the face is detected in the first frame and dlib's face detector was used [11]. After that, in the second step, the facial ROI in successive frames is tracked by the trackers to be compared. Also, tracked ROI is divided into skin mask areas by threshold-based skin segmentation methods [8, 9]. It should be noted in this paper that a comparative analysis is performed on the essential image processing algorithms for rPPG, not signal processing for extracting rPPG signals. The third step is signal processing to extract the rPPG signal from the average value of the chrominance signal in the skin area obtained by image processing. The chrominance-based method is suitable for realistic HCI situations where angle changes among camera, skin, and light sources or illumination variations are frequent. By separating luminance and chrominance, chrominance-based rPPG measurement is useful for separating the rPPG signal from diffuse reflections with noise [1]. In signal processing, rPPG extraction was performed using a chrominance signal, and the signal was detrending for eliminating high-frequency components and noise mitigation, and FIR bandpass filtering to obtain the frequency band of 42–240 bpm [1]. Finally, in step 4, the heart rate is calculated by transforming the signal into a frequency domain signal through the discrete Fourier transform.

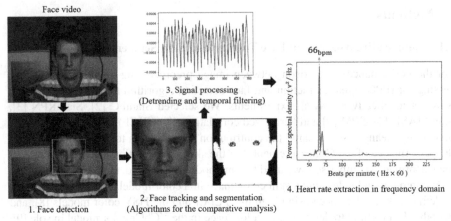

Fig. 1. Performance comparison between rPPG methods using trackers with various backbone. The radius of the circle represents the number of model parameters in the backbone network. The larger radius, the more model parameters.

2.3 Pulse Rate Detection Dataset (PURE Dataset)

The PURE dataset [7] is a public dataset designed to evaluate the performance of various rPPG methods. Image sequences in this dataset were captured with an RGB camera in 30 Hz frame rate at 640×480 resolution. A contact PPG sensor was used for the reference heart rate data. We used *pulox CMS50E* sensor providing pulse wave with a sampling rate of 60 Hz. This dataset was collected in six different environments with ten participants. These six setups consist of steady, talking, slow translation, fast translation, small rotation, and medium rotation. This rPPG public dataset with the subject's motion was suitable for evaluating the performance of the tracker, and a total of one minute of facial video was image sequence data for rPPG extraction.

2.4 Performance Evaluation Metric

To evaluate the performance of rPPG methods using different tracking algorithms and skin segmentation methods, Pearson correlation coefficient (PCC) and signal-to-noise ratio (SNR) were used. PCC shows the correlation between the reference heart rate and that measured by rPPG. We calculated a time series heart rate profile of 10 to 12 approximately, by applying a 10-s sliding window with a 5-s overlap to a 1-min long video for PCC calculation. SNR was computed from the ratio that a summation of the power of bandwidth corresponding to the reference heart rate in the frequency domain and a summation of the power of the remaining bandwidth, which is the noise signal [1]. Figure 2 shows the power spectral density of the signal bandwidths (each 5 bins width) corresponding to the ground truth used in the SNR calculation and the remaining bandwidth corresponding to the noise components. The equation for calculating SNR based on the template window of Fig. 2 is as follows.

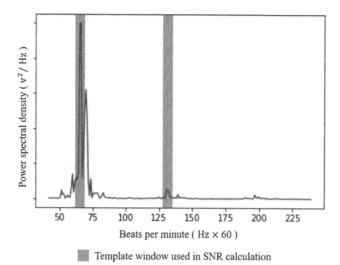

Fig. 2. Power spectral density of signal sections (in the template window) and noise sections (in the other windows) used for SNR metric

$$SNR = 10 \log_{10} \left(\frac{\sum_{42}^{240} \left(U_t(f) \hat{S}(f) \right)^2}{\sum_{42}^{240} ((1 - U_t(f)) \hat{S}(f))^2} \right) \qquad (1)$$

Where, $U_t(f)$ is a binary function representing the template window where the signal component exists. $\hat{S}(f)$ is power spectral density function of the rPPG signal. The bandwidth at which the SNR value is calculated by temporal filtering is 42–240 bpm, and the SNR metric is calculated by Eq. (1) [1]. As the final metric for processing speed measurement, fps according to the rPPG measurement method using different tracker and segmentation methods are calculated.

3   Results

We first conducted the analysis between the skin segmentation method by defining a skin color space in the YCbCr color model [9] and the skin segmentation method by defining the skin color space in the YCbCr and HSV color models [8]. Table 1 shows the performance comparison between the two on the total PURE dataset. In the case using both HSV and YCbCr, noise was generated by skin masks which is likely to be sensitive to changes in motion and lighting. On the other hand, skin segmentation using YCbCr model in sole showed higher PCC and SNR since the skin mask was more stable than skin segmentation using the HSV and YCbCr model. Therefore, the skin segmentation method using the YCbCr model [9] was superior in the correlation between the reference data and the rPPG signal and the quality of the rPPG signal.

Table 1. Comparison of rPPG accuracy between the two skin segmentation methods.

Tracker	HSV&YCbCr [8]		YCbCr [9]	
	PCC	SNR	PCC	SNR
KCF [6]	0.64	18.77	**0.77**	**21.36**
SiamFC [2]	0.65	18.39	**0.8**	**21.28**
SiamRPN [3]	0.65	19.19	**0.79**	**21.53**
SPM [4]	0.65	18.91	**0.79**	**21.1**

*The bold fonts indicate higher performances, and all Siamese tracker's backbones are Resnet 50.

Next, we conducted the analysis between an rPPG method using the tracked ROI without skin segmentation and an rPPG method using the tracked ROI in conjunction with skin segmentation. Table 2 shows the result in the fast-speed head translation situations, and Table 3 shows the result in a medium-speed head rotation situations. According to Tables 2 and 3, each rPPG method to which different tracker was applied shows a significant improvement in performance when skin segmentation was additionally used. Besides, in the case of the rPPG method without skin segmentation, changes in PCC and SNR were observed depending on the type and performance of the tracker, but the rPPG method with skin segmentation shows rPPG accuracy that is no longer dependent on the tracker. However, KCF [6] showed lower PCC than Siamese trackers in facial rotation environment despite using skin segmentation. This corresponds to the fact that the KCF tracker is known to be vulnerable to object tracking with object deformation compared to Siamese trackers.

Table 2. Performance comparison between rPPG methods according to the use of skin segmentation in fast head translation situations.

Tracker	Without skin segmentation			With skin segmentation		
	PCC	SNR	FPS	PCC	SNR	FPS
KCF [6]	0.46	8.74	**77.2**	**0.84**	**21.11**	76.06
SiamFC [2]	0.5	9.88	**47.5**	**0.84**	**20.43**	46.66
SiamRPN [3]	0.45	8.66	**87.05**	**0.83**	**21.27**	86.9
SPM [4]	0.48	10.72	**77.25**	**0.83**	**20.94**	76.75
SiamMASK [5]	Unifying approach			0.58	10.97	50.97

*The bold fonts indicate higher performances, and all Siamese tracker's backbones are Resnet 50.

Table 3. Performance comparison between rPPG methods according to the use of skin segmentation in medium head rotation situations.

Tracker	Without skin segmentation		With skin segmentation	
	PCC	SNR	PCC	SNR
KCF [6]	0.41	8.47	**0.64**	**22.65**
SiamFC [2]	0.53	13.5	**0.79**	**22.76**
SiamRPN [3]	0.51	10.62	**0.84**	**23.79**
SPM [4]	0.64	16.03	**0.79**	**21.76**
SiamMASK [5]	Unifying approach		0.39	5.56

*The bold fonts indicate higher performances, and all Siamese tracker's backbones are Resnet 50.

Table 4 and Table 5 show the performance comparison results of rPPG trackers for steady and talk situations, respectively. Unlike the previous Tables 2 and 3, in the steady situation of Table 4, there is no clear performance improvement depending on whether skin segmentation is performed. Most of them show a slight performance improvement, but in the case of KCF, PCC decreases after skin segmentation. It can be seen that skin segmentation is effectively separating the rPPG signal from diffuse reflections mixed with motion noise in a situation where subject motion exists as in Tables 2 and 3. On the other hand, it was confirmed that only a slight improvement in performance was shown in a steady state without movement. In addition, as shown in Table 5, when there is no movement and a non-rigid change in the skin area continuously occurs due to talk, it was confirmed that the rPPG performance was very poor. Even if skin segmentation was performed, it was confirmed that overall performance was low in all trackers. Through Table 4 and Table 5, it was confirmed that skin segmentation plays an important role in improving rPPG measurement performance in a realistic HCI situation where movement occurs frequently. In addition, when a continuous non-rigid change occurs in the skin area such as talk, the rPPG accuracy was significantly reduced.

Table 4. Performance comparison between rPPG methods according to the use of skin segmentation in steady situations.

Tracker	Without skin segmentation		With skin segmentation	
	PCC	SNR	PCC	SNR
KCF [6]	**0.89**	25.98	0.88	**26.96**
SiamFC [2]	0.73	24.87	**0.87**	**26.83**
SiamRPN [3]	**0.88**	20.87	**0.88**	**26.66**
SPM [4]	0.89	22.61	**0.90**	**26.78**
SiamMASK [5]	Unifying approach		0.71	18.62

*The bold fonts indicate higher performances, and all Siamese tracker's backbones are Resnet 50.

Table 5. Performance comparison between rPPG methods according to the use of skin segmentation in talk situations.

Tracker	Without skin segmentation		With skin segmentation	
	PCC	SNR	PCC	SNR
KCF [6]	0.47	10.35	**0.54**	**11.75**
SiamFC [2]	0.45	9.67	**0.60**	12.03
SiamRPN [3]	0.45	7.33	**0.51**	11.67
SPM [4]	0.26	8.37	**0.53**	11.63
SiamMASK [5]	Unifying approach		0.38	3.58

*The bold fonts indicate higher performances, and all Siamese tracker's backbones are Resnet 50.

Finally, we performed the experiment based on the observation that there is no performance difference between the Siamese tracker-based rPPG methods with skin segmentation and that the KCF tracker is vulnerable to the rPPG measurement environment with head rotation. In the final experiment, results using trackers with various backbone networks were compared, and skin segmentation using YCbCr model was used. Figure 3 shows the performance comparison result between rPPG methods using a tracker with less convolutional layers than Resnet 50 as a backbone network. In Fig. 3 (a), the horizontal axis represents the speed of the rPPG method, and the vertical axis represents the rPPG accuracy. Also, the size of the circle represents the number of model parameters. Figure 3 (a) shows the final experimental results, and even if the rPPG method using a tracker that is more than twice as fast is used, only a little performance loss occurs. The loss of rPPG accuracy was less than 0.05 in the PCC, and there was very little difference in beat per minute of heart rate. This phenomenon also appeared in Fig. 3 (b), which is a performance evaluation based on SNR and processing speed. In addition, even the Resnet 18-based SiamRPN model outperformed trackers with several deeper layers in SNR metric.

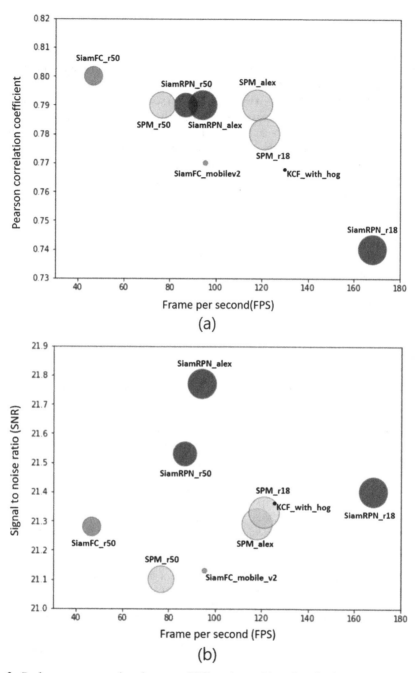

Fig. 3. Performance comparison between rPPG trackers with various backbone. The radius of the circle represents the number of model parameters in the backbone network. The larger radius, the more model parameters. (a) Performance evaluation based on PCC and processing speed. (b) Performance evaluation based on SNR and processing speed.

4 Conclusion

In this paper, a comparative analysis of object tracking and skin segmentation methods for chrominance-based rPPG method [1] was reported. Through performance comparison experiments, it was confirmed that the problem that the rPPG accuracy was dependent on the tracking algorithm can be solved by using the skin segmentation method. In addition, it was confirmed that rPPG accuracy was significantly improved by skin segmentation in realistic HCI situations with head translation and rotation. Also, through comparative analysis, it was confirmed that there was no difference in rPPG accuracy between the Siamese tracker-based methods with skin segmentation. Based on these observations, we conducted a performance comparison of various Siamese trackers, and it was verified through experiments that rPPG measurement with only a 0.05 loss of PCC is possible even with the two times faster rPPG method.

Acknowledgements. This work is supported by Samsung Research Funding Center of Samsung Electronics under project number SRFC-TC1603-52. Also, this research is supported by Ministy of Culture, Sports and Tourism and Korea Creative Content Agency (Project Number: R2020040186). Also, this work was financially supported by a Grant (2018000210004) from the Ministry of Environment, Republic of Korea.

References

1. De Haan, G., Jeanne, V.: Robust pulse rate from chrominance-based rPPG. IEEE Trans. Biomed. Eng. **60**(10), 2878–2886 (2013)
2. Bertinetto, L., Valmadre, J., Henriques, J.F., Vedaldi, A., Torr, Philip H.S.: Fully-Convolutional siamese networks for object tracking. In: Hua, G., Jégou, H. (eds.) ECCV 2016. LNCS, vol. 9914, pp. 850–865. Springer, Cham (2016). https://doi.org/10.1007/978-3-319-48881-3_56
3. Li, B., Yan, J., Wu, W., Zhu, Z., Hu, X.: High performance visual tracking with siamese region proposal network. In: Proceedings of the IEEE Conference on Computer Vision and Pattern Recognition, pp. 8971–8980 (2018)
4. Wang, G., Luo, C., Xiong, Z., Zeng, W.: Spm-tracker: Series-parallel matching for real-time visual object tracking. In: Proceedings of the IEEE Conference on Computer Vision and Pattern Recognition, pp. 3643–3652 (2019)
5. Wang, Q., Zhang, L., Bertinetto, L., Hu, W., Torr, P.H.: Fast online object tracking and segmentation: a unifying approach. In: Proceedings of the IEEE Conference on Computer Vision and Pattern Recognition, pp. 1328–1338 (2019)
6. Henriques, J.F., Caseiro, R., Martins, P., Batista, J.: High-speed tracking with kernelized correlation filters. IEEE Trans. Pattern Analysis Mach. Intell **37**(3), 583–596 (2014)
7. Stricker, R., Müller, S., Gross, H. M.: Non-contact video-based pulse rate measurement on a mobile service robot. In: The 23rd IEEE International Symposium on Robot and Human Interactive Communication, pp. 1056–1062. IEEE (2014)
8. Dahmani, D., Cheref, M., Larabi, S.: Zero-sum game theory model for segmenting skin regions. Image and Vis. Comput. **99**, 103925 (2020)
9. Phung, S.L., Bouzerdoum, A., Chai, D.: A novel skin color model in ycbcr color space and its application to human face detection. In: Proceedings. International Conference on Image Processing, vol. 1, pp. I-I. IEEE (2002)

10. Zhao, C., Mei, P., Xu, S., Li, Y., Feng, Y.: Performance evaluation of visual object detection and tracking algorithms used in remote photoplethysmography. In: Proceedings of the IEEE International Conference on Computer Vision Workshops (2019)
11. King, D.E.: Dlib-ml: A machine learning toolkit. J. Mach. Learn. Res. **10**, 1755–1758 (2009)

Fuzzy-PID-Based Improvement Controller
for CNC Feed Servo System

Nguyen Huu Cuong[1], Trung Trong Nguyen[2], and Tran Viet Thang[2(✉)]

[1] Can Tho University, Can Tho, Vietnam
[2] Vietnam Research Institute of Electronics, Informatics and Automation, Hanoi, Vietnam
tvthang74@gmail.com

Abstract. The feed servo system is one of the most important mechatronics parts in CNC machine tools. However, it is very difficult to model its control system accurately because of time-varying parameters in continuous-time nonlinear system and load disturbances. This paper introduces the control methods on feed servo system aimed to promote higher precision for CNC milling machine. The main objective of our research is focus on the improvement of Fuzzy PID (FPID) controller applied to feed servo system of CNC milling machine. The performances between the developed FPID controller and conventional PID controller have been compared and evaluated on MATLAB/SIMULINK simulation. The simulation results of AC servo system with the FPID controller show that this control method not only has more robustness but also has a better adaptation capability. Furthermore, the proposed control method has been tested in the TI developed board with one horsepower Yaskawa AC servo motor. The experimental results demonstrate that the accuracy of the motion precision of CNC milling machine's axis using the proposed control system with one horsepower AC motor can achieve about 8–10 µm.

Keywords: Fuzzy PID controller · Fuzzy logic controller · CNC machine tool · Feed servo system

1 Introduction

The PID controller is widely used in industrial process control and achieved good control effect duo to its sample structure and easy control. In fact, the feed servo system model is not only related to the friction characteristics, but also considered the disturbance torque factors, especially taken the mechanical system damping, inertia, stiffness and other parameters into account [1], and the PID parameters is very difficult to get better. In order that the servo system can have good accuracy and control performance in more running mode, more recently, intelligent control and conventional PID control methods are combined to design the position controller of ball-screw servo system. Wang and et al. [2] applied a Kalman filter to overcome the influence of measurement noise and control noise in the CNC servo control system. Zhou [3] established a two-input and three-output of a single variable speed loop self-tuning fuzzy controller which get better

© Springer Nature Switzerland AG 2021
M. Singh et al. (Eds.): IHCI 2020, LNCS 12616, pp. 38–46, 2021.
https://doi.org/10.1007/978-3-030-68452-5_4

control performance than normal PID control. Shu [4] presented a servo system of NC machine tool can be controlled by means of PID controller combined with the fuzzy logic which is designed fro application to position controller.

Fuzzy control has its advantage of strong robustness, less sensitive to the parameters of the controlled object and small overshoots. This study proposes a fuzzy control method is combined with the traditional PID control to apply to CNC servo system. This controller not only has flexibility and adaptability advantages of fuzzy control, but also has high precision characteristics of the PID control. The controller parameters can be adjusted automatically in real time for the feed servo system which has a nonlinear, time varying and random.

2 Design of Feed Servo System with Ball Screw

2.1 Ball-Screw Selection

Computation experiments are carried out simulating the ball screw feed drive using an inertial model of a commercial version BIF 4010-10 [5]; $X = 1711$ mm; $Y = 1112$ mm; $Z = 902$ mm (Fig. 1).

Inertial moment of the ball-screw:

$J_{sx} = 1.97 \times 10^{-2} \times 1711 = 3370.67 \times 10^{-2}$ (kgcm^2) $= 33.7 \times 10^{-4}$ (kgm^2)

$J_{sy} = 1.97 \times 10^{-2} \times 1112 = 21.9 \times 10^{-4}$ (kgm^2)

$J_{sz} = 1.97 \times 10^{-2} \times 902 = 17.77 \times 10^{-4}$ (kgm^2)

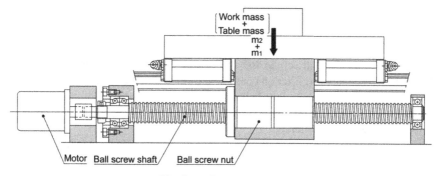

Fig. 1. Ball-screw system

Input parameters of ball screw

- X-axis (Horizontal): total mass: 905 kg, movement length: 800 mm, $V_{max} = 10$ m/min;
- Y-axis (Horizontal): total mass: 1040 kg; movement length: 500 mm; $V_{max} = 10$ m/min;
- Z-axis (Vertical): total mass: 445 kg; movement length: 400 mm; $V_{max} = 10$ m/min

2.2 Motor Selection

Using the same motor for 3 axes X, Y, and Z: AC-SERVO [6]; Model: SGMG-44A2AB; Rated Speed: 1500 rpm; Max Speed: 3000 rpm; Power: 4.4 kW (5.9HP); Rated Torque: 28.4 Nm; 290 kgcm; Peak Torque: 71.1 Nm; Rated Current: 32.8 A; Max Current: 84 A; Torque Constant: 0.91 Nm/A; Moment of Inertial: $67.5*10-4$ (kgm^2)

Axial load in horizontal ball-screw:

- Friction torque due to an external load

$$T_1 = \frac{F_a \times Ph}{2\pi \times \eta} = \frac{(\mu Mg + f) \times Ph}{2\pi \times \eta}$$

Required torque for acceleration:

$$T_2 = \left[M \left(\frac{Ph}{2\pi} \right)^2 + J_s + J_m + J_c \right] \ddot{\theta}$$

In horizontal the friction always opposite direction.
Motor torque required as follow:

$$\tau_{eu} = T_1 + T_2 = \frac{(\mu Mg + f) \times Ph}{2\pi \times \eta} + \left[M \left(\frac{Ph}{2\pi} \right)^2 + J_s + J_m + J_c \right] \ddot{\theta}$$

$$\ddot{\theta} = \frac{\tau_e - \frac{(\mu Mg + f) \times Ph}{2\pi \times \eta}}{M \left(\frac{Ph}{2\pi} \right)^2 + J_s + J_m + J_c}$$

where F_a: Force of the ball screw; Ph: Ball screw lead; η: Ball screw efficiency; f: Guide surface resistance; μ: Friction coefficient of the guide surface; M: Table mass + Work mass; J_s: initial moment of the ball-screw; J_m: initial moment of motor; J_C: initial moment of the coupling; τ_e: Electromagnetic torque of the motor.

Calculation with parameters:

$Ph = 0.01$ m; $\eta = 0.9$; $f = 15$ N; $\mu = 0.1$; $J_m = 67.5 \times 10^{-4}$ kgm^2; $J_c = 0$; $M_x = 905$ kg; $J_{sx} = 33.7 \times 10^{-4}$ kgm^2; $M_y = 1040$ kg; $J_{sy} = 21.9 \times 10^{-4}$ kgm^2

Movement equation for X-axis:

$$\ddot{\theta}_x = \frac{\tau_e - 1.5973}{7.3 \times 10^{-4} + 101.2 \times 10^{-4}} = \frac{\tau_e - 1.5973}{108.5} \times 10^4$$

Movement equation for Y-axis:

$$\ddot{\theta}_y = \frac{\tau_e - 1.5973}{8.385 \times 10^{-4} + 89.4 \times 10^{-4}} = \frac{\tau_e - 1.831}{97.785} \times 10^4$$

Axial load in vertical ball-screw
Friction torque due to an external load

– During upward uniform motion

$$T_1 = \frac{F_{a2} \times Ph}{2\pi \times \eta} = \frac{(Mg + f) \times Ph}{2\pi \times \eta}$$

– During downward uniform motion

$$T_2 = \frac{F_{a5} \times Ph}{2\pi \times \eta} = \frac{(-Mg + f) \times Ph}{2\pi \times \eta}$$

• Required torque for acceleration

$$T_3 = \left[\left(M((Ph)/2\pi)^2 + J_s + J_m + J_c\right)\ddot{\theta}\right]$$

Motor torque required as follow:

– Upward motion:

$$\tau_{eu} = T_1 + T_3 = \frac{(Mg + f) \times Ph}{2\pi \times \eta} + \left[M\left(\frac{Ph}{2\pi}\right)^2 + J_s + J_m + J_c\right]\ddot{\theta}$$

$$\ddot{\theta} = \frac{\tau_e - \frac{(Mg + f) \times Ph}{2\pi \times \eta}}{M\left(\frac{Ph}{2\pi}\right)^2 + J_s + J_m + J_c}$$

– Downward motion:

$$\tau_{eu} = T_2 + T_3 = \frac{(-Mg + f) \times Ph}{2\pi \times \eta_{t_e} + \frac{(Mg-f)\times Ph}{2\pi \times \eta}} + \left[M\left(\frac{Ph}{2\pi}\right)^2 + J_s + J_m + J_c\right]\ddot{\theta}$$

$$\ddot{\theta} = \frac{\tau_e + \frac{(Mg-f)\times Ph}{2\pi \times \eta}}{M\left(\frac{Ph}{2\pi}\right)^2 + J_s + J_m + J_c}$$

Calculation with parameters:
    $Ph = 0.01$ m; $\eta = 0.9; f = 15$ N; $\mu = 0.1; J_m = 67.5 \times 10^{-4}$ kgm^2; $J_c = 0; M_z = 445$ kg; $J_{sz} = 17.77 \times 10^{-4}$ kgm^2

– Upward motion:

$$\ddot{\theta} = \frac{\tau_e - \frac{(Mg + f) \times Ph}{2\pi \times \eta}}{M\left(\frac{Ph}{2\pi}\right)^2 + J_s + J_m + J_c} = \frac{\tau_e - \frac{(445 \times 9.81 + 15) \times 0.01}{2\pi \times 0.9}}{455\left(\frac{0.01}{2\pi}\right)^2 + 17.77 \times 10^{-4} + 67.5 \times 10^{-4} + 0}$$

$$\ddot{\theta} = \frac{\tau_e - 7.746}{11.5 \times 10^{-4} + 85.27 \times 10^{-4}} = \frac{\tau_e - 7.746}{96.77} \times 10^4$$

– Downward motion:

$$\ddot{\theta} = \frac{\tau_e + \frac{(Mg - f) \times Ph}{2\pi \times \eta}}{M\left(\frac{Ph}{2\pi}\right)^2 + J_s + J_m + J_c} = \frac{\tau_e + 7.693}{455\left(\frac{0.01}{2\pi}\right)^2 + 17.77 \times 10^{-4} + 67.5 \times 10^{-4} + 0} = \frac{\tau_e + 7.693}{96.77} \times 10^4$$

3   Modeling of Feed Servo Motor with Ball-Screw

Model of motor of the servo drive system adopts permanent magnet synchronous machine (PMSM) [7]. In the MATLAB/Simulink environment establishing and combining the ball-screw transmission model, vector-control simulation model, PID controller simulation model, PMSM model and PWM inverter simulation model as shown in Fig. 2.

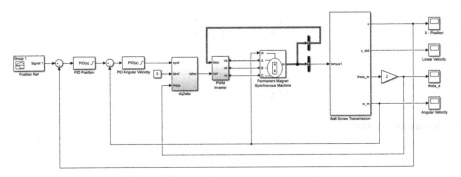

Fig. 2. MATLAB/Simulink model of the ball-screw feed servo system

The model of the ball-screw feed servo system is simulated on MALAB/Simulink environment with the motor and PID controller parameters as follows:

– Stator resistance: $R_s = 0.0918\ \Omega$
– Inductances: $L_d = 0.0009262$ H; $L_q = 0.001024$ H
– Flux linkage: $\lambda = 0.1688$
– Pole pairs: $p = 2$
– PID position: $K_p = 7000$; $K_i = 10$; $K\text{-}_d = 10$
– PID angular velocity: $K_p = 10$; $K_i = 1$; $K_d = 0.5$

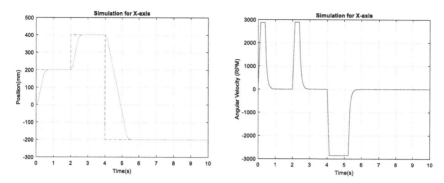

Fig. 3. Simulative results

Simulation results of position and velocity of X-axis are shown as in Fig. 3.

In order to highlight the efficiency of the proposed Fuzzy-PID (Fig. 4), a comparison of the Fuzzy-PID and PID controllers is performed as Fig. 5. The parameters of the PID position and Fuzzy-PID controllers are selected respectively as follows:

- PID position controller: $K_p = 2000$; $K_i = 5$; $K_d = 5$
- Fuzzy-PID position controller: $K_{pmin} = 1000$; $K_{pmax} = 10000$; $K_{imin} = 1$; $K_{imax} = 10$

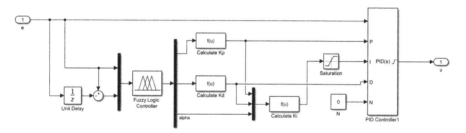

Fig. 4. Fuzzy-PID controller

The result of the simulation as Fig. 6 shows that the Fuzzy-PID controller has some advantages, such as the setting time is better and selection of controller parameters becomes easier.

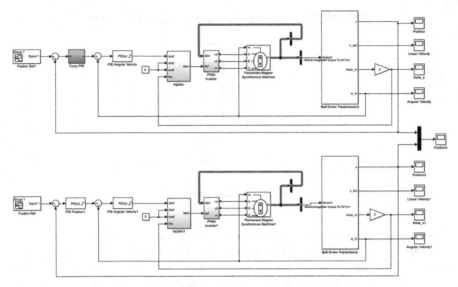

Fig. 5. Comparison of the Fuzzy-PID and PID controllers

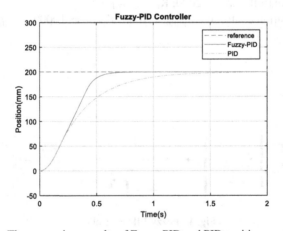

Fig. 6. The comparison results of Fuzzy-PID and PID position controllers

4 Experimental Results

Using the simulation results the values of K_p, K_d, and K_i parameters of the PID controller is calculated and applied to a real feed servo system for experiment. The real system is shown as Fig. 7 which consists of a PMSM, a PWM inverter, and a PID controller.

The experiment is carried out in no-load condition with the PID controller parameters are listed as in Table 1.

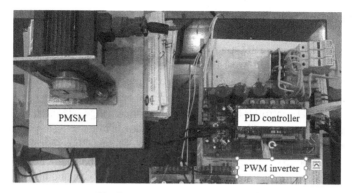

Fig. 7. Experimental servo system

Table 1. PID controller parameters

Velocity controller		Position controller	
K_p	3	K_p	1.5
K_i	0.001	K_i	0.0001
K_d	0.001	K_d	10

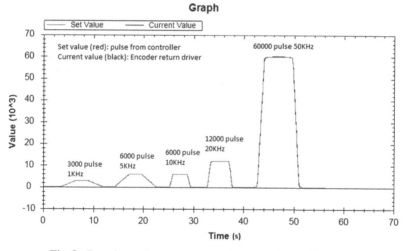

Fig. 8. Experimental result of real feed servo system with PMSM

Figure 8 shows the result of experiment where the red is set value from the controller which includes some specific values (3000 pulse 1 kHz; 6000 pulse 5 kHz; 6000 pulse 10 kHz; 12000 pulse 20 kHz; and 60000 pulse 50 kHz) and the black is current returned value from encoder, respectively.

5 Conclusions

This paper proposes a fuzzy-PID controller to control the speed and velocity of CNC feed servo system. The designed fuzzy-PID controller is applied to the MATLAB/Simulink model of CNC feed servo system considering the electromechanical coupling effects including mechanical feed link. The controller can be online auto tuning the K_p, K_d, and K_i parameters of PID controller by measuring error and change rate of error. The fuzzy-PID controller is compared with a traditional PID controller on the system response. The simulation and experimental results show that the use of fuzzy-PID controller can reduce the difficulty of model construction; the fuzzy-PID controller has higher robustness than conventional PID controller for a wide range, and rapid response and high accuracy.

References

1. George, C.: Dynamic definition of machine tool feed drive models in advanced machine Tools. In: Proceedings of 3rd International Conference on the Industry 4.0 Model for Advanced Manufacturing AMP 2018. pp. 115–138 (2018)
2. Wang, W., Zhao, P.: Application of kalman filter in the cnc servo control system. Procedia Eng. **7**, 442–446 (2010)
3. Zhou, R.: Modeling and simulation of fuzzy self-tuning control of the feed servo system. J. WUT **33**(1), 47–50 (2011)
4. Shu, D.: Application of the fuzzy controller to the servo system of nc machine tool. Techn. Autom. Appl. **29**(6), 33–36 (2010)
5. BIF 4010–10, Detail Specifications Product Information, THL. https://tech.thk.com/en/pro ducts/get_all_attributes.php?id=2045 Accessed 09 Feb. 2020
6. Rotary Servo Motors, Yaskawa. https://www.yaskawa.com/products/motion/sigma-7-servo-products/rotary-servo-motors Accessed 09 Feb. 2020
7. Permanent Magnet Synchronous Machine, MATLAB-Simulink, Mathworks. https://www.mat hworks.com/help/physmod/sps/ug/permanent-magnet-synchronous-machine.html Accessed 09 Feb. 2020

Plastic Optical Fiber Sensors Based on in-Line Micro-holes: A Review

Hyejin Seo and Jaehee Park[✉] [iD]

Keimyung University, Daegu 42601, South Korea
jpark@kmu.ac.kr

Abstract. The sensors fabricated with plastic optical fiber (POF) having in-line micro-holes have recently been interested in because of their merits such as simple structure, low cost, high tensile strength, and easy handing. In-line micro-holes were fabricated using a low-cost drilling machine with a bias and a microdrill bit. Refractive index, liquid level, and respiration rate measurements have been performed with these POF sensors. These measurement results show that these POF sensors can be used in field because they have sufficient sensitivity and are unaffected by ambient noise. In this document, a review of the current state of the POF sensors including in-line micro-holes is presented.

Keywords: Plastic optical fiber · Sensor · In-line Micro-hole

1 Introduction

In spite of developing almost at the same time, the success of glass optical fiber has been brilliant over plastic optical fibers (POFs). However, POFs have made an excellent substitute to glass optical fibers (GOFs) owing to recent technological advances of POFs. Furthermore, POFs are expected to be chosen as the next optical fibers for making advanced optical fiber sensors in future [1].

POF sensors have most of the merits of GOF sensors, including rapid response, remote sensing capability, small size, and immunity to electromagnetic interference [2]. Moreover, POF sensors also offer some advantages over GOF sensors, such as easy handling, high elastic strain limits, high flexibility in bending, high fracture toughness, and high sensitivity to strain [3]. Thus, various types of POF sensors have already been developed due to these advantages [4]. However, these POF sensors have a little bit low measurement sensitivity because the measurement method of most POF sensors are based on intensity-modulation. Therefore, researches have been interested in developing the POF sensors with high measurement sensitivity.

The optical fiber with in-line micro-holes have been making increasing interest [5]. They are fabricated by making micro-holes in single mode glass fiber using chemical etching and femtosecond laser processing. The accurate alignment is not required because the desired alignment is obtained after fabricating micro-holes in glass fiber. Several GOF devices having in-line micro-holes have been produced and shown good performance [6]. Yet, the expensive equipment is needed for making in-line micro-holes.

© Springer Nature Switzerland AG 2021
M. Singh et al. (Eds.): IHCI 2020, LNCS 12616, pp. 47–52, 2021.
https://doi.org/10.1007/978-3-030-68452-5_5

Recently, the development of POF sensors based on in-line micro-holes fabricated using an inexpensive equipment has been reported. All of them had sufficient sensitivity to measure physical parameters. This paper reviews recent studies on the POF sensors including in-line micro-holes.

2 Theory

The characteristics of in-line micro-holes based POF sensor are theoretically investigated. A multimode POF have a lot of modes in the visible wavelength range because of about mm sized core. Therefore, the theoretical analysis has to be performed from geometrical optics perspective.

A schematic diagram of a POF including one in-line circular micro-hole is shown in Fig. 1. If there are no micro-hole in the POF and rays pass through P point with less angle to the optical(fiber) axis than $\theta_B (= \tan(\frac{a}{L}))$, they travel without loss where a and $L(= a \tan\theta_C)$ are radius of the core and the distance of point P from the hole center, respectively. θ_c is the angle of incidence at the core-cladding interface where the refraction angle is 90°(critical angle). If there are one in-line micro-hole and rays pass through P point when the angle to the optical axis is less than θ_B and greater than θ_A (the angle between a tangent ray passing through P point and the optical axis), they propagate without loss. When rays pass through P point with an angle smaller than θ_A, rays experiences loss because of the refraction at the boundary if they pass through the hole and arrive with an angle less than θ_C at the interface. Otherwise, the rays guide into the fiber without loss. The transmittance (T $= \frac{Optical power after passing through an in-fiber hole}{optical power before arriving at an in-fiber hole}$) [7] of the POF sensor having one in-line micro-hole is obtained as

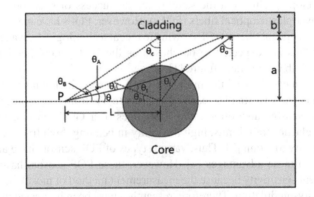

Fig. 1. Schematic diagram of POF having one in-line circular micro-hole [1]

$$T = \frac{\left[cos^{-1}\left(\frac{n_L}{n_O}\right) - sin^{-1}\left(\frac{r}{L}\right) + tan\left(\frac{r sin\theta_{h,max}}{L - r cos\theta_{h,max}}\right)\right]}{cos^{-1}\left(\frac{n_L}{n_O}\right)} \tag{1}$$

$$\theta_{h,max} = 2sin^{-1}\left(\frac{n_O}{n}sin\theta_{i,max}\right) - \theta_{i,max} - \frac{\pi}{2} + \theta_C \tag{2}$$

$$\theta_i = \theta_h + \tan\left(\frac{r\sin\theta_h}{L - r\cos\theta_h}\right) \qquad (3)$$

where r and n are the hole radius and the refractive index of the material in an in-fiber micro-hole, respectively. θ_h is the angle measured from the micro-hole center between the optical axis and the point where rays reach the micro-hole, and θ_i is the incident angle at the core-hole boundary. n_L and n_O are the refractive index of cladding and the core, respectively. Equations show that the transmittance depends on the micro-hole radius and the refractive index of the material in the in-line micro-hole.

3 POF Sensors

The POF sensors including in-line micro-holes have been developed to measure refractive indices, water level, and respiration rate. The measurement method of these POF sensors has been based on the intensity modulation.

3.1 POF Sensor Fabrication

The fabrication setup for making in-line micro-holes is shown in Fig. 2. The fabrication setup consisted of an inexpensive drilling machine (SMC HD-280), a microdrill bit (NEO Technical System), a power meter (PM, Thorlabs PM100D), and a 630 nm laser diode (LD). In-line micro-holes in the POF were produced while monitoring transmission optical power. The POF used to make various types of sensors had a 1.48 mm core diameter and core and cladding refractive indices were 1.49 and 1.41, respectively. Figure 3 shows in-fiber micro-holes in POF produced using the experiment setup

3.2 POF Refractive Index Sensor

The POF refractive index sensor having an in-line micro-hole [7] was studied. This sensor consisted of a POF with one 0.7 mm diameter micro-hole. Three liquids (water (n = 1.33), 99% glycerol (n = 1.47), and n-dodecane (n = 1.42)) were used to analysis the characteristics of this POF sensor. The values of all refractive index are measured at 589.3 nm. When a liquid of the lowest refractive index was applied to the micro-hole using a micro pipette, the POF sensor transmittance was obtained. The reference optical power for obtaining sensor transmittance was optical power measured before

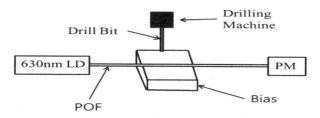

Fig. 2. In-line micro-hole fabrication setup [8].

Fig. 3. In-line micro-holes in the POF produced using an inexpensive setup.

drilling. All transmittances were obtained as the liquids were put on the micro-hole in sequence from the lowest refractive index to the highest. The experimental transmittance data obtained at about 670 nm and theoretical transmittance calculated by Eq. (1) are shown in Fig. 4. As expected, the POF sensor transmission increased as the refractive index increased. Figure 4 indicates that the experimental and analytical results were in excellent agreement. The sensor sensitivity was about 13.4 dB/RIU. The measured refractive index range was 1.33–1.42.

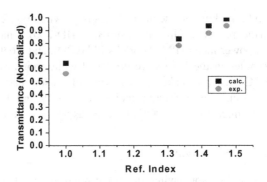

Fig. 4. Transmittance of a POF refractive index sensor [7].

High sensitivity POF refractive index sensor [9] was produced with the POF with three cascaded in-line micro-holes of 0.7 mm diameter. The sensitivity of this sensor was about 43.8 dB/RIU. This sensitivity is about three times that of single micro-hole sensor. This result demonstrates that the sensitivity is proportionnate to the number of in-line micro-hole.

3.3 POF Level Sensor

The POF sensor with in-line micro-holes for measuring water level [1] has been studied. The POF sensor for measuring water level was the POF including eight in-line micro-holes of 0.9 mm diameter. The micro-hole space was around 5 cm. Figure 5 shows experiment setup composed of a 633 nm wavelength laser diode, a 55 Cm high measuring cylinder, and a power meter (Thorlabs PM100D). The POF sensor for measuring

water level sensor were analyzed experimentally by putting water into the cylinder. The experimental transmittance data (solid line) and theoretical transmittance data (dotted line) calculated using transmittance equation are shown in Fig. 5. The transmittance equation was $T^{(8-n)}T_1^n$ where T_1 was the sensor transmittance when one in-line micro-hole of the sensor was in water and n was in-fiber micro-hole number submerged in water. The transmittance was proportional to micro-hole number immersed in water as expected. Because refractive index (RI) of water was closer to the RI of POF core than the RI of air, transmittance increase occurred. Figure 5 shows that the experimental value and theoretical value were in excellent agreement.

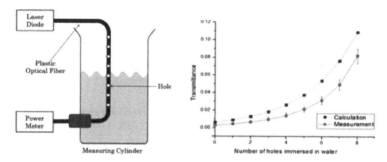

Fig. 5. Experiment setup and transmittance of POF Respiration Sensor [1].

A simple structure POF sensor including multiple in-line micro-holes for monitoring respiration [10] was investigated. This sensor was the POF with several in-line micro-holes made along the POF which stitched to an abdomen-band. The respiration signal was obtained with the optical power change induced by bending of the POF induced by the variations of abdomen circumference. The variations of abdomen circumference was made by patient's respiration. The proposed respiration POF sensor (Fig. 6) consisted of the POF with several in-line micro-holes in central part, plastic buckle, and an elastic fabric abdomen-band. Several types of respiration sensors were produced with POFs including different number and different space of 0.7 mm diameter in-line micro-holes. The sensitivity of the POF sensor for monitoring respiration was proportionate to the micro-hole number. It was also in proportion to micro-hole spacing. However, from the sensitivity and strength perspective, the POF respiration sensor having 7 holes spaced at 15 mm had the best performance for respiration monitoring. Respiration signals obtained from the POF respiration monitoring sensor (top waveform) and a MP150WSW Biopac medical device for respiration measurement (bottom waveform) are shown in Fig. 6. This figure indicates that the performance of POF respiration sensor is similar with the performance of the Biopac respiration measurement device.

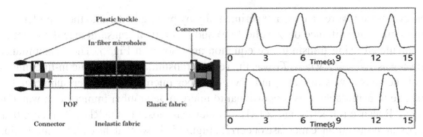

Fig. 6. Configuration of the POF respiration sensor and respiration signals [1].

4 Conclusions

In this document, the POF sensors including in-line micro-holes were reviewed. All of these sensors showed sufficient sensitivity and strength to use in field. The development of different POF sensors with the same structure is expected.

Acknowledgements. : This work was supported by a National Research Foundation of Korea (NRF) grant funded by the Korea government (MIST), No. 2018R1D1A1B07048066.

References

1. Park, J., Park, Y., Shin, D.: Plastic optical fiber sensor based on in-fiber microholes for level measurement. Jpn. J. Appl. Phys. **54**, 028002 (2015)
2. Park, J.: Plastic optical fiber sensor for measuring driver-gripping force. Opt. Eng. **50**(2), 020501 (2011)
3. Peter, K.: Polymer optical fiber sensors-a review. Smart Mater. Struct. **20**(1), 013002 (2011)
4. Zhou, Q., Kritz, D., Bonnell, L., Sigle, G.: Porous plastic optical fiber sensor for ammonia measurement. App. Opt. **28**(11), 2022–2025 (1989)
5. Lai, Y., Zhou, K., Zhang, L., Bennion, I.: Microchannels in conventional single-mode fibers. Opt. Lett. **31**(17), 2559–2561 (2006)
6. Yuan, L., Huang, J., Lan, H., Wang, H., Jiang, L., Xial, H.: All-in-fiber optofludic sensor fabricated by femtosecond laser assisted chemical etching. Opt. Lett. **39**(8), 2316–2358 (2014)
7. Shin, J., Park, J.: Plastic optical fiber refractive index sensor employing an in-line submillimeter hole. IEEE Photonics Technol. Lett. **25**(19), 1882–1884 (2013)
8. Seo, H., Park, J.: Plastic optical fiber sensor based on in-fiber rectangular hole for mercury detection in water. Sen. Mater. **32**(6), 2117–2125 (2020)
9. Shin, J., Park, J.: High-sensitivity refractive index sensors based on in-line holes in plastic optical fiber. Microwave Opt. Technol. Lett. **57**(4), 918–921 (2015)
10. Ahn, D., Park, Y., Shin, J., Lee, J., Park, J.: Plastic optical fiber respiration sensor. Microwave Opt. Technol. Lett. **61**, 120 (2019)

Long-Distance Real-Time Rolling Shutter Optical Camera Communication Using MFSK Modulation Technique

Md Habibur Rahman(ID), Mohammad Abrar Shakil Sejan(ID),
and Wan-Young Chung$^{(\boxtimes)}$(ID)

Department of Electronic Engineering, Pukyong National Univeristy,
Busan 48513, South Korea
wychung@pknu.ac.kr

Abstract. Optical camera communication (OCC) is the subset of a visible light communication. Due to some practical challenges for data transmission, most of the existing OCC systems fail to provide long-distance real-time communication performance. In this study, we proposed and demonstrated a long-distance real-time OCC system that utilizes the multiple frequency shift on-off keying modulation technique on the transmitter side. Moreover, to observe the performance of the proposed modulation scheme, we compare it to the others for the OCC system with respect to the bit-error-rate and signal to noise ratio. Besides, the rolling shutter mechanism of a smartphone embedded CMOS image sensor used to enhance the data rate and communication distance. The system performance shows that it can successfully communicate 7 m distance between the light-emitting diode as a transmitter and smartphone image sensor as a receiver. The proposed system can be applied to VLC based indoor positioning system, smart indoor lighting, autonomous vehicle system in outdoor, and indoor optical wireless healthcare monitoring system.

Keywords: Optical camera communication · Visible light communication · Light-emitting diode · Multiple frequency shift on-off keying · Smartphone CMOS image sensor

1 Introduction

Due to the high demand for wireless communication technology, visible light communication (VLC) has emerged as a promising solution for the optical wireless communication system (OWC). It makes a significant advantage in the various field such as indoor localization, unlicensed broad bandwidth spectrum, electromagnetic interference-free communication. Moreover, OWC is the alternative and complementary choice to the existing radio frequency communication. OWC

This work was supported by the National Research Foundation of Korea(NRF) grant funded by the Korea government(MSIT) (No. 2020R1A4A1019463).

M. Singh et al. (Eds.): IHCI 2020, LNCS 12616, pp. 53–62, 2021.
https://doi.org/10.1007/978-3-030-68452-5_6

can be divided into four main categories, such as visible light communication, free-space optical communications, light-fidelity, and optical camera communication (OCC) [1].

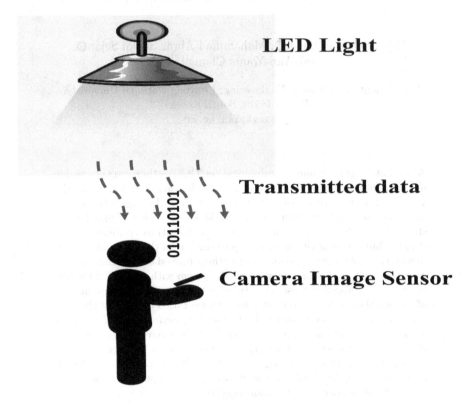

Fig. 1. Optical camera communication scheme between transmitter and receiver.

However, each technology uses visible light-medium to transmits the data or information to the receiver. OCC is the new and subset field of visible light communication as shown in Fig. 1. The OCC technology that used the optical camera image sensor as a receiver. Therefore, the researcher classifies the image sensor-based OCC system mainly into two different types: high frame rate camera (HFR) OCC and low frame rate camera (LFR) OCC [2]. Most of the smartphone and mirror-less camera are provides the LFR performance. Therefore, this study we mainly attention only LFR camera based-OCC system with long distance communication. Using the LFR camera of a non-flickering OCC system can be achieved by the two-communication modulation scheme, such as an under-sampled modulation scheme [3] and rolling shutter-based scheme [4].

In recent years there have many OCC systems has proposed that has various kinds of application such as VLC based wireless healthcare system [5,6], visible light indoor positioning system [7], vehicle to vehicle communication and etc.

A lower data rate and longer communication distance proposed using an under-sampled modulation scheme in [3]. A higher data rate and lower communication distance between transmitter and receiver using a rolling shutter camera-based proposed in [4,8]. The short communications range using photo-diode as a receiver and multiple input multiple outputs OWC has presented in [9]. A commercial digital camera (Canon EOS 700D) and 1024-QAM OCC system using duel LED demonstrated in [10]. In [10], the system provides a lower communication distance with 1.5 m, and it performed off-line processing in MATLAB. The study [11] represents, smartphone camera based-non-line of sight with a short distance OCC system where it can not provide the smartphone real-time processing. Color-shift keying based with 8.64 kbps data rate and 4 cm communication distance OCC system demonstrated in [12]. OCC system with 1.5 m communication distance using a color intensity modulation scheme, which requires a computer for processing in [13].

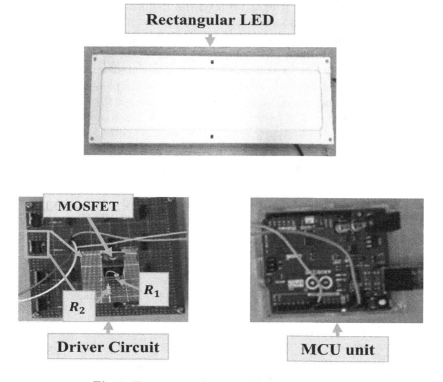

Fig. 2. Transmitter design with its components.

In this study, we demonstrated a real-time long-distance OCC system using a rolling shutter embedded smartphone complementary metal-oxide-semiconductor (CMOS) camera image sensor. A flicker-free multiple frequency shift on-off keying (MFSK) modulation technique is used to enhance data rate and minimize the bit error rate. The experimental results show that the proposed system can communicate almost 7 m distance between the transmitter and smartphone image sensor receiver.

The rest of this study is organized as : Sect. 2 descried the system design. The details of the experimental setup and performance evaluation provided in Sect. 3. Finally, concluding, and future work remarks in Sect. 4.

2 System Architecture

The proposed OCC system architecture is shown in Fig. 3. The proposed system consists of two main sections: the OCC transmitter and the OCC receiver. The transmitter consists of a rectangular-shaped light-emitting diode (LED) with 20 W power; a high-speed switching LED driver consists of a metal-oxide-semiconductor field-effect transistor (MOSFET) and a microcontroller (MCU) ATmega328p. Figure 2 shows the transmitter design with its components. To control the flickering effect of light we used the frequency shift-keying modulation scheme. The standard visual flickering happened if the LED on-off state observed by the human eye is above 200 Hz [14]. The LED light keeps at the higher frequencies than the smartphone camera can detect to encode the modulate the data. Finally, the LED to transmit the modulated data to the optical channel. In contrast, the receiver side, (Samsung S8) smartphone, CMOS image sensor is used. The modulated signal transmitted through the optical channel to the image sensor via on-off keying. After receiving the signal, it processes the

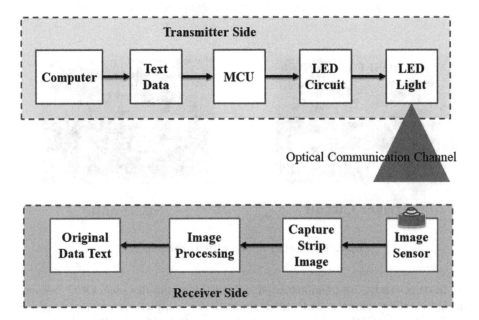

Fig. 3. The proposed system architecture with a transmitter and a receiver.

frame with the help of the OpenCV library for image processing. The rolling shutter camera captures the dark and bright strip and measures the strip width for demodulation. At the end of the demodulation and successful decoding, the original data is shown on display.

2.1 Rolling Shutter OCC Encoding

In the transmitter section, the modulation mechanism is performed in the wireless system to transmit the information through a defined band frequency. The MFSK modulation technique is used to encode the data. For the case of a rolling shutter modulation scheme, the light signal transmitted by used to create a square wave in its modulation, as shown in Fig. 4(a).

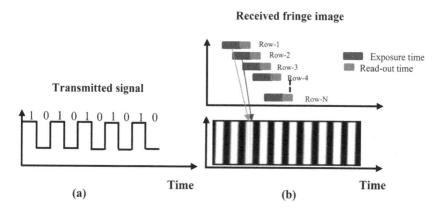

Fig. 4. Rolling shutter mechanism of receiving image; (a) Transmitted signal (b) Received fringe image

A square shaped wave of a certain frequency transmitted by the light that generates in a dark and bright strip pattern in the captured image. Also, the different frequencies make the different dark and bright strip width. Those different frequencies of the square wave to represent the different symbols of the data signal and the image sensor demodulate those data signals if it can measure the width of the dark and bright strip width. In this OCC system, we used an 8-FSK modulation scheme in which the encoding data signal represents three bits of every consecutive frequency.

2.2 Rolling Shutter OCC Decoding

There have two types of camera one is the global shutter and the rolling shutter. In a global shutter, the captured frame exposed at one time and rolling shutter it captured using row-by-row exposure. The image sensor quality and characteristics vary for the smartphone to smartphone. On the receiver side, we used a smartphone embedded CMOS image sensor that can work rolling shutter mode to scan every pixel value row by row and capture a fringe image from the LED on-off state, as shown in Fig. 4(b). In the real-time OCC system requires a fast mode of processing of the received image information from the transmitter. Real-time processing should be in the individual frame duration of the camera.

If the frame rate of the camera is 1/frame rate, then the time limit for each frame in real-time is the frame rate. To demodulate the data from light signal, we measured the width of the strip w, which is generated by different transmitted frequencies f_n $(n = 1,2,\ldots)$ from the transmitter LED light. The width of the strip mostly depends on the modulated frequencies. Also, the number of the strip on the image plane depends on the communication distance between the transmitter LED and the smartphone image sensor. The strip width refers to the number of pixels occupied by the dark and bright strip in the received image on the image plane. If the square waveform of signal frequency is f_g, then the duration of the complete cycle will be $1/(2f_g)$ seconds for a couple of dark and bright stripes. Therefore, the width of the strip w is defined as:

$$w = \frac{1}{2f_g t_r} \tag{1}$$

$$f_g = \frac{1}{2w t_r} \tag{2}$$

where t_r is the read-out duration which is usually an unknown parameter of each camera.

3   Experimental Results Evaluation

The experimental test has conducted in terms of the real-time decoding performance rate and bit error rate. The experimental testbed setup is shown in Fig. 5, where the distance from the LED transmitter and smartphone image sensor was 7 m. We developed the camera application to modify some parameters on the android studio platform for the purposes of real-time decoding. To perform the image processing inside the camera application, OpenCV 320 library was added. The important parameters of the transmitter and receiver section are shown in Table 1.

3.1   Decoding Accuracy Evaluation

The decoding accuracy of the OOC system relies on the distance between the transmitter and receiver; the transmitter LED size, and the intensity of the light. To test the decoding accuracy's performance, we performed an experiment by measuring the accuracy rate with respect to the distance, as shown in Fig. 6. From the test, we can see that the decoding rate is almost 100% until the distance 4 m, and after that, it decreased to a maximum of 6.45 m. The image sensor could not decode properly after the maximum distance due to the reduced number of strips in the captured image with increasing distance.

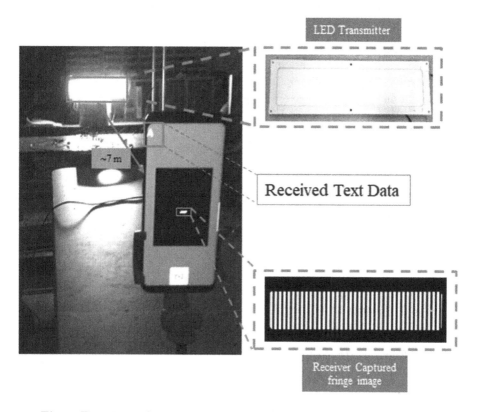

Fig. 5. Experimental setup with a transmitter and a receiver arrangement.

Table 1. Transmitter and receiver parameters.

Section	Parameters	Values
Transmitter	LED Shape	Rectangular
	LED Power	20 W
	Microcontroller	ATmega328p
	Modulation Scheme	MFSK
Receiver	Camera Frame Rate	20 fps
	Camera Shutter Speed	1/16000 s
	Sensor Size	1/3.6 in
	Image Resolution	600 × 800
	Camera2 API	API 24
	Image Processing Library	OpenCV

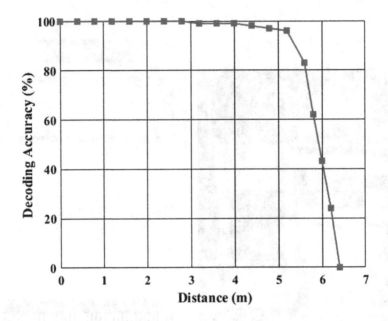

Fig. 6. Performance of decoding accuracy rate with respect to distance.

3.2 Bit-Error Rate Analysis

The performance of the bit-error-rate of the system is considered an indoor wireless Rician fading channel. Because MFSK modulation is very simple for the communication over this channel, and it is known [15] where the bit-error-rate is defined as:

$$L_s = \frac{1}{M_s} \sum_{n=2}^{M_s} (-1)^n \binom{M}{n} exp\left[-\left(1-\frac{1}{n}\right)L\right] \tag{3}$$

where M_s is the modulation order, and L is the signal-to-noise instantaneous values. Figure 7 shows that the comparison figure between the different modulation order and signal to noise ratio of MFSK modulation scheme. Where we can see that if the modulation order is growing up, then the bit-error-rate is probability is decreasing. So that the bit-error-rate becomes lower at the symbol duration increase in the higher modulation index.

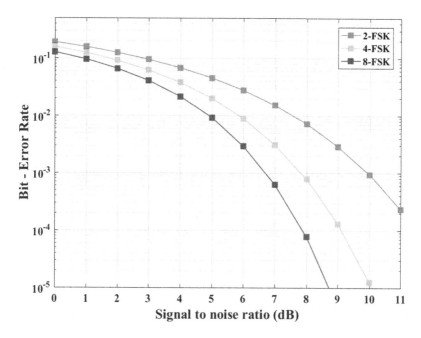

Fig. 7. Bit-error rates for the different MFSKs.

4 Conclusion

In this study, we proposed and demonstrated the long-distance real-time OCC system that utilizes the rolling shutter mechanism of a smartphone embedded CMOS image sensor. The MFSK modulation scheme is used to enhance the data rate and transmission the light information through the optical channel. The experimental results imply that the system can communicate up to 7 m distance between the transmitter and smartphone image sensor receiver. In the future, we will apply this system in the VLC based indoor localization system and optical wireless healthcare monitoring system.

References

1. Saeed, N., Guo, S., Parkm, K.-H., Al-Naffouri, T.-Y., Alouini, M.-S.: Optical camera communications: survey, use cases, challenges, and future trends. Phys. Commun., **37**, 100800 (2019). https://doi.org/10.1016/j.phycom.2019.100900
2. Rachim, V.P., Chung, W.-Y.: Multilevel intensity-modulation for rolling shutter-based optical camera communication. IEEE Photonics Technol. Lett. **30**(10), 903–906 (2018)
3. Roberts, R.D.: Undersampled frequency shift ON-OFF keying (UFSOOK) for camera communication (CamCom). In: Proceedings 22nd Wireless Optical Communucations Conference, Chongqing, China, pp. 645–648 (2013)

4. Danakis, C., Afgani, M., Povey, G., Underwood, I., Haas, H.: Using a CMOS camera sensor for visible light communication, In: Proceedings of the IEEE Globecom Workshops, pp. 1244–1248. Anaheim, CA, USA (2012)
5. Sejan, M.A.S., Chung, W.-Y.: Lightweight multi-hop VLC using compression and data-dependent multiple pulse modulation. Opt. Express **28**(13), 19531–19549 (2020)
6. Sejan, M.A.S., Rahman, M.H., Chung, W.-Y.: MPPM based bi-directional long range visible light communication for indoor particulate matter monitoring. In: Proceedings 2020 IEEE 3rd International Conference Computer Communications Engineering Technology (CCET), Beijing, China, pp. 263–266 (2020)
7. Rahman, M.H., Sejan, M.A.S., Kim, J.-J., Chung, W.-Y.: Reduced tilting effect of smartphone CMOS image sensor in visible light indoor positioning. Electronics **9**(10), 1635 (2020)
8. Chow, C.-W., Chen, C.-Y., Chen, S.-H.: Enhancement of signal performance in LED visible light communication using mobile phone camera. IEEE Photon. J. **7**(5), 1–7 (2015)
9. Zeng, L., et al.: High data rate multiple input multiple output (MIMO) optical wireless communications using white LED lighting. IEEE J. Sel. Areas Commun. **27**(9), 1654–1662 (2009)
10. Luo, P., et al.: Experimental demonstration of a 1024-QAM optical camera communication system. IEEE Photonics Technol. Lett. **28**(2), 139–142 (2016)
11. Lain, J., Jhan, F., Yang, Z.: Non-line-of-sight optical camera communication in a heterogeneous reflective background. IEEE Photon. J. **11**(1), 1–8 (2019)
12. Chen, H.W., et al.: Color-shift keying for optical camera communication using a rolling shutter mode. IEEE Photon. J. **11**(2), 1–8 (2019)
13. Tian, P., Huang, W., Xu, Z.: Design and experimental demonstration of a real-time 95 kbps optical camera communication system. In: Proceedings of the Sixth International Conference WCSP, pp. 23–25. Hefei, China (2014)
14. Berman, S.M., Greenhouse, D.S., Bailey, I.L., Clear, R.D., Raasch, T.W.: Human electro retinogram responses to video displays, fluorescent lighting, and other high frequency sources. Optom. Vis. Sci. **68**(8), 645–662 (1991)
15. Arthurs, E., Dym, H.: On the optimum detection of digital signals in the presence of white gaussian noise–a geometric interpretation and a study of three basic data transmission systems. IRE Trans. Commun. Syst. **10**(4), 336–372 (1962)

A Systematic Review of Augmented Reality in Multimedia Learning Outcomes in Education

Hafizul Fahri Hanafi[1]([✉]), Mohd Helmy Abd Wahab[2], Abu Zarrin Selamat[3], Abdul Halim Masnan[4], and Miftachul Huda[5]

[1] Department of Computing, Faculty of Art, Computing and Creative Industry, Universiti Pendidikan Sultan Idris, Malim, Malaysia
hafizul@fskik.upsi.edu.my
[2] Department of Electronic Engineering (Computer Engineering), Faculty of Electrical and Electronic Engineering, University Tun Hussein Onn, Parit Raja, Malaysia
helmy@uthm.edu.my
[3] Department of Moral, Civic and Character Building Studies, Universiti Pendidikan Sultan Idris, Malim, Malaysia
zarrin@fsk.upsi.edu.my
[4] Department of Early Childhood Education, Faculty of Human Development, Universiti Pendidikan Sultan Idris, Malim, Malaysia
abdul.halim@fpm.upsi.edu.my
[5] Department of Islamic Studies, Universiti Pendidikan Sultan Idris, Universiti Pendidikan Sultan Idris, Malim, Malaysia
miftachul@fsk.upsi.edu.my

Abstract. This research involved a systematic and thematic review of literature of studies involving the use of Augmented Reality (AR) in the educational context. Specifically, the review was performed by searching and identifying relevant and recent articles on several leading online databases, which were published from 2014 to 2020. The review yielded 40 articles that discuss various applications of this novel technology. Based on the analysis of their contents, 15 articles are related to multimedia learning involving augmented reality in primary classroom settings. On the other hand, 25 articles discuss studies involving the use of AR in university settings. In terms of content, 18 articles articulate the applications of AR from the explorative perspective, while another 18 articles elaborate the use of AR books in education. By contrast, only 4 articles highlight the use of game-based AR applications in education. The findings of the content analysis carried out suggest that student learning can be improved in AR-enabled learning environments with multimedia elements that provide high interactivity and immersion in which not only students can see and visualize learning objects and contents but also they can readily interact with such three-dimensional visual objects. Arguably, in exploring or navigating and interacting in such environments, students can enhance their cognitive and spatial skills as well. Overall, the above findings provide a greater insight into the understanding of AR characteristics that have a profound impact on the teaching and learning process.

Keywords: Augmented reality · Interactivity · Immersion · Multimedia elements · Learning environments · Thematic review

© Springer Nature Switzerland AG 2021
M. Singh et al. (Eds.): IHCI 2020, LNCS 12616, pp. 63–72, 2021.
https://doi.org/10.1007/978-3-030-68452-5_7

1 Introduction

Over recent years, Augmented Reality (AR) has become a popular educational research topic [1–3] that focuses on the potentials of AR in the teaching and learning process. The main factor that contributes to the wide acceptance of AR are its cost-effectiveness in creating highly immersive and enjoyable learning environments in which students can learn more efficaciously [4, 5]. Specifically, teaching and learning can be carried out in real-world environments containing embedded virtual contents to support for formal and informal learning [6, 7]. Several researchers have also pointed out that AR can significantly help improve learning in various contexts, particularly in conceptual understanding, inquiry learning, technology, and engineering [8–10]. Furthermore, the deployment of multimedia contents on internet resources during lessons enables lecturers and teachers to perform their respective tasks more flexibly and conveniently [11] to explain theoretical aspects and increases students' motivation in invigorating settings [12–14]. With the use of mobile devices, AR learning contents can be made more accessible to students transcending temporal and geographical barriers [15, 16]. To date, several researchers have begun conducting studies that focus on learning assisted by AR in many disciplines and fields [17, 18]. In this study, a systematic review of the current literature of the applications of AR in various educational contexts was carried out o highlight its advantages and disadvantages and the challenges in its implementation. The findings of this undertaking helped the researchers to determine the main research objective and formulate appropriate research questions that focused on the instructional design of AR mobile learning with the use of relevant multimedia elements.

2 Related Work

In the current literature, many educational benefits accorded by AR have been highlighted in many educational contexts [19–21] in a number of disciplines and fields [22]. In particular, many researchers have conducted several systematic reviews of the applications of AR in a diverse range of important fields, which have identified a host of educational benefits and potentials [23, 24]. The same reviews have also identified several barriers in the implementation of this technology in educational contexts [25]. In such reviews, the main focus was on the use of AR in multimedia context learning [26] and on the insight into how AR could serve as an assistive tool for students to learn more efficiently and effectively [27]. In this regard, according to Erbas et al. [28], most students perceive the use of AR in learning to be both highly useful and enjoyable, signifying the educational potential of this novel technology [29]. Interestingly, most of the studies in the current literature were carried out based on a qualitative approach that helped highlight the impact of AR on students' performance, motivation, attitudes, and interest in learning [30–32]. Given these revelations, it can be reasonably argued that the implementation of AR technology in educational settings can further enhance the teaching and learning process through interactive and immersive learning environments, benefitting both teachers and students [33, 34]. Table 1 summarizes some of the studies of the applications of AR in multimedia education in classroom settings.

Table 1. The objectives and results of AR studies in multimedia education

Authors	Type of Literature Review	Aim of the study	Result of the study
Zhang et al. [24]	Integrative review	The main objective of this research is to manipulate AR-based objects applied in a class environment	The result suggests that AR is able create innovative learning objects that can enhance students' learning performance and motivation
Radu [35]	Integrative review	This aim of this research is to analyse the positive and negative impacts of educational Augmented Reality on student learning	The overall result of this research indicates that physical tasks performed by students in AR-based collaborative learning environments can help improve their motivation, cognitive skills, and spatial abilities
Arulanand et al. [36]	Semi-systematic review	This main aim of the research is to provide a comprehensive use of Augmented Reality and multimedia application on three-dimensional printing to promote self-learning activities	This research finding signifies that the use of this novel technology can stimulate creative thinking among students as they learn to develop learning objects by themselves
Kara et al. [37]	Semi-systematic review	The main aim of this study is to articulate the values in technology in education, notably on early childhood education by comparing a smart toy and conventional learning using print media	The research finding underscores the importance of smart AR toys to attract young children's attention to learn in enjoyable learning environments in the classroom
Chen [38]	Review papers and systematic reviews	This primary aim of the study is to analyse the advantages of an AR-based scaffolding video in helping students in the learning of English language	This research finding suggests that such a novel AR learning tool is able to enhance students' motivation, satisfaction, and learning achievement, indicating a major advantage of using augmented reality for educational purposes

(*continued*)

Table 1. (*continued*)

Authors	Type of Literature Review	Aim of the study	Result of the study
Maas et al. [39]	Semi-systematic review	This research aims to highlight the positive impacts of appropriate versions of Virtual Reality (VR) and Augmented Reality (AR) on the teaching and learning of general education. Additionally, the study aims to explore the educational benefits of Mixed Reality (MR) in instructional education	This research finding indicates that practitioners should exercise extreme caution in choosing a suitable platform for the deployment of AR, VR, and MR technologies for educational purposes
Chen [40]	Review papers	This research focuses on the use of AR technology on early childhood education of young children in acquiring sound reading skills in comparison to a traditional approach that relies on textbooks	The research finding suggests that the use of AR learning tools can help young children to learn to read more efficaciously than learning based on conventional methods, as the former can make learning more attractive by allowing students to learn using interactive educational games and interactive embedded videos
Buchner et al. [41]	Semi-systematic review	This main aim of this research is implement a multimedia and instructional model using an AR book as an educational medium in learning	The research finding indicates that multimedia elements of AR books can help improve students' motivation and interest to learn
Martin et al. [42]	Integrative review	The primary aim of this study is to examine the effects of a novel game-based AR on the learning of digital circuit in a mobile learning environment	The research finding suggests that engineering students will be able to gain a firm understanding of the principles of digital circuit. The researchers also caution that teachers need to possess sufficient programming experiences to help them teach with greater efficacy

Figure 1 highlights the flow of the steps taken by the researchers in carrying out the systematic review of the current literature of studies of Augmented Reality in multimedia learning in education. This review helped reveal the positive impacts of multimedia Augmented Reality technology in helping students to achieve learning outcomes in a diverse range of educational settings [43–45]. More importantly, the review helped highlight the potential of improving learning efficacy by capitalizing on mobile devices, notably smartphones, to provide mobile learning with the use of interactive multimedia elements [46, 47]. The following section provides a detailed account of the systematic review carried out by the researchers by focussing on collecting appropriate data, which is essential in the early stages of a systematic review process [48, 49].

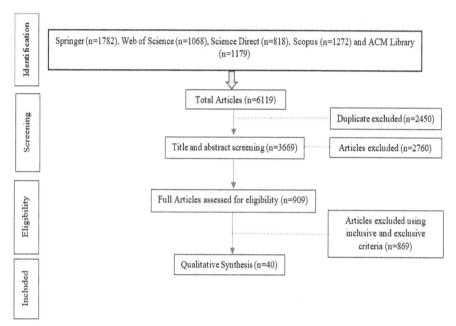

Fig. 1. The flow diagram of the systematic review of literature of studies of Augmented Reality in multimedia contexts

3   Method

3.1   The Manuscripts Selection

The selection of relevant academic articles and was made by searching several online research databases. The researchers conducted such a process on 20 August 2020 by applying appropriate keywords or strings of phrases, such as "Augmented Reality" AND Education AND Learning AND Multimedia. Such a query search helped the researchers to identify and select relevant documents that were both appropriate and current. Specifically, the online databases searched were several leading online publications, namely

Springer, ISI Web of Science, ScienceDirect, Scopus, and ACM Digital library. The selection process of articles was inter-coded through the agreement and disagreement between two different coders that helped determine the veracity of such articles based on several inclusion and exclusion criteria. Ultimately, this filtering process yielded 40 articles that were deemed relevant and current according to such criteria as summarized in Table 2.

Table 2. The inclusion and exclusion criteria in selecting relevant articles relating to Augmented Reality in the Multimedia learning context

Inclusion Criteria	Exclusion Criteria
The articles in respective databases have to be published within a period from January 2014 to January 2020.	The studies only focus on "Augmented Reality" and "Virtual Reality."
The articles have to be peer-reviewed prior to publication.	The contents of the articles do not include teaching and learning in a multimedia context.
The articles must be related to empirical research design, multimedia, and the utilization of augmented reality in education.	The area dealt with in related studies is not related to the application of Augmented Reality.

The following subsection will analyse the manuscripts based on the content analysis to deduce the textual information.

3.2 Analysis of the Manuscripts

In this study, the researchers carried out a thematic analysis of contents and contexts of selected papers or articles. Even though such an analysis is quite simple, the results it yield can be significant for research [50, 51]. In principle, the systematic categorization process [52] of a large amount of data consisting of textual information can help determine patterns [53] and trends of words, as depicted in Fig. 2. The reliability of the selected articles was deemed high based on the calculated value of Cohen's Kappa at 0.94, which is regarded extremely high [54]. In this research, the thematic analysis was implemented based on a qualitative methodology. Figure 2 shows the characteristic of the thematic and content analysis.

The analysis process

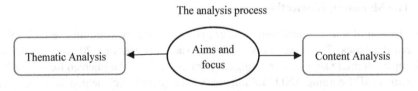

Fig. 2. The characteristic of the thematic and content analysis

The analysis process was performed by the researchers in determining a group based on the categorical analysis of sub-categories. Later, research questions were grouped based on the domains and characteristics of the criteria of the study. This content analysis is also suitable for multi-faceted criteria based on the AR classification process [55, 56]. In the following stage of the systematic review process, emerging classifications were refined to reveal the appropriate sub-categories [57]. Arguably, other researchers can manually code studies depending on the sub-categories that have been defined earlier. For this type of research, iterative discussions among researchers are essential.

4   Result

As highlighted, this study involved a systematic review of the current literature of studies of the applications of AR in education. Specifically, the review was performed by searching and identifying relevant and recent articles on several leading online databases, which were published from 2014 to 2020. The review yielded 40 articles that discuss various application of the novel technology. Based on the analysis of their contents, 15 articles are related to multimedia learning involving augmented reality in primary classroom settings. On the other hand, 25 articles discuss studies involving the use of AR in university settings. In terms of content, 18 articles articulate the applications of AR from the explorative perspective, while another 18 articles elaborate the use of AR books in education. By contrast, only 4 articles highlight the use of game-based AR in education. Interestingly, 81% of the researchers used image-based AR technology. The above findings helped the researchers to address both the first and second research questions of this study. The findings of studies discussed in the selected articles suggest that specific instructional models are needed based on appropriate learning contexts [58, 59]. For instance, Salar et al. [60] discovered that students' focus and attention were influenced by a particular AR-learning approach used in a classroom setting. In addition, Zafar et al. [61] argues that students can learn more efficaciously in a AR-enabled learning environment because they can visualize learning objects that help improve cognition.

5   Discussion

In this study, the researchers used a qualitative research methodology involving a systematic and thematic review of literature. The process helped highlight descriptive accounts of relevant papers or articles that discuss studies of the applications of AR in various learning contexts. In particular, such a review enabled the researchers to examine the characteristics of AR that could help students to learn in learning environments that are engaging, entertaining, and motivating, which ultimately can lead to enhanced learning through which students gain better understanding of learning contents.

References

1. Kesim, M., Ozarslan, Y.: Augmented reality in education: current technologies and the potential for education. Procedia - Soc. Behav. Sci. **47**, 297–302 (2012). https://doi.org/10.1016/j.sbspro.2012.06.654

2. Georgiou, Y., Kyza, E.A.: Bridging narrative and locality in mobile-based augmented reality educational activities: effects of semantic coupling on students' immersion and learning gains. Int. J. Hum. Comput. Stud. **56**, 102546 (2020)

3. Staccini, P.: Serious games, simulations, and virtual patients. In: Digital Innovations in Healthcare Education and Training. pp. 17–27. Elsevier (2020)

4. Gargrish, S., Mantri, A., Kaur, D.P.: Augmented reality-based learning environment to enhance teaching-learning experience in geometry education. Procedia Comput. Sci. **172**, 1039–1046 (2020)

5. Ibáñez, M.B., Portillo, A.U., Cabada, R.Z., Barrón, M.L.: Impact of augmented reality technology on academic achievement and motivation of students from public and private Mexican schools. A case study in a middle-school geometry course. Comput. Educ. **145**, 103734 (2020)

6. Laine, T.H., Nygren, E., Dirin, A., Suk, H.-J.: Science Spots AR: a platform for science learning games with augmented reality. Educ. Technol. Res. Dev. **64**(3), 507–531 (2016). https://doi.org/10.1007/s11423-015-9419-0

7. Zhu, L., Cao, Q., Cai, Y.: Development of augmented reality serious games with a vibrotactile feedback jacket. Virtual Real. Intell. Hardw. **2**, 454–470 (2020)

8. Dunleavy, M., Dede, C., Mitchell, R.: Affordances and limitations of immersive participatory augmented reality simulations for teaching and learning. J. Sci. Educ. Technol. **18**, 7–22 (2009)

9. Bujak, K.R., Radu, I., Catrambone, R., MacIntyre, B., Zheng, R., Golubski, G.: A psychological perspective on augmented reality in the mathematics classroom. Comput. Educ. **68**, 536–544 (2013)

10. Flores-Bascuñana, M., Diago, P.D., Villena-Taranilla, R., Yáñez, D.F.: On augmented reality for the learning of 3D-geometric contents: a preliminary exploratory study with 6-grade primary students. Educ. Sci. **10**, 4 (2020)

11. Sampaio, D., Almeida, P.: Pedagogical strategies for the integration of Augmented Reality in ICT teaching and learning processes. Procedia Comput. Sci. **100**, 894–899 (2016)

12. Midak, L.Y., Kravets, I.V., Kuzyshyn, O.V., Pahomov, J.D., Lutsyshyn, V.M.: Augmented reality technology within studying natural subjects in primary school (2020)

13. Ducasse, J.: Augmented reality for outdoor environmental education. In: Augmented Reality in Education. pp. 329–352. Springer (2020)

14. Rossing, J.P., Miller, W., Cecil, A.K., Stamper, S.E.: iLearning: The future of higher education? Student perceptions on learning with mobile tablets (2012)

15. Kiryakova, G.: The Immersive Power of Augmented Reality. In: Human-Computer Interaction. IntechOpen (2020)

16. Radosavljevic, S., Radosavljevic, V., Grgurovic, B.: The potential of implementing augmented reality into vocational higher education through mobile learning. Interact. Learn. Environ. **28**, 404–418 (2020)

17. Akçayır, M., Akçayır, G., Pektaş, H.M., Ocak, M.A.: Augmented reality in science laboratories: The effects of augmented reality on university students' laboratory skills and attitudes toward science laboratories. Comput. Human Behav. **57**, 334–342 (2016)

18. Sucihati, R., Malyan, A.B.J., Cofriyanti, E.: Augmented reality in the registration flow for enterance eximination at state polytechnic of sriwijaya based on Android. In: Journal of Physics: Conference Series, p. 12128 (2020)

19. Habig, S.: Who can benefit from augmented reality in chemistry? Sex differences in solving stereochemistry problems using augmented reality. Br. J. Educ. Technol. **51**, 629–644 (2020)

20. Fan, M., Antle, A.N., Warren, J.L.: Augmented reality for early language learning: a systematic review of augmented reality application design, instructional strategies, and evaluation outcomes. J. Educ. Comput. Res. 0735633120927489 (2020)

21. Alzahrani, N.M.: Augmented reality: a systematic review of its benefits and challenges in e-learning contexts (2020). https://doi.org/10.3390/app10165660

22. Costa, M.C., Manso, A., Patrício, J.: Design of a Mobile Augmented Reality Platform with game-based learning purposes. Information **11**, 127 (2020)
23. Pigueiras, J., Ruiz-Zafra, A., Maciel, R.: An augmented reality-based mlearning approach to enhance learning and teaching: a case of study in Guadalajara. In: International Conference in Methodologies and intelligent Systems for Techhnology Enhanced Learning. pp. 174–184. Springer (2020)
24. Zhang, Z., Li, Z., Han, M., Su, Z., Li, W., Pan, Z.: An augmented reality-based multimedia environment for experimental education. Multimedia Tools Appl. **3**, 1–16 (2020). https://doi. org/10.1007/s11042-020-09684-x
25. Nechypurenko, P.P., Stoliarenko, V.G., Starova, T.V, Selivanova, T.V, Markova, O.M., Modlo, Y.O., Shmeltser, E.O.: Development and implementation of educational resources in chemistry with elements of augmented reality. (2020)
26. Bektas, K.: Toward a pervasive gaze-contingent assistance system: attention and context-awareness in augmented reality. In: ACM Symposium on Eye Tracking Research and Applications. pp. 1–3 (2020)
27. Priestnall, G.: Augmented reality. Geogr. Educ. Digit. World Link. Theory Pract. **39** (2020)
28. Erbas, C., Demirer, V.: The effects of augmented reality on students' academic achievement and motivation in a biology course. J. Comput. Assist. Learn. **35**, 450–458 (2019)
29. Chang, S.-C., Hwang, G.-J.: Impacts of an augmented reality-based flipped learning guiding approach on students' scientific project performance and perceptions. Comput. Educ. **125**, 226–239 (2018)
30. Hanafi, H.F., Said, C.S., Wahab, M.H., Samsuddin, K.: Improving students' motivation in learning ICT course with the use of a mobile augmented reality learning environment. IOP Conf. Ser. Mater. Sci. Eng. **226**, 012114 (2017). https://doi.org/10.1088/1757-899X/226/1/012114
31. Sahin, D., Yilmaz, R.M.: The effect of Augmented Reality Technology on middle school students' achievements and attitudes towards science education. Comput. Educ. **144**, 103710 (2020)
32. Tuli, N., Mantri, A.: experience fleming's rule in electromagnetism using augmented reality: analyzing impact on students learning. Procedia Comput. Sci. **172**, 660–668 (2020)
33. Muangpoon, T., Osgouei, R.H., Escobar-Castillejos, D., Kontovounisios, C., Bello, F.: Augmented reality system for digital rectal examination training and assessment: system validation. J. Med. Internet Res. **22**, e18637 (2020)
34. Triepels, C.P.R., Smeets, C.F.A., Notten, K.J.B., Kruitwagen, R.F.P.M., Futterer, J.J., Vergeldt, T.F.M., Van Kuijk, S.M.J.: Does three-dimensional anatomy improve student understanding? Clin. Anat. **33**, 25–33 (2020)
35. Radu, I.: Augmented reality in education: a meta-review and cross-media analysis. Pers. Ubiquitous Comput. **18**(6), 1533–1543 (2014). https://doi.org/10.1007/s00779-013-0747-y
36. Arulanand, N., Babu, A.R., Rajesh, P.K.: Enriched learning experience using augmented reality framework in engineering education. Procedia Comput. Sci. **172**, 937–942 (2020)
37. Kara, N., Cagiltay, K.: Smart toys for preschool children: a design and development research. Electron. Commer. Res. Appl. **39**, 100909 (2020)
38. Chen, C.: AR videos as scaffolding to foster students' learning achievements and motivation in EFL learning. Br. J. Educ. Technol. **51**, 657–672 (2020)
39. Maas, M.J., Hughes, J.M.: Virtual, augmented and mixed reality in K–12 education: a review of the literature. Technol. Pedagog. Educ. **29**, 231–249 (2020)
40. Chen, L., Yang, X., Wang, B., Shu, Y., He, H.: Research on augmented reality system for childhood education reading. In: 2018 12th IEEE International Conference on Anti-counterfeiting, Security, and Identification (ASID), pp. 236–239. IEEE (2018)

41. Buchner, J., Jeghiazaryan, A.: Work-in-progress–the ari 2 ve model for augmented reality books. In: 2020 6th International Conference of the Immersive Learning Research Network (iLRN), pp. 287–290. IEEE (2020)
42. Martin, S., Parra, G., Cubillo, J., Quintana, B., Gil, R., Perez, C., Castro, M.: Design of an augmented reality system for immersive learning of digital electronic. In: 2020 XIV Technologies Applied to Electronics Teaching Conference (TAEE). pp. 1–6. IEEE (2020)
43. Syawaludin, A., Gunarhadi, G., Rintayati, P.: Enhancing elementary school students' abstract reasoning in science learning through augmented reality-based interactive multimedia. J. Pendidik. IPA Indones. **8**, 288–297 (2019)
44. Hanafi, H.F. bin, Said, C.S., Ariffin, A.H., Zainuddin, N.A., Samsuddin, K.: Using a collaborative Mobile Augmented Reality learning application (CoMARLA) to improve Improve Student Learning. IOP Conf. Ser. Mater. Sci. Eng. 160, 012111 (2016). https://doi.org/10.1088/1757-899X/160/1/012111
45. İbili, E., Çat, M., Resnyansky, D., Şahin, S., Billinghurst, M.: An assessment of geometry teaching supported with augmented reality teaching materials to enhance students' 3D geometry thinking skills. Int. J. Math. Educ. Sci. Technol. **51**, 224–246 (2020)
46. Morris, N.P., Lambe, J.: Multimedia interactive ebooks in laboratory bioscience education. High. Educ. Pedagog. **2**, 28–42 (2017)
47. Videnovik, M., Trajkovik, V., Kiønig, L.V., Vold, T.: Increasing quality of learning experience using augmented reality educational games. Multimedia Tools and Applications **6**, 23861–23885 (2020). https://doi.org/10.1007/s11042-020-09046-7
48. Meline, T.: Selecting studies for systematic review: Inclusion and exclusion criteria. Contemporary Issues in Communication Science and Disorders. ASHA. 33 (2006)
49. Baragash, R.S., Al-Samarraie, H., Alzahrani, A.I., Alfarraj, O.: Augmented reality in special education: a meta-analysis of single-subject design studies. Eur. J. Spec. Needs Educ. **35**, 382–397 (2020)
50. Sandelowski, M., Leeman, J.: Writing usable qualitative health research findings. Qual. Health Res. **22**, 1404–1413 (2012)
51. Vaismoradi, M., Turunen, H., Bondas, T.: Content analysis and thematic analysis: Implications for conducting a qualitative descriptive study. Nurs. Health Sci. **15**, 398–405 (2013)
52. Mayring, P.: Qualitative content analysis. A companion to Qual. Res. **1**, 159–176 (2004)
53. Morgan, D.L.: Qualitative content analysis: a guide to paths not taken. Qual. Health Res. **3**, 112–121 (1993)
54. Primer, A.P.: Quantitative methods in psychology. Psychol. Bull. **112** (1992)
55. Struck, H.N.: International E-Conference on Advances in Engineering, Technology and Management-ICETM 2020 S
56. Pfeiffer, T.: A Context-Aware Assistance Framework for Implicit Interaction with an Augmented Human
57. Donyavi, Z., Asadi, S.: Diverse training dataset generation based on a multi-objective optimization for semi-Supervised classification. Pattern Recognit. **108**, 107543 (2020)
58. Inglese, T., Korkut, S.: Modeling the instructional design of a language training for professional purposes, using augmented reality. In: Dornberger, R. (ed.) New Trends in Business Information Systems and Technology. SSDC, vol. 294, pp. 205–222. Springer, Cham (2021). https://doi.org/10.1007/978-3-030-48332-6_14
59. Kanivets, O.V, Kanivets, I.M., Kononets, N.V, Gorda, T.M., Shmeltser, E.O.: Development of mobile applications of augmented reality for projects with projection drawings (2020)
60. Salar, R., Arici, F., Caliklar, S., Yilmaz, R.M.: A model for augmented reality immersion experiences of university students studying in science education. J. Sci. Educ. Technol. **29**(2), 257–271 (2020). https://doi.org/10.1007/s10956-019-09810-x
61. Zafar, S., Zachar, J.J.: Evaluation of HoloHuman augmented reality application as a novel educational tool in dentistry. Eur. J. Dent. Educ. **24**, 259–265 (2020)

RFID Technology for UOH Health Care Center

Velugumetla Siddhi Chaithanya[1], M. Nagamani[1], Venkateswara Juturi[2]([✉]),
C. Satyanarayana Prasad[3], and B. Sitaram[4]

[1] UOH, SCIS, Gachibowli, Hyderabad, India
aumsiddhi369@gmail.com, selmt.2015@gmail.com
[2] American International group, Newyork, USA
sagarjuturi@gmail.com
[3] TCS, Gachibowli, Hyderabad, India
satyanarayanaprasad.c@gmail.com
[4] Technical Head Netpeach Technologies Pvt Ltd., Hyderabad, India
rburle@netpeach.com
http://www.uohyd.ac.in

Abstract. RFID is the emerging technology in the human achievement in terms of security and authentication. The contact less communication with intervention of digital media enables this technology for sophisticated use in every human computer interaction. The passive tag based RFID enables a cost effective, less maintenance system for information transfer and user identification. This paper is to automate the identification of users(patients) at health care centers to ensure proper health maintenance and drug distribution. It can be applied to reduce the crowd and maintain social distance by developing good interactive system at health care centers.

Keywords: Passive Tags · RFID (Radio Frequency Identification) ·
Contact less communication

1 Introduction

Physical contact less communication is demanding present pandemic period. Eliminating human direct interaction and sophisticated automatic Human Computer Interaction is possible with the Technology using Radio Frequency Identification(RFID).

1.1 Introduction to RFID

RFID(Radio Frequency Identification) is a technology is based on radio frequency waves for efficient way of communication which is functioning of an already existing systems. It contains two main parts, RFID tag and a reader device [1].

Supported by University Of Hyderabad.

In general terms, Radio Frequency Identification systems consist of an RFID tag (typically many tags) and an receiver(transceiver in passive tag technology and RFID reader in active tag technology). The receiver emits a field of electromagnetic waves from an antenna, which are identified by the tag if that waves frequency matches the tags frequency. The absorbed energy is used to power the tag's microchip and a signal that includes the tag identification number is sent back to the receiver [1].

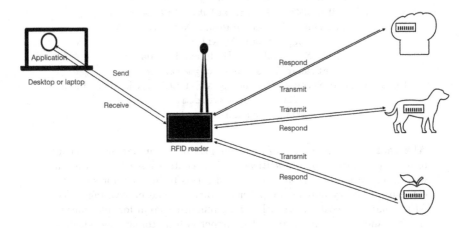

Fig. 1. Basic RFID system

1.2   RFID Tags

The RFID tag and antenna are the communication points. Tags are source to communicate data. Tags are categories two types based on the source of energy initiated.

– Active Tags:External source of energy provided
– Passive Tags: Self sufficient to energies when communication establish

Active RFID Tags: Active RFID systems contains sensors associated with batteries that provides external energy source to various access points throughout an area (like a building) and transfer data to the receiver. These systems are commonly used for real time location tracking [7]. There are three types of Active Tags. Beaconing active RFID, Transponding active RFID, Intelligent active RFID.

Passive RFID Tags: Passive RFID systems does not use any battery, it responds upon initiative taken by some other source here in this case transceiver. These transceivers sends electromagnetic waves, if the frequency of this waves matches the frequency of RFID tag for which it is programmed for, then it responds accordingly by sending back the necessary information [7].

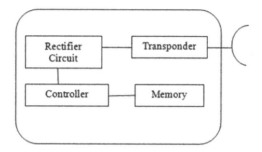

Fig. 2. Formal RFID tag chip

Transponder: It receives the radio wave from the reader and sends the feedback signal back to the reader, if frequency of the electromagnetic waves matches for which tag is programmed for [5].

Rectifier Circuit: It is used to store energy from the radio wave and store across the capacitor and then energy is used as a supply for the controller as well as the memory element inside the RFID tag [5].

Memory: Memory inside the passive tag. It stores the information or data provided, it can hold up to 2KB data [5].

Beaconing RFID: In this type of RFID, for every some particular intervals of time a message is sent back to the receiver. This message is encrypted with static key, so it is easy for the receiver to decrypt it. It operates at a maximum frequency of 900 MHz [6].

Transponding Active RFID: There is a Transponding active RFID, in which a combination of both active and passive RFID. It takes the passive RFID component, and it uses a tag that doesn't transmit the information without the intervention of receiver. The only similarity in Transponding active RFID compared to Active RFID is that, they both contain battery [6].

Intelligent Active RFID: paginationIt is a combination of both beaconing and transponding active RFID. An intelligent active RFID doesn't continuously sends information back to the user, instead it wakes up for every particular interval time and scans its location to find the nearest receiver and sends the information to that receiver [6].

2   Methodology

RFID uses electromagnetic waves particularly radio waves to keep track of a certain object [18]. For example, now a days dogs are being attached with RFID tags so that whenever a dog is lost or theft, based on the intensity of the signals(whenever we are close to it the intensity is high) that receiver is showing up we move in that direction and trace the object. The same methodology is used in this research work for automation of health center interaction between staff and user like appointment booking with available doctor, pharmacy to check availability of medicines received medicine acknowledge and its recording, laboratory etc. To identify a patient, he will be given ID numbers with a unique tag associated with his/her university authenticated id card to ease of use for many purpose like health center or as per required field. For the Health Center he/she must to use the RFID tag(ID) and then according to the appointment or medical history the patient should have to follow the next step for doctor and after the prescription provided by the doctor, the medical store must know automatically medicine name from the database, then issued medicine details should be automatically update by default. The laboratory system admin must notify the details about the report of the person visited either patient have to collect any report or not. if patient have to collect any report from the lab system must be notify about the status of the report either report is completed or awaited. This will help patient without intervention of the lab assistant he can get the report.

Proposed system follows the passive RFID tag based authentication system with support of Mobile application to check the availability of doctor through the duty roaster, and get appointment online, once the confirmation given from the doctor then only user will visit the health care. Once the user reaches the antenna receives the tag information and notify the doctor and patient so that waiting time is reduced and immediately user can consult the doctor by eliminating physical verification of user authenticity by staff.

3   Block Diagrams

3.1   Block Diagram

In the below block diagram 3 [5], the instruction that needs to be executed is sent via centralised database which contains all the information regarding patients history and does the provided instruction [5].

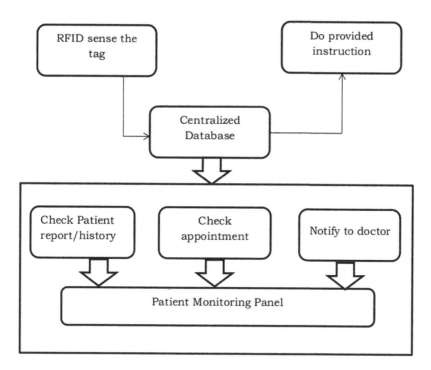

Fig. 3. Block diagram of proposed RFID based patient activity tracking in UOH health care system[5].

3.2   Activity Diagram

The below activity diagram explains the way the entire maintenance less system works. Whenever user visits the health care, the user id is been checked for the appointment, if there is an appointment then doctor checks the patient and updates medical history and laboratory details and manage database accordingly [5].

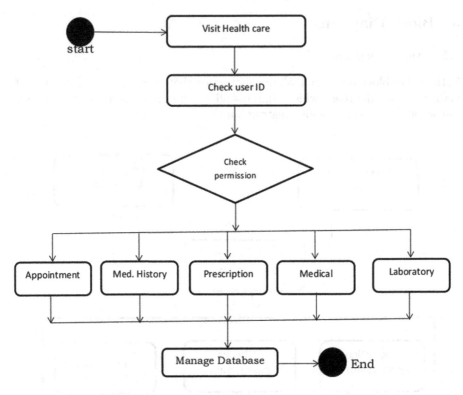

Fig. 4. Activity diagram for UoH campus healthcare centre system with RFID [5]

4   Implementation

4.1   Micro Strip Patch Antenna

The Micro strip Patch Antenna is a light, thin metallic strip placed above the ground plane, in between there is a dielectric material used as an electric insulator. It is used band microwave wireless applications [17].

The Antenna has been designed with the help of IE3D platform. To design antenna the first part that need to be done is to estimate the antenna's length(L) and width(W). To estimate the length and width the following equations were used [17],

$$W = c/(2f_0\sqrt{(\epsilon_r + 1)/2} \tag{1}$$

$$\epsilon_{eff} = (\epsilon_r + 1/2) + (\epsilon_r - 1/2)[1 + 12(h/w)]^{-1/2} \tag{2}$$

$$L_{eff} = c/2f_0\sqrt{\epsilon_{eff}} \tag{3}$$

$$\Delta_L = 0.412h * (\epsilon_{eff} + 0.3) * (w/h + 0.264)/(\epsilon_{eff} - 0.258) * (w/h + 0.8) \tag{4}$$

$$L = L_{eff} - 2\Delta_L, h < 0.05c/f_0 \tag{5}$$

Parameters are used in Above Equation are

- f_0 : Resonance Frequency.
- W : patch width.
- L : patch length.
- h : thickness.
- ϵ_r : Relative Permitivity(dielectric constant) for dielectric substrate.
- c : speed of the light which is given as $3 * 10^8$.
- Δ_L : thickness of the material.

The Characteristics that should be Considered While Choosing an Antenna

- *Antenna Radiation Patterns* describes, how RF field will be generated if you use that particular antenna. It also describes how we need to position our antenna in order to get better results. It is used to determine many properties like antenna gain [12].
- *Antenna Gain* defines how efficiently the antenna can convert the input source into radio waves(transmitting antenna) and how efficiently the antenna can convert radio waves into electrical signals(receiving antenna) [15].
- *Directivity* is the antenna gain in a particular direction commonly called as directive gain. The higher the directivity, the more concentrated is the beam radiated by an antenna[14].
- *Polarization* of antenna describes how we should place our antenna in order to get better results. This is because an horizontally polarized antenna can receive and transmit horizontally polarized waves and vertically polarized antenna can receive vertically polarized waves [13].
- The *Efficiency* of an antenna is, out of the total power radiated from the antenna(transmitting) how much power is delivered to the antenna(receiving). Efficiency of the antenna is directly proportional to the power delivered to the antenna. If an antenna delivers more power than it means that efficiency of that antenna is high and if it delivers less power than its efficiency is low [16].
- *Bandwidth* is the range of frequencies that antenna can operate correctly. Like, consider an example of radio station, we tune to specific frequency in order to go to a particular station [14].

4.2 Designing of Antenna

Consider

- f_0 = 2 GHZ
- ϵ_r = 1 (dielectric constant of air)
- c = $3 * 10^8$

Calculate W, L, H from Eqs. 1 and 5

- $L = 74\,\text{mm}$
- $W = 75\,\text{mm}$
- $H = 1\,\text{mm}$

By using the IE3D design system and considering the above values as parameters we design the antenna. IE3D is used to define and design antenna in simulation environment [19] and also a verification platform that delivers the modeling accuracy for the combined needs of high-frequency circuit design and signal integrity engineers across multiple design domains [11].

5 Results

The present work mainly focusing the antenna design for the proposed application using the RFID system communication distance used for user authentication process in context of UoH campus health care system.

5.1 Radiation Pattern of the Designed Antenna

The below picture shows the radiation pattern for the designed antenna [18,19], whenever the object equipped with RFID is near to the RFID receiver the color is thick, as it moves away from the antenna the thickness of the color goes on decreasing.

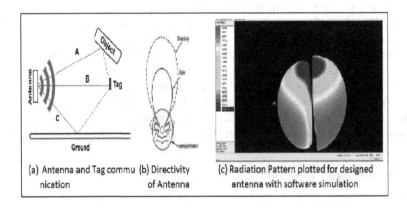

(a) Antenna and Tag commu (b) Directivity (c) Radiation Pattern plotted for designed
    nication   of Antenna   antenna with software simulation

Fig. 5. Radiation pattern of designed antenna [14]

5.2 GUI

The RFID technology based system interface to University Of Hyderabad campus Health centre and its User communication explained here. The user who wants to visit the health centre will do book their slot through the online mobile application by finding the availability of the concern doctor and his appointment so that the crowd at the health centre is reduced and possible to maintain the social distance with minimal number of the people in the Centre. The GUI shown in Figure No. 6 explains the way the patient is given appointment by for doctor for his consultancy. Whenever the users RFID tag is scanned, it is redirected to page which asks for users token number, once the token number is been entered, the status of the patient along with medical history appears, if the token is verified then he will be given the doctors appointment [5] with user authentication verification in secured manner [20].

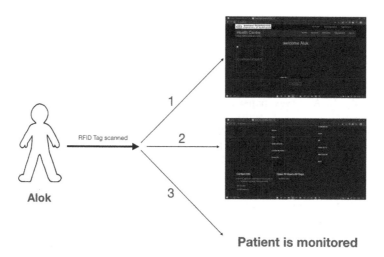

Fig. 6. How the patients are given appointment and monitored with GUI

6 Conclusion and Future Scope

Main objective of this research work is to build an automation of health care organization using RFID. It can be efficiently used by any Health care organization to reduce Man power, consume time and cost. The research work done in this paper mainly focuses on how well the hospital systems can be improved using RFID technology for eliminating physical contact in verifying the authenticity of user. Existing health care center uses human intervention for verifying the authenticity and information access through supporting staff. Human interactive work most of the time causes error scope delay and time consuming which causes difficult for students and employees for physical verification. Instead if we use technology and system which can store and monitor the patients health condition effectively than it will be easy for the doctor to assist the patient efficiently [13]. The designed antenna of RFID and tag communication performance measure to reach the expectation of proposed system implementation in real time various applications to eliminate Human to Human communication and incorporate Human to machine interaction for better quality of system functioning.

Acknowledgements. We would like to acknowledge Univesity of Hydeerabad Registrar Mr. Sardhar Singh and Capt Dr. Ravindra from health care who gave opportunity to work on health care system and supported us to understand the entire system functionality. We would also like to acknowledge the students of Mtech Artificial Intelligence and Information Technology who did the ground work by going into the health center of UOH and getting the details required for the project. We would also like to acknowledge the MCA students named Alok Kumar Sinha and Saumya Chitransi Chowdary who helped us to complete the paper successful.

References

1. Davinder Parkash, C., Preet, K.: The RFID Technology and its Applications: A Review. ISSN 2249–684X (2012)
2. Christoph, J.: A survey paper on radio frequency identification (RFID) Trends (2010)
3. Santhi, L., Lakshmi, S., Sakthivel, R.: RFID Technology and Its Applications With Reference To Academic Libraries ISSN 2278–0661
4. Moutaz, H., Anna, S.: RFID applications and adoptions in healthcare. In: HCIST Conference : International Conference on Health and Social Care Information and System Technologies (HCist) (2018)
5. Alok Kumar, S., Saumya, C., Nagamani, M.: Swatch UOH Campus through RFID Technology, Master of Technology application project report, SCIS, UoH (2020)
6. Sparkfun homepage https://learn.sparkfun.com/. Accessed 21 Sept 2020
7. RFID Journal. https://www.rfidjournal.com/. Accessed 21 Sept 2020
8. Antenna theory home page. http://www.antenna-theory.com. Accessed 20 Oct 2020
9. Electronic notes home page. http://www.electronics-notes.com. Accessed 20 Oct 2020
10. George, R., Vasilis, K.: Rfid in pervasive computing: State-of-the-art and outlook. https://www.academia.edu. Accessed 20 Oct 2020
11. https://www.rfidglobal.net. Accessed 20 Oct 2020
12. Free patents homepage. https://www.freepatentsonline.com/
13. Rebecca Angeles, Potential Consumer Responses to RFID Product Item Tagging and Emergent Privacy Issues, https://www.irma-international.org/
14. MobileMark Homepage. https://www.mobilemark.com/
15. everythingrf Homepage. https://www.everythingrf.com/
16. Grand Stream Homepage. http://www.grandstream.com/

17. Zhi Ning, C., Xianming, Q.: Antennas for RFID applications https://ieeexplore.ieee.org/Xplore/home.jsp
18. Ari, J., Christof, P.: RFID Security and Privacy. https://link.springer.com
19. Nikitin, P., Rao, K.V.S.: Antennas and propogation in UHF RFID systems. In: IEEE International Conference on RFID (frequency Identification), IEEE RFID, pp. 277–288 (2008) https://doi.org/10.1109/RFID.4519363
20. Reynolds, M., Weigand, C.: Design considerations for embedded software-defined RFID readers: Emerging Technology. pp. 14–15 (2015)

An IOT Based Smart Drain Monitoring System with Alert Messages

Samiha Sultana[1], Ananya Rahaman[1], Anita Mahmud Jhara[1], Akash Chandra Paul[1], and Jia Uddin[2(✉)]

[1] Department of Computer Science and Engineering, BRAC University, Dhaka, Bangladesh
samihasultana100@gmail.com, ananyarahman95@gmail.com,
anitajhora14@gmail.com, akash103691@gmail.com
[2] Department of Technology Studies, Woosong University, Daejeon, South Korea
jia.uddin@wsu.ac.kr

Abstract. Everyone has the right to live in a healthy environment. Flooding due to obstructed drain is a very common phenomenon in Bangladesh and many other developing countries that leads to unhygienic surroundings. Because of this, many health issues appear when the air gets poisoned due to sewage gas. One of the key causes of increasingly growing Aedes mosquitoes is the stagnant water on the road for too long. It is difficult to track manually, and the issues become apparent only when they are already clogged and the whole region is filled with water. A warning system that sends the sensed data to the managing authorities using GSM techniques and IOT is generated to prevent such incidents even before it could impact the public is proposed in this paper. MQ135 is used for sewage gas, an ultrasonic sensor for sewage distance, and a water level sensor is used to keep track of water flow. If the level reaches a certain threshold, it will send a text message using GSM to the authority to report the issues mentioning which areas should be fixed with a location using GPS, and authority and general people will also be able to keep track of real-time data via an online website which is implemented using NodeMCU. It is possible to adjust the threshold values according to the user's choice. As it ensures that the authority can get notified in time and also shows live data, the system can help people in the community live healthier lives.

Keywords: Drain water · GSM SIM808 · IoT · Alert system · MQ135 · Ultrasonic sensor · Water sensor · GPS module · Adafruit IO

1 Introduction

A poor drainage system can lead to many diseases like campylobacteriosis, typhoid, hepatitis A, gastroenteritis. Diseases like these can spread through direct touch or even through indirect contact. It can also spread if people step on the dirty sewage or drain water as they can carry the germs from one area to another. Insects, dogs, mice, cockroaches, and other living organisms also participate in spreading [1]. Solid waste and water staying in a place for too long are one of the factors for the alarming rise of dengue [2]. In Bangladesh, a total of 7,450 dengue patients were recorded in 2018 [3]. Good

© Springer Nature Switzerland AG 2021
M. Singh et al. (Eds.): IHCI 2020, LNCS 12616, pp. 84–95, 2021.
https://doi.org/10.1007/978-3-030-68452-5_9

Health and well-being is the third goal and Clean Water and Sanitation is the sixth goal of Sustainable Development Goal [4]. If drainage water is not cleaned for too long, it can harm the environment and make the area smelly and hard to live in. Due to clogged drain, even after slight rainfall, the streets get filled with water. The drain gets clogged because some people are throwing plastic bags, plastic bottles, etc. in drains [5]. So, it has become a necessity to have a way to monitor if the drain is clogged and has made the whole area filled with dirty water and if the area is filled with harmful sewage gas that can cause asthma. Research says if the concentration of sewage gas rises above 0.0005 parts per million then a very bad smell of rotten eggs gets detected which makes our daily life difficult [6]. In 1972, 90% of people were under extreme poverty lines which gradually decreased to 9% in 2018. So, now people can afford more and want to spend for a better living. A teenage girl faced death as she fell into an open manhole in Bogra of Bangladesh and many other deaths took place due to having an open manhole. A low-cost solution like keeping a manhole lid open which causes the death of many is not acceptable [8].

Therefore, we came up with a solution that keeps in mind the safety of general people that can monitor the real-time data through a website which shows conditions of the drainage system and to have an alert system to make sure if the condition gets hazardous, the authority gets to know about it and clean the manhole and drain in that area to have a better drainage system. The sewage gas level is tracked using an MQ135 sensor and if the threshold is above 400 ppm, it sends the authority message "Need to clean at latitude, longitude: CO2 Level High". For sewage distance, an ultrasonic sensor is used and if the threshold is above 20 cm, it sends the reason that "Sewage level high" with the location. For water level, when the value is above 500, it sends the reason "Water level High". An extra feature is used, if all the three components are high it simply sends the location and says "Sewage High, CO2 High, water high". The latitude and longitude can be used to find the location using Google Maps.

The objective of this paper is to provide a way for a healthier environment for everyone at a reasonably affordable cost.

The rest of the paper is organized as follows: Sect. 2 describes other existing solutions and their background and why the new solutions are more convenient. Section 3 explains the proposed model. Section 4 shows the results achieved from experiments and Sect. 5 concludes the paper and explains the limitations and future work.

2 Literature Review

This section reviews some of the existing systems that also tried to come with a solution to these problems that comes with a poor drainage system.

In a paper by M et al. (2019) the idea of IoT, GSM, and water flow sensor was used to keep track of water level but they did not use any gas sensor or any method to keep track of sewage gas level. However, even if the water level is not too high at this moment still too much sewage builds up can increase the level of harmful gases which can make the area filled up with rotten egg smell [9]. In another paper by Rjeily et al. (2017), a solution of monitoring real-time data continuously was explained but did not have a way of sending an alarm if it suddenly exceeds a certain level. For a solution like this, a person needs to always keep monitoring [7]. This research lacked in the sector of not having any automatic way to let the humans know an emergency has occurred and thus they might miss out on the opportunity to update the drainage problem before it becomes inconvenient for people to live in that area [10]. An explanation about a solution using GPS and GSM was given but did not apply IoT to monitor real-life data that we are using. In addition to that, no use of tracking sewage gas levels was seen in a paper by M et al. (2016) [11]. Authors R et al. (2019) suggested Raspberry Pi, Arduino board, ultrasonic sensor, gas sensor, temperature sensors were used. However, it makes the system complicated and costly for ordinary people even though the features provided are similar to this research [12]. The BMP 280 Ultrasonic Sensor, Ultrasonic Sensor HC-SR04, SW-420 Vibration Sensor, Arduino [17], and GSM 300 are used in the paper by Nataraja et al. (2018). These sensors are deployed in the manholes and linked directly to the microcontroller-based Arduino Uno, and if the manhole cover is ever opened, an alarm is used to alert localities and send messages to the authorities concerned [13]. A paper by SK et al. (2014) suggested IoT-based underground drainage and manhole monitoring system, model. They used LCD display to show warning messages. A buzzer warning sensor was used to alert when a manhole lid is open. Temperature detection sensors were used to detect the condition of the manhole lid but they did not display any real-time data and did not provide any solution to the problem of a situation when anyone is not near the LCD during the emergency [14]. However, in another paper by Gowthaman et al. (2018) IOT, Magnetic Float Sensor, Arduino is used to for detecting the sewage level and sending messages and mails but did not provide any solution to the dangerous gases like CO_2, methane, etc. which are produced from sewage which makes the environment harmful for people [15]. In a paper by Viswanadh et al. (2019) NodeMCU, Flow Sensor, Blynk Application (Blynk app, Blynk server, Blynk libraries) were used to monitor the underground drainage system but did not provide any real-time data [16].

After reviewing the related works, we noted that a system to get alert remotely if a sudden threshold is crossed while not in the office is also needed in addition to the presence of real-time data. Real-time data is needed to keep track even when it has not reached the threshold yet which helps in detecting the levels of the elements produced in the drain. Moreover, by real-time data, it can be predicted that which elements level has not exceeded the threshold but will exceed soon and can therefore take safety precautions according to that. The system proposed in this paper is affordable and convenient as it uses components that are low of cost and easy to get. The solution to the lacking of existing models is provided in this as it shows the real-time data of the levels of elements such as gases, water, sewage, etc. and also if any of these elements exceeds the threshold value it will alert by sending messages along with the location.

3 Proposed Model

3.1 Components Used

As we want real-time data for viewing the live situation of water level, sewage gas level, sewage level, that is why our proposed model is built using NodeMCU for showing the real-time data on the monitor using Adafruit IO. Moreover, for making more user friendly and useful for the environment and human being, GSMsim808 and GPS have been used. Arduino Mega has been used as the processing unit by connecting various sensors, GSM, GPS for showing the present state to ensure whether the sewage level crossed the threshold values. We use Arduino Mega because we need many ports for processing. Arduino Nano and Arduino Uno do not have that many ports as Arduino Mega. We can also use raspberry pi for our processing but the cost of raspberry pi is much higher than our processors (Arduino Mega and NodeMCU) to make the system more affordable.

It is possible to give public access for the website so that anyone gets to see the graph if they want and private access option is also possible if security reasons or maintenance issue arises. The feed can be shared with other Adafruit account holders and thus it is very convenient to do maintenance work. If the connection is lost somehow then the graph will not show. If we want to store the graph we have to store it manually by downloading CSV files or data that can be stored on the website up to 1 KB. To store up to 1 KB feed history should be turned on otherwise only the last value received will be stored. Besides, if the graph is not shown then we have to assume that something is wrong and the authority needs to fix this.

3.2 Circuit Details

Here, we have displayed the input section, microprocessor, and output section in the block diagram shown in Fig. 1. The input portion of our project includes the Ultrasonic sonar sensor, the MQ135, and the water level sensor. All sensors here take the analog value and sends the analog value for processing to the microprocessor. For our processing, we used two microprocessors: Arduino Mega is one and NodeMCU is another. Since we used a Water Level Sensor, MQ135, Ultrasonic Sensors, GSM, GPS, the processing requires a massive amount of memory, so we are using Arduino Mega. As we need to

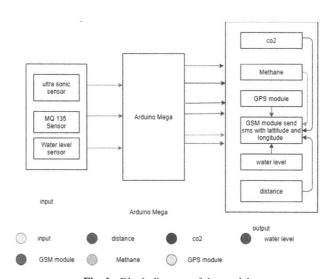

Fig. 1. Block diagram of the model

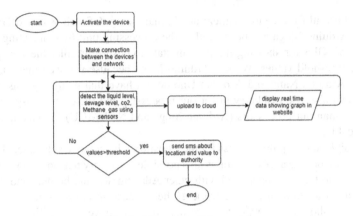

Fig. 2. Flowchart of the model

send real-time data to the server, we have used NodeMCU for IoT (Internet of Things) purposes. For MQ135, if the value reaches the threshold after receiving the value from Arduino, then a message will be sent by GSM to the authority whom we want to notify with latitude and longitude. For the case of water level sensor and sonar, the same is achieved, i.e. when the value is greater than the threshold, a text with the location will be sent. Besides, all the data will be shown in real-time on the Adafruit website so that the authority can see the draining state. To ensure transparency, the website will also be available to the local public. It can be visible from the graphs whether the value is rising or not.

Figure 2 shows the flowchart. The system starts its work after taking inputs from sensors. We have a sonar sensor for sewage distance detection, MQ135 sensor for gas detection, and a water sensor is used for detecting the level of water. Sonar is an active sensor and it has a trigger pin and echo pin. After taking input from the sensors, our system will check the water level, sewage level, hazardous gas level (CH4, CO2, CO) with sensors. If the value which is detected by the sensors exceeds the threshold value, then it sends an SMS with a GSM module to the authority with the exact location (latitude and longitude). Our system has one more feature that it uploads the real-time data to the cloud so that authority can check the situation live and the sonar value and gas sensor value is shown by graphs and the water level is shown by an alert button. The alert button will appear red if the water level crosses the threshold or otherwise it will be green.

3.3 Implementation of the System

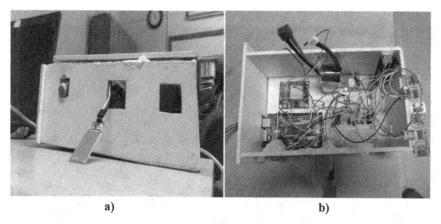

Fig. 3. a) Outside view on the left b) Inside view on the right side

For hardware implementation, a box is made to keep internal wires inside and only the input sensors and antenna of GPS outside as shown in Fig. 3. The input sensors used are MQ135, water level sensor, and Ultrasonic Sonar Sensor. Here, two sonar sensors are used for accuracy. Using one sensor the system faces some difficulty taking both readings accurately at the same time with Arduino Mega and NodeMCU.

We initially interfaced GSM SIM808 with Arduino Mega and digital pins 10, 11 as RX and TX are the pins used for the interface. The digital pin 9 for the trig pin and digital pin 12 for the echo pin are then used for interfacing with Arduino and sonar. Analog pin 1 of the Arduino Mega is used for the water sensor and analog pin 0 is used for the MQ135. For Arduino, we give a baud rate of 9600. With sensors and NodeMCU, we have different interfaces. For MQ135 we used the analog pin in NodeMCU. NodeMCU has just 1 analog pin. Then for the sonar sensor, D2 is used as a trig pin, and D1 is used as an echo pin. For the water level sensors, we used D3 for interfacing with NodeMCU. For NodeMCU we also used Arduino IDE for coding but here we used the baud rate 115200. We first need an account in Adafruit for NodeMCU and also need external Wi-Fi access to access the Adafruit website.

4 Experimental Analysis

4.1 Graphs

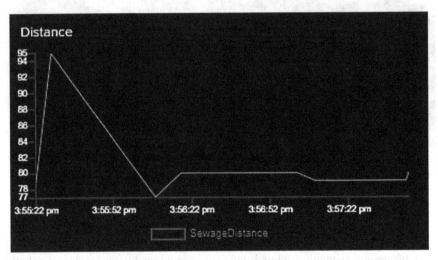

Fig. 4. Graph of sewage distance

Here, the graph in Fig. 4 shows the sewage distance of the drainage system on LIVE after every 30 s. We experimentally used our hands to create obstacles. We tested 7 or 8 times and got real-time data update on our server.

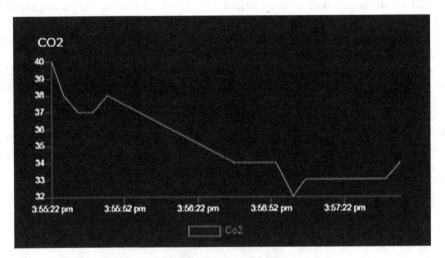

Fig. 5. Graph of MQ135

Here, the graph in Fig. 5 is showing the CO2 level of the drainage system on LIVE after every 30 s. We experimentally used aerosol gas to increase the level of CO2 so that our sensor can detect a higher level of CO_2 and can send an update to our server. We tested 7 or 8 times and got some real-time data update on our server. The curve of the graph is the CO_2 level of the environment (Fig. 6).

Fig. 6. Water level sensor alert system (Color figure online)

Here, Green Alert is used to show the water level notification for Low-Level water. On the other hand, Red Alert is used to show the High-Level notification of water. When the water is less than 500 then the button is green and if the level crosses 500 then the button is red.

The sensors were tested for a longer period for observation as well. The graphs for both MQ135 and Ultrasonic sonar sensors are shown in Fig. 7.

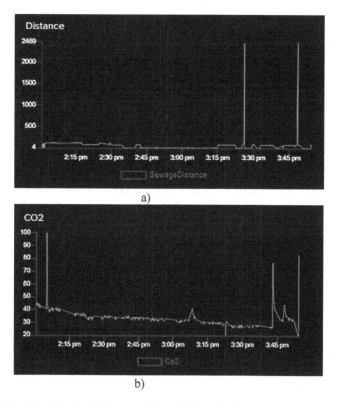

Fig. 7. a) Graph of ultrasound above b) Graph of mq135 values below

4.2 Data Table

Here, we are using a water sensor. If the water level distance less than 500 m, then our GSM will never send any SMS or update. However, for change in water level, it will send a notification to the server that "Water Empty"/"Water Low"/"Water Medium"/"Water High". If the value is more than 500 then it will notify that the water level is high and will send an SMS to the authority declaring the fact that the threshold is crossed (Table 1).

Table 1. Dataset for water level

Distance	Notification	SMS
0 < value <= 100	Water Empty	No
100 < value <= 300	Water Low	No
300 < value <= 500	Water Medium	No
Value > 500	Water High	Yes

If the level of CO2 changes, then GSM will notify the local server about the state of CO2, i.e. if low, medium, or high. If the value is less than or equal to 400 then the CO2 level is normal, GSM will not send any SMS. But if the level is more than 400, then GSM will send SMS to the authority (Table 2).

Table 2. Dataset for Hazardous Gas

Level	Notification	SMS
0 < value <= 100	Gas level low	No
100 < value <= 200	Gas level low	No
200 < value <= 400	Gas level medium	No
Value > 400	Gas level high	Yes

If the sewage level is zero, GSM will not send any SMS and it will show the zero level to our server. But if the sewage level is more than 0 and less than 20 then GSM will send the SMS to the authority (Table 3).

Moreover, if the sewage level, CO2 level, and water level distance are high at the same time then GSM will send a notification to the local server "Need to clean everything" and the SMS that "Sewage High, CO2 High, water high" will be sent.

The SMS includes the latitude and longitude and the specified problem as shown in Fig. 8. Using the latitude and longitude, the exact location can be found by using Google Maps.

Table 3. Dataset for sewage level

Sewage level	Notification	SMS
Value > 30	Empty	No
20 < value <= 30	Close Distance	No
0 < value <= 20	Need to clean sewage	Yes

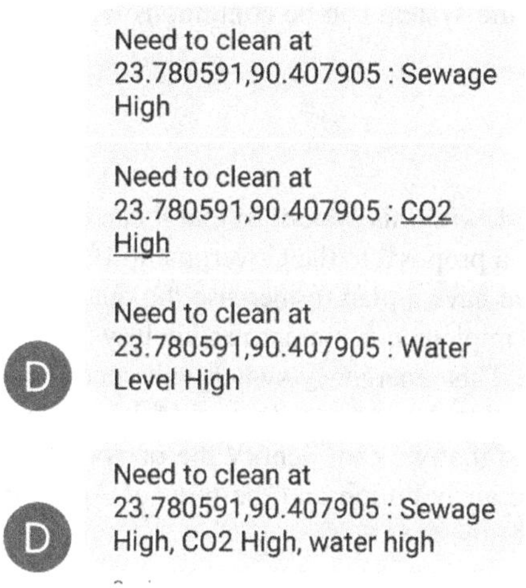

Need to clean at
23.780591,90.407905 : Sewage
High

Need to clean at
23.780591,90.407905 : CO2
High

Need to clean at
23.780591,90.407905 : Water
Level High

Need to clean at
23.780591,90.407905 : Sewage
High, CO2 High, water high

Fig. 8. Sample SMS

5   Conclusion

The poor drainage system is unhygienic and alarming for our environment. It carries a vast amount of harmful gases like CO_2, CO, CH_4, NH_3, etc. We have designed the device to control the overflow of drain water. In an overpopulated country like Bangladesh, it is a common scenario where drains are filled and blocked with plastics. As a result, the vast amount of sewage and harmful gas is increasing day by day. But authority does not get any alert about the blockage. Sometimes the dirty water comes out from the drain and blocks the entire road with this drain water. People face hustle to walk or move for these situations. That is why we have designed this device to relieve human suffering. We used GPS to get the actual location of the drain. Moreover, we used here Wi-Fi-based NodeMCU so that the nearest authority can get the daily update of the water level, sewage gas, and sewage distance of the drain. Before the overflow of the drain water, the authority can take action to clean the drain. The local people also have read-only access to the website which increases credibility and answerability and thus it can be a path to solve corruption problems as well. We have implemented this system so that only the authority can get control of the system. Because of security purposes, we are not giving full control access to the local people. The local people can only have read access to the website according to our plan.

5.1 Limitation

Although we designed the proposed system for our educational purpose and tried to design this on a limited budget, this system works accurately but it has some limitations. Firstly, we have used a water level sensor which we have to replace every 3 years because the sensor can get damaged due to rust water and harsh environment. Secondly, a biometric switch to press for local people to contact authority does not exist in our system. Another limitation is that the server used can store only up to 1 KB data. Lastly, a LiPo battery was used in our model which is rechargeable but a power system will be more convenient so that the system can be continuous without frequent replacing and charging.

5.2 Future Work Plan

In the future, we will work with this system so that it can overcome its limitations. We have a plan of presenting a proposal to the Government. If we can get a handsome fund from the Government, we have a plan to increase the quality of power source, server, water sensor quality, and implement a biometric switch system that will be used to press for emergency purposes. This emergency switch will send a message to the authority if our device has any functional problems. To avoid false messages or updates, we will use a biometric switch so that we can identify the person. This proposed system has the potential to solve the air pollution and the birth of Aedes mosquitoes and can be extended to have more features.

References

1. How a Working Drainage System Saves You. https://benjaminfranklinplumbingfortworth.com/plumbing-blog/working-drainage-system-saves-diseases/. Accessed 21 Jan 2020
2. The Sun Daily: Poor drainage contributing factor to rise in dengue cases in Kelantan (2014). https://www.thesundaily.my/archive/1143446-MRARCH267960. Accessed 2 Feb 2020
3. Kamruzzaman, M.: Dengue reaches 'epidemic proportions' in Bangladesh. https://www.aa.com.tr/en/asia-pacific/dengue-reaches-epidemic-proportions-in-bangladesh/1540832. Accessed 19 Jan 2020.
4. #Envision2030: 17 goals to transform the world for persons with disabilities. https://www.un.org/development/desa/disabilities/envision2030.html. Accessed 12 Jan 2020
5. Mishu: Drainage Problem in Dhaka City and Needs More Awareness Program (2008). https://bd-halchall.blogspot.com/2008/01/drainage-problem-in-dhaka-city-and.html. Accessed 2 Mar 2020
6. Health Problems Caused By Sewer Gas. https://www.burtonplumbingco.com/health-problems-caused-by-sewer-gas/. Accessed 13 Jan 2020
7. Rahman, A., Rahman, T., Ghani, N.H., Hossain, S., Uddin, J.: IoT based patient monitoring system using ECG sensor. In: International Conference on Robotics, Electrical and Signal Processing Techniques, Dhaka, Bangladesh, pp. 378–382 (2019)
8. The Daily Star: One more death in manhole (2017). https://www.thedailystar.net/backpage/one-more-death-manhole-1374478. Accessed 16 Jan 2020
9. Gunasekaran, M., et al.: IoT-enabled underground drainage monitoring ssystem using water flow sensor. Int. Res. J. Eng. Technol. 6(3), 2427–2430 (2019)

10. Rjeily, Y., et al.: Smart system for urban sewage: feedback on the use of smart sensors. In: 2017 Sensors Networks Smart and Emerging Technologies. IEEE, Beirut (2017)
11. Bhuvanyadevi, M., et al.: Uunderground gutter drainage monitoring using GSM. IJARMATE, 174–181 (2016)
12. Chandraprabha, R., et al.: Smart real time manhole monitoring system. Int. Res. J. Eng. Technol. **6**(7), 934–938 (2019)
13. Nataraja, N., et al.: Secure Manhole Monitoring System Employing Sensors and GSM Techniques. 3rd IEEE International Conference on Recent Trends in Electronics, Information & Communication Technology. IEEE, Bangalore (2018).
14. Muragesh, S.K., Rao, S.: Automated Internet of Things for underground drainage and manhole monitoring system for metropolitan cities. Int. J. Inf. Comput. Technol. **4**(12), 1211–1220 (2014)
15. Gowthaman, G., et al.: Sewage level maintenance using IoT. Int. J. Mech. Eng. Technol. **9**(2), 389–397 (2018)
16. Viswanadh, K., et al.: Under ground drainage monitoring system using IoT. JETIR **6**(4), 21–26 (2019)
17. Muntasir Rahman, A.M., Hossain, M.R., Mehdi, M.Q., Alam Nirob, E., Uddin, J.: An automated zebra crossing using Arduino-UNO. In: International Conference on Computer, Communication, Chemical, Material and Electronic Engineering, Rajshahi, pp. 1–4 (2018)

Design and Implementation of a Safety Device for Emergency Situations Using Arduino and GSM Module

Samiha Sultana[1], Ananya Rahaman[1], Akash Chandra Paul[1], Evea Zerin Khan[1], and Jia Uddin[2(✉)]

[1] Department of Computer Science and Engineering, BRAC University, Dhaka, Bangladesh
samihasultana100@gmail.com, ananyarahman95@gmail.com,
akash103691@gmail.com, evea.zerin@gmail.com
[2] Department of Technology Studies, Woosong University, Daejeon, South Korea
jia.uddin@wsu.ac.kr

Abstract. Everyone is entitled to live the way they want and to feel happy, but there is a lot of verbal abuse and aggression that people face which is a huge challenge to everyone. It is a general urge to tell our close ones about our current position in emergencies and warn them about the fact that we expect danger. In the event of sudden abduction or sudden illness in an uncertain place, we can feel powerless. A way is therefore built to save lives from possible dangerous circumstances in which a person is at risk of being abducted, robbed, or raped. In the model, two voice inputs are detected-"Danger" and "Help" and also a switch button input is taken. As the voice recognition model is trained that way, the device can only recognize the user's voice. It sends the text to the police and parents for the "Danger" word input or push-button input. It also sends the location to the parents for the "Help" word. On the other hand, when the pulse rate is not within a certain range, the hospital and relatives are informed. The trained words, mobile numbers, BPM range used can be modified according to the user's preference. The model is designed to send SMS containing latitude and longitude which can be used to find the exact position on Google Maps. Using our model, individuals can make their lives free of fear as it will serve as a supporting hand in crisis situations.

Keywords: Safety · Emergencies · Arduino uno · Voice Recognition Module VR3 · Push-button sensor · GPS module · GSM Sim808 module

1 Introduction

In this modern world, everybody has the right to feel safe and to roam anywhere they want to go, but people still feel caged because of fear of multiple types of violent crime. In Bangladesh, 94% of women have been exposed to sexual assault while traveling on public transport. Often, the mode of harassment is physical and often verbal, but the incidence of harassment is troubling [1]. Speakers at a discussion that took place at the National Museum of Bangladesh said that the risky environment for women at

M. Singh et al. (Eds.): IHCI 2020, LNCS 12616, pp. 96–107, 2021.
https://doi.org/10.1007/978-3-030-68452-5_10

work and other institutions is a major obstacle to Bangladesh's ambition to become a middle-income economy [2]. In 2020, between January to September 975 rape cases were reported which is very alarming [19]. Women's protection has become a global concern, it is not just limited to Bangladesh. India recorded an average of 87 rapes per day in 2019 and violence against women last year rose 7.3% over 2018, as said by the National Crime Records Bureau [3]. A woman traveling with her three children from Lahore to Gujranwala at night on September 9 of 2020, was robbed and raped by three men at gunpoint after her car ran out of fuel on Pakistan's Lahore-Sialkot Motorway [4]. Fear of rape is greater than fear of other crimes if we compare women's fear of facing different forms of violence. Women fear rape because they know there is a possibility of getting raped and that the consequences appear to be serious [5]. Even children are not safe from becoming victims. A girl who was just 7 years old was found raped in Feni of Bangladesh in 2019 [6]. In addition to rape, children are at high risk of being abducted. Children under the age of 13 years old have more fear-related emotions and are more concerned about their safety than teenagers have [7]. In the capital of Bangladesh, Dhaka alone, figures from 2018 indicate that 82 people have been kidnapped and many more all over Bangladesh [8]. Men are not exempted from being the victims of murder cases either. A blogger called Avijit Roy was killed and there are more victims like these [9]. In certain cases, because of the lack of clues, bodies are discovered very late. A man missing for the last seven days has been discovered dead in the Khilgaon area of Dhaka. The body was found attached to a pillar underwater, with a cable. After a spectator first noticed pieces of his body surfacing from the water, police retrieved the body [10]. In another scenario, an old woman named Geneva Chambers was found dead at her home after about three years of her death [13]. Elderly people are at high risk of dying from a heart attack if symptoms are not detected early. A research study says the cost of heart disease care is projected to rise from 69% in 2012 to 80% in 2030 for people who are above the age of 65 [12]. A typical heart rate may range from 70 to 100 beats per minute (BPM). Higher or lower heart rates may indicate abnormality [11].

As crime rises at an unprecedented rate, the impetus for this research has emerged. When going out, women do not feel comfortable. The parents are concerned about the child playing, fearing that the child is going to be lost or abducted. Family members are unsure of how many days the man of the house is gone and whether he will ever return. The elderly with heart problems die every day, but if it is regularly monitored, they might get saved at the right time. This analysis aims to come up with a plausible solution. The proposed model uses a voice module to record the voice and it can detect only the voice it was trained. So, other people talking will not be a big issue and will not start the device. The "Danger" word input sends the message to the parents and police while the word input "Help" only sends location feedback to the parents. The system used a push-button system for critical situations where the victim finds talking more harmful or impossible. Lastly, the pulse sensor input is used to keep track of the health and potential danger of heart attack. The mobile number, pulse range, word inputs can be modified according to the preference of the customer using the device. The latitude and longitude sent as text can be used to find the exact location using Google Maps.

Therefore, the purpose of this paper is to suggest an approach that will provide a proactive response to emergencies through the use of the few affordable sensors and modules available.

The rest of the paper is organized as follows: Sect. 22 describes other existing solutions and their background and why the new solutions are more convenient. Section 3 explains the proposed model. Section 4 shows the results achieved from experiments and Sect. 5 concludes the paper and explains the limitations and future work.

2   Literature Review

This section addresses relevant previous work on the design of women's safety equipment and some initiatives that have introduced similar SOS equipment.

The concept of making protection just a press button away was used in a paper by Edward et al. (2018). The paper focuses on the use of a women's protection mechanism based on GSM, in times of danger or untoward circumstances, using location sending systems. Predefined mobile numbers are sent to an SMS containing the latitude and longitude coordinates via the GPRS, informing them of the danger and the victim's location. The drawback of this project is that a location accuracy up to a range of 100 m to 300 m can be detected by GPRS, so there is a potential for improvement in the access to the accurate location [14]. In another paper by Bansode et al. (2015), the concept of being able to use voice to solve daily problems was used. However, it was again incomplete as it concentrated on a certain category of individuals i.e. for blind people only and just one sensor gave the features available [15]. A solution for child protection by using face recognition for bus services was explained in the paper by Sanam and Sawant (2016), but that is too expensive and turns out not to be practical for everyday use but is certainly a very good paper and well thought of child security [16]. Authors Sharmila et al. (2020) suggested a system that had two different transmitter and receiver units. The transmitter module consists of copper thread that can be stitched together in female clothing that is not visible to the perpetrator from the outside. Between these threads, there will be a connection flow that is continuously controlled by the comparator. The GSM module makes an emergency call when the link flow is disconnected by breaking the copper string and sends a warning message along with the victim's current location to the predefined numbers, and the RF transmitter sends the signal to the transmitter module's receiver. When the receiver receives a signal from the transmitter, the buzzer is switched on. The person can tear the thread to send warning messages in emergency cases. The idea seems to be fresh and modern, but it can only be used once. Once the copper wire is broken, it is important to repair the device again [17]. A research paper by Srinivasan et al. (2020) suggested a heart rate analyzer that is attached to the fingertip of the user which counts the pulse every 30 s and the signal is provided to the microcontroller. The proposed system includes a pulse sensor and this system is tested with different data of heart rate. This system allows medical people to keep a track of the health status of the patient through GSM. However, the system has three levels: Normal, low, and high. They send a message for all the little changes which is very redundant [18].

After analyzing the related works, we observed that a GSM along with a GPRS system is not sufficient for identifying the accurate location of a person. So we emphasize the use of a GPS module for accurate identification of a person's location. Our proposed system uses a voice recognition module that can be used to reach out for help and alternatively push-button can also be used for an emergency text. Moreover, heart rate irregularities can be detected from a person's fingertip and can automatically send an SMS to the required authority to help the person on time.

3 Proposed Model

3.1 Components Used

We will start by listing the hardware components that we have used and will show how they are used in our system. We used an Arduino Uno microcontroller [20, 21], Voice Recognition module, GSM module, GPS module, switch, buck converter, pulse sensor, and battery. Arduino Uno has been used as the processing unit by connecting various sensors, GSM, and GPS modules. We have used Arduino Uno as we are using many sensors and the ports available are sufficient for our desired outcome.

The Arduino board is directly connected to the battery via Vin pin. The battery source is connected to a buck converter which converts 9 V to 5 V. The buck module is used to power GSM, GPS, voice recognition module, and other components in our project. The voice module, switch, and pulse sensors are used to take inputs from humans in our system. If any of the input values cross a certain threshold value, the coordinates input from GPS are taken and the system sends SMS using the GSM module. These components are described below, and the datasets are later shown in the experimental analysis section.

We used the VR3 Voice Recognition module developed by the Elec house. It can store up to 80 voice commands. The analog voice input is taken using the microphone which is sensitive to only one user which means only one person whose voice was recorded can be detected. Anyone speaking the same words cannot be accepted by this module which puts this module at a higher advantage than other modules. On the other hand, 7 voice commands are effectively detected by this module at the same time.

The switch is used as an added safety function that initiates the whole system. It is an active-low device.

We have used the pulse sensor manufactured by Pulsesensor.com. It is used to measure the heartbeats per minute (BPM) of a person.

GSM SIM808 is used to send SMS to parents and police and GPS is used to collect the latitude and longitude of the victim to send in the text.

3.2 Circuit Details

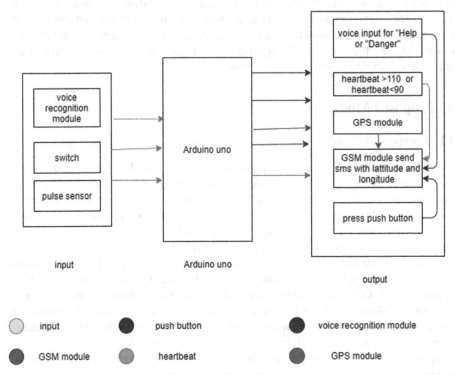

Fig. 1. The block diagram

Here, in the block diagram shown in Fig. 1, we have shown the input section, micro-processor, and output section. In our project, the input section contains a Push-button, Voice recognition Module, Pulse sensor. Here, the voice module and pulse sensor take an analog value and send that analog value to the microprocessor for processing. Arduino Uno was used because we have used five components and the processing needs a huge memory as well as many ports.

For the voice recognition module, after getting the value from Arduino, if the command is "Help", then a message will be sent by GSM to the parents only. On the other hand, if the command is "Danger", then a message will be sent by GSM to the parents and police, whom we want to inform the latitude and longitude. The same is done for the case of a pulse sensor, i.e. if the value is greater or less than the two specified threshold values, then an SMS will be sent with the location. If the heartbeat is abnormal for example if the heartbeat is greater than 110 or less than 90 BPM, then our device will send an SMS to the relatives of that person as well as hospitals.

Figure 2 shows the flowchart. Our project starts its work after taking inputs from sensors.

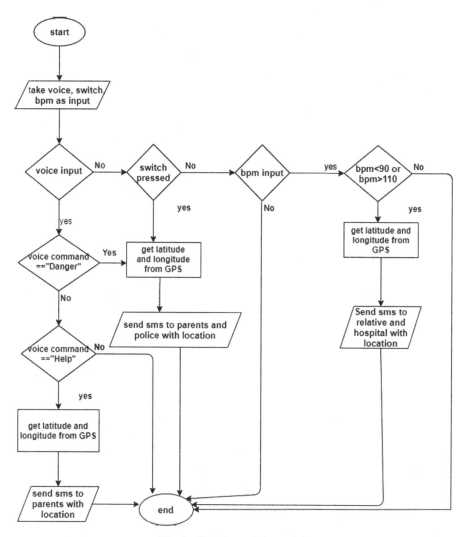

Fig. 2. Flowchart of the model

As we want to ensure safety in any possible situations, we have considered some scenarios as given below:

Case 1: For women, children, or old people our model is an ideal project. If they face any kind of kidnapping situation, the person can use his/her voice saying commands like "Help" or "Danger" which will be detected by the microphone. If the command is "Help", the voice recognition module sends the command to Arduino as input. Then the Arduino sends the command to GSM. GPS supplies the current location and sends the location to the GSM module. At last, the GSM sends SMS to the parents or guardians containing the current location where he/she is staying. If the victim gives the command "Danger", the GSM module will send SMS to both the parents or guardians and the police station.

Case 2: We are also using a push-button switch for the emergency case. If the victim is unable to give a voice command then he can use a switch, which will send the command to Arduino as input. The Arduino will trigger GPS for the current location. Then the GPS will send a command to the GSM module and at last, GSM will send SMS to the parents and police.

Case 3: We are also using another sensor named pulse sensor. This sensor can read the BPM (beats per minute) of the person in contact. This device will be attached to a hand or finger and it can continuously read the BPM of the person. If the BPM crosses the normal range of heartbeat (90 to 110), the Arduino will send a command to take immediate steps and it will send SMS to the relatives and hospital with location to take care of the person. Heart rate can vary from person to person; if a person does physical exercise or uses stairs then the heart rate rises. It also varies due to the different ages of the users. The threshold value can be customized according to the user's normal heart rate range.

3.3 Hardware Implementation

For hardware implementation, we made a box to keep internal wires inside and only the input sensors and antenna of GPS outside. The input sensors used are voice module, switch, and pulse sensor. The voice module is kept inside but the microphone is outside the box to take the voice input and also to make the system more high-level for general users. We kept the switch outside because people need to press the button in case of an emergency. If we keep the switch inside the box then no one can press the button and it will not be useful. The pulse sensor is also placed outside to take the pulse input. Figure 3 shows the full hardware implementation of our project.

Fig. 3. Hardware Implementation

4   Experimental Analysis

Table 1. Dataset for voice command

Voice Command	Send SMS	To whom
Yes "Help" not "Danger"	Yes	Parents
Not "Help" yes "Danger"	Yes	Parents and Police
Yes "Help" yes "Danger"	Yes	Parents and Police
Not "Help" not "Danger"	No	None

Here, we used the voice recognition module VR3 to collect our voice command dataset.

Case 1: If the person using the device says "Help", then the message is sent to parents that she/he is at that particular location so that they can get alert. This command is used when someone is suspected of being followed, but not under life-threatening risk (Table 1).

Case 2: However, if the situation portrays a life-threatening risk that police involvement is needed then the voice input of the "Danger" word is needed. An example of such a situation is when someone is facing severe physical or verbal assault. The "danger" word input will send a message to both parents and the police. In our system, the voice command of only a particular person is accepted as trained. Therefore, random noises will not create a problem.

Case 3: If a person says both "Help" and "Danger", it will send a message to both parents and police.

Case 4: If there is no required input, i.e. the person has neither said "Help" nor "Danger", then no text is sent.

Table 2. Dataset for switch command

Switch Command	Send SMS	To whom
Press switch	Yes	Parents and Police
Not press switch	No	None

In situations where voice command is not possible, but the person needs urgent help can use the switch button (Table 2).

Case 1: If the switch button is pressed then SMS is sent to both parents and police. For example, if a taxi takes the wrong turn and even after asking him many times the car does not stop or if he/she is unable to speak due to any impairment, then using the switch command is more appropriate.

Case 2: If there is no button pressing then no SMS is sent.

Table 3. Dataset for abnormal heart rate situation

Normal Heart Rate (in BPM)	Heartbeat Found (in BPM)	Send SMS and Location
90–110	88	Yes
90–110	85	Yes
90–110	89	Yes
90–110	100	No
90–110	105	No
90–110	110	No
90–110	120	Yes
90–110	115	Yes
90–110	90	No

In our model, the pulse sensor is used to collect a dataset of heart rate. We have considered that 90–110 BPM as normal heart rate. Thus, when the heart rate is above 110 BPM, a text is sent as that indicates that the person is either sick or being chased by someone. On the other hand, a heartbeat under 90 BPM also represents sickness. The threshold values are chosen for experiment and can be changed (Table 3).

Case 1: If the heart rate is between 90–110 BPM, no SMS is sent.

Case 2: If the heart rate is above 110 BPM, an SMS is sent to relatives and the person (doctor) who needs the information.

Case 3: If the heart rate is below 90 BPM, an SMS is sent to relatives and the person (doctor) who needs the information.

The latitude and longitude can be used to find the exact location using Google Maps as shown in Fig. 4.

Fig. 4. Text message and Google Map location

5 Conclusion

Since the world is transitioning to a livelier lifestyle, having a safe system for people has become an issue of primary importance. For instance, in this paper, we have discussed the methodology of safety devices for women, children, men, and elderly people as well. This device can also help in keeping track of the regular heartbeat of an individual. This device is using a voice recognition module, push-button, and a pulse sensor. The arrangements of this device have been designed to send an emergency message through a voice command to the concerned authority or parents of the person. Moreover, those who cannot speak due to disability or an untoward situation can use the push button sensor. Besides this, GPS will provide the latitude and longitude to detect the exact location of the person and GSM will send the message containing the location. In summary, this device has the potential to ensure the safety of human lives from the increasing rate of crimes and health issues.

5.1 Limitation

The device is effective for security purposes but in the device, there are some technical limitations. Firstly, we planned to use Arduino Nano to make the device small in size. But we found there were fewer ports and the ports were not enough to connect the four modules. If we used an Arduino Nano, the device would be smaller in size and every component will stay in a shorter space. For this, there will be a lower possibility of components getting disconnected from the ports. Secondly, we used the LiPo battery which is rechargeable, long-lasting, and suitable for research purposes but a lightweight compact size rechargeable battery is more suitable for regular use. Lastly, the device cannot identify who is pressing the switch as a normal push-button was used.

5.2 Future Work Plan

In the near future, we plan to improvise the design of our system. A more sensitive voice recognition module could result in a much more consistent analog signal, which is a challenge even after using the latest voice recognition module. We intend to use more modified versions of the module for voice recognition. In the limitations, we have mentioned the size constraint of the device, hence, we would like to make the device smaller in size by using a more compact sized rechargeable battery. We also have a plan of presenting our proposal to the Government to make our proposed model marketable. If we can get funding from the Government, we will be able to improvise the system by adding a biometric switch. IoT can be also implemented to keep a track of heart rate where graphs can be shared with doctors. Our project can be further extended to make it much easier to wear and carry outside, which would be suitable for its sole purpose to provide a better sense of security for everyone.

References

1. Kabir, R.: Study: 94% women victims of sexual harassment in public transport (2018). https://www.dhakatribune.com/bangladesh/crime/2018/03/07/study-94-women-vic tims-sexual-harassment-public-transport
2. The Daily Star: Safe workplace for women must (2019). https://www.thedailystar.net/city/news/safe-workplace-women-must-1733869
3. Sharma, S.: At least 87 rapes per day in India: report (2020). https://asiatimes.com/2020/09/india-rape-outrage-highlights-womens-plight/
4. Ahmed, R.: In Pakistan, women stage nationwide protest in response to shocking rape incident (2020). https://globalvoices.org/2020/09/23/in-pakistan-women-stage-nationwide-protest-in-response-to-shocking-rape-incident/
5. O'Donovan, A., et al.: Antecedents to women's fear of rape. Behaviour Change. **24**(3), 135–145 (2007)
6. The Daily Star: Man arrested for raping a 7-yr-old girl (2019). https://www.thedailystar.net/frontpage/news/man-arrested-raping-7-yr-old-girl-1745563
7. Wilson, B., et al.: Children's and parents' fright reactions to kidnapping stories in the news. Commun. Monographs. **72**(1), 46–70 (2005)
8. Bangladesh Police. https://www.police.gov.bd/en/crime_statistic/year/2018
9. Hussain, M.: Bangladesh criticized for slow progress in blogger murders (2017). https://www.voanews.com/east-asia-pacific/bangladesh-criticized-slow-progress-blogger-murders
10. The Daily Star: Man missing for 7 days found dead (2019). https://www.thedailystar.net/city/news/man-missing-7-days-found-dead-1817161
11. Alsulami, M. et al.: The use of smart watches to monitor heart rates in elderly people: a complementary approach. In: 2016 IEEE/ACS 13th International Conference of Computer Systems and Applications (AICCSA). IEEE, Agadir (2016)
12. Heidenreich, P., et al.: Forecasting the impact of heart failure in the United States. Circul. Heart Failure. **6**(3), 606–619 (2013)

13. Morel, L.: Reclusive woman may have been dead in Largo home for three years (2013). https://bit.ly/3en6mpb

14. Edward, S., et al.: GSM based women's safety device. Int. J. Pure Appl. Math. **119**(15), 915–920 (2018)

15. Bansode, M., et al.: Voice recognition and voice navigation for blind using GPS. IJIREEICE. **3**(4), 91–94 (2015)

16. Sanam, A., Sawant, S.: Safety system for school children transportation. In: 2016 International Conference on Inventive Corporation Technologies (ICICT). IEEE, Coimbatore (2016)

17. Sharmila, R., et al.: Women safety thread. Int. J. Eng. Tech. Res. **9**(5), 167–170 (2020)

18. Srinivasan, P., et al.: Heart beat sensor using fingertip through Arduino. J. Crit. Rev. **7**(7), 1058–1060 (2020)

19. Ellis-Petersen, H.: Bangladesh approves death penalty for rape after protests (2020). https://www.theguardian.com/world/2020/oct/12/bangladesh-approves-death-penalty-for-after-pro tests

20. Alam, A. I., Rahman, M., Afroz, S., Alam, M., Uddin, J., Alam, M. A.: IoT enabled smart bicycle safety system. In: Joint 7th Internation Conference on Informatics, Electronics & Vision and 2nd Int. Conference on Imaging, Vision & Pattern Recognition, Japan, pp. 374–378 (2018)

21. Rahman, F., Ritun, I. J., Ahmed Biplob, M. R., Farhin, N., Uddin, J.: Automated aeroponics system for indoor farming using Arduino. In: Joint 7th Int. Conference on Informatics, Electronics & Vision and 2nd Inernational. Conference on Imaging, Vision & Pattern Recognition, Japan, pp. 374–378 (2018)

Design and Implementation of an IoT System for Predicting Aqua Fisheries Using Arduino and KNN

Md. Monirul Islam[1], Jia Uddin[2(✉)], Mohammod Abul Kashem[1], Fazly Rabbi[3], and Md. Waliul Hasnat[1]

[1] Department of Computer Science and Engineering, Dhaka University of Science and Technology, Gazipur-1700, Dhaka, Bangladesh
monir.duet.cse@gmail.com, drkashem11@duet.ac.bd,
waliulbd@yahoo.com
[2] Department of Technology Studies, Woosong University, Daejeon, South Korea
jia.uddin@wsu.ac.kr
[3] Department of Statistics, Jahangirnagar University, Savar Dhaka-1342, Bangladesh
fazlyrabbi77@gmail.com

Abstract. This paper presents an Internet of Things (IoT) system using K Nearest Neighbors Machine Learning Model for selection fish species by analyzing a fish data set. For storing real time data from used sensors, we used a cloud server. We make a dynamic website for giving information of various fish species living in an aquatic environment. This website is connected with cloud server; anyone can easily watch it on a web application. Therefore, they can easily decide what should follow the next step, which kinds of fish are surviving in the water. For constructing the proposed IoT system, we utilized 5 sensors including mq7, ph, temperature, ultrasonic and turbidity. These sensors are connected with an Arduino Uno. The real time data of water environment using sensor is obtained in the cloud server as a csv format file. In this study, we have utilized a server of thingspeak. The end user of fish farming can monitor easily remotely using the proposed IoT system.

Keywords: Internet of things (IoT) · Prediction based aqua fishing system · Water sensing network (WSN)

1 Introduction

Fish is vital to nutritious count calories in numerous regions over the world. Fish and angle items are recognized not as it were as a few of the most beneficial foods on the planet but too as a few of the slightest impactful on the normal environment. For these reasons, they are imperative for national, territorial and worldwide nourishment security and sustenance procedures and have a huge portion to play in changing nourishment frameworks and disposing of starvation and malnutrition. Fish have given approximately 3.3 billion individuals with nearly 20% of their normal per capita admissions of creature protein. In 2017, fish accounted for almost 17% of add up to creature protein and 7% of all proteins expended globally [1]. World aquaculture production of cultivated oceanic creatures has been ove whelmed by Asia with an 89% share within the final two decades or so. An important part of IoT is used in so many agricultural equipment's, sensors and devices to monitor the smart cultivation system [2]. Therefore, we can help a fish farmer with the help of IoT as modern technology and solve all kinds of fish related problems and which water is suitable for fishes. IoT based prediction model for fisheries can be implemented at minimum cost hereby maximizing user satisfaction. Thus the main aim of IoT based prediction model for fisheries is its cost-effectiveness, real data perception than the Bio floc technique [3, 4].

In our study, a machine learning model is approached for analyzing dataset to survive fish species in an aquatic environment and a prediction based model for end user has been implemented occupying the Internet of Things (IoT). This system can directly send data collected from used sensors to IoT cloud storage. The whole work is done by connecting sensors with Arduino Uno and Ethernet shield with the Internet. Data have been processed in Arduino Uno which uses processing software Weka [5], and thingspeak cloud storage [6] is used for showing graphically form and getting real time data. An automatic notification will be sent to the website and authorized user in case of abnormalities found in the water's environment. This system is hazard free and cost-effective for remote users as they do not need to visit the ponds regularly. Farmers also can gain fish related information by visiting this website.

The remainder of the paper is sorted out as follows. Part 2 shows the literature review. In Sect. 3, proposed architecture is stated. The flowchart and cloud server is depicted in part 4. Finally, we conclude the paper in part 5.

2 Literature Review

Research in aquaculture is an input to extend stabilized production. In the last decade, various scientists have made sustained efforts that resulted in the development of recent production technologies that have revolutionized farm production. A system is shown using IoT devices for aqua culture in [7, 17]. A. Ramya [8] proposed an IoT system for smart fish farming. W. Sung [9] developed a system for monitoring fish aquaculture based on WSN. D. Prangchumpol [10] developed a model of mobile application for fish feeding. T. Joseph, et al. [11] dictates a fish farming system using IoT devices. Authors [12] utilized an IoT system for smart fish farming. An IoT based model is proposed for aquaculture in authors [13]. Table 1 shows the details related tasks of some recent years.

But all these authors do not make any website for perception data using machine learning algorithms to remote monitoring. The main aim of the system is too remote monitoring of the fish farming system by using the various sensors to reduce the risks. By using these sensors all the work is automated and it will even be easy to watch the fish farming remotely from other locations. The proposed system is successfully developed and tested to examine the effectiveness and reliability of the system. This is a low-cost smart system that can be used for water environment monitoring so that users can take precautions at the proper time for fishes.

Table 1. Related task

Paper	Used sensor	Accuracy	Dynamic website
MM Islam [7]	Temperature, Turbidity, CO, pH, water level	No prediction	No
A. Ramya [8]	pH, Ultrasonic sensor	No prediction	No
W. Sung [9]	Temperature, pH, Dissolved sensor	No prediction	No
D. Prangchumpol [10]	pH sensor	No prediction	No
T. Joseph et al. [11]	DO, pH, Turbidity and Temperature sensor	No prediction	No
Authors [12]	Temperature, pH and Oxygen sensor	No prediction	No
Authors [13]	Temperature, water level, turbidity, pH sensor	95.31%	No
Proposed Model	Temperature, Turbidity, CO, pH, Ultrasonic sensor	100%	Yes

3   System Architecture of a Prediction Model of Aqua Fisheries Using IoT Devices

The architecture of the proposed system is divided into two parts. One is hardware setup and second is machine learning model for analyzing dataset of aquatic environment.

3.1   Hardware architecture

Figure 1 shows the block diagram of hardware architecture which mainly consists of three parts. They are - the Water Sensing Network (WSN), IoT cloud, and GUI. The components used in the water Sensing network are PH Sensor, Temperature sensor, Turbidity sensor, Ultrasonic sensor, MQ7, Ethernet shield and Arduino Uno [18].

Water Sensing Network (WSN)
Water Sensing Network is set for assembling physiological various sensors and Arduino microcontrollers. In our system, various sensors have been used to collect data from the water environment over long hours. The water environment data gathered from sensors are transmitted to the IoT cloud and website. Our embedded system is shown in Fig. 2.

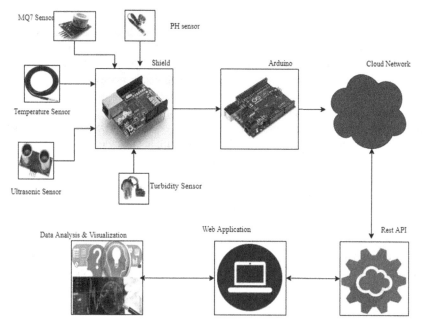

Fig. 1. Proposed Architecture of the IoT system for aqua fisheries

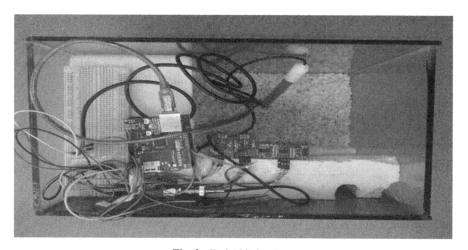

Fig. 2. Embedded system

PH Sensor

A PH meter may be a logical instrument that measures the hydrogen-ion movement in water-based solutions, indicating its causticity or alkalinity communicated as PH. We use PH sensor, model PH sensor module probe BNC for Arduino (03SEN14). In our project, we use this sensor for test the soil acidity and water quality.

Water Temperature sensor

In this system, we use a water temperature sensor whose model is DS18B20. It is capable to measure the temperature from 55 °C To 125 °C with accuracy of ±5.

Turbidity sensor

The Arduino turbidity sensor recognizes water quality by measuring level of turbidity. It is used in this system to find out the suspended particles in water.

Ultrasonic Sensor

In our examination, we have utilized an Ultrasonic Sensor to decide the separation of an obstacle from the sensor. The fundamental head of ultrasonic separation estimation depends on ECHO.

MQ7 Sensor

The MQ7 is an easy to-utilize Carbon Monoxide (CO) sensor reasonable for detecting CO fixations noticeable all around. It can distinguish CO-gas focuses somewhere in the range of 20 to 2000 ppm.

Ethernet Shield

The Ethernet Shield is microcontroller board based on the ATmega328 which is used in this system for allowing Arduino board to interface with the internet.

Arduino Uno

Arduino Uno is a microcontroller board on ATmega328P (datasheet). It is used in this system for reading data signals from various sensors.

IoT Cloud

With the aid of the IoT cloud in the Aqua fisheries monitoring system, we can store data, modify data and all the fisheries data are saved here. It can moreover show the graphical representation of this system. We use ThingSpeak cloud server in our study [6].

Graphical User Interface (GUI)

In our work, we make a dynamic website. By using this website, people can easily know about their pond's water component condition. They can monitor their pond, fish from any corner of the world. Figure 3 shows the dynamic website sample.

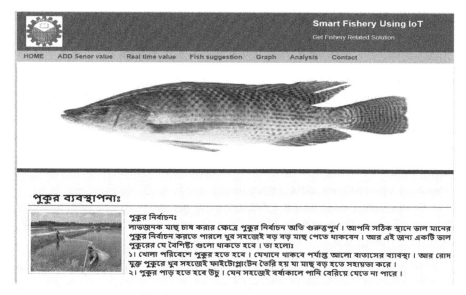

Fig. 3. Home page of dynamic website.

3.2 Machine Learning Model Analysis

Figure 4 shows the block diagram for analyzing dataset using machine learning model for fish survival prediction in an aquatic environment. There are 4 sections in the block diagram. First part is importing. In importing section, we upload the dataset.

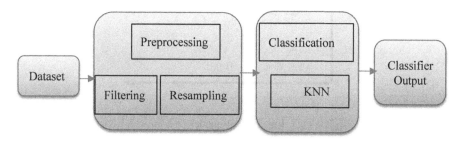

Fig. 4. Block diagram of machine learning model

Data Description

The data used in this study involving the factor of fish farming data extracted from the University of Dhaka, Faculty of fisheries, Dhaka, Bangladesh. There are 88 instances of 6 attributes. The dataset is divided into 2 groups. One is water environment factors and another is fish species.

Water Environment Factors: We applied ph, depth, CO, temperature and turbidity as water environment characteristics in our study.

Fish species: In our dataset, we used complete 8 fish species as the target variable. The fish species in our dataset are katla, prawn, rui, koi, sing, pangas, tilapia, silver carp.

Preprocessing

In the preprocessing part, we separated our dataset utilizing a resampling alternative for watching the current connection of in-positions and characteristics of the dataset. In the quality choice window, we can check missing, interesting and unmistakable estimation of each ascribes.

Classification

In classification section, we utilized KNN model.

K- Nearest Neighbor (KNN) is a distance-based non-parametric algorithm. K-NN algorithm calculates the K number of neighboring data point distances, and finds the best K for the dataset. KNN model is frequently utilized for Regression additionally concerning Classification yet generally it's utilized for the Classification issues [14, 15]. KNN is perceived as apathetic learner calculation since it doesn't gain from the preparation set promptly rather it stores the dataset and at the hour of order, it plays out an activity on the dataset.

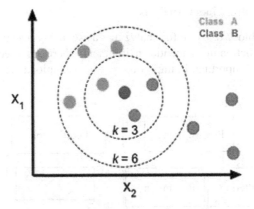

Fig. 5. KNN model

Suppose your input data is located at the star sign in the Fig. 5.

- If you set the value of K to 3 (three neighbors) then you see that your two neighbors are Class B and one neighbor is Class A. Since most of the neighbors are class B, we will assume that our data is class B in the case of K's value 3.
- If you set the value of K to 6 (six neighbors) then you see that two of your neighbors are of class B and four of your neighbors are of class A. Since most of the neighbors are class A, we will assume that our data is class A in the case of K's value.

Diverse separation capacities are utilized to quantify the separations; a famous decision is a Euclidean separation given by Eq. 1.

$$d(x, \acute{x}) = \sqrt{(x_1 - \acute{x}_1)^2 + \cdots + (x_n - \acute{x}_n)^2} \tag{1}$$

More subtleties, given a positive whole number K, a groundbreaking perception x, and a comparability metric d, the KNN calculation plays out the accompanying advances by Eq. 2.

- It goes through the whole dataset figuring d among x and each preparation perception. We'll call the K focuses in the preparation information that are nearest to x the set A. Note that K is ordinarily odd to stop tie circumstances.
- It at that point assesses the restrictive likelihood for each class, that is, the portion of focuses in A with that given class mark. (Note I(x) is that the marker work which assesses to 1 when the contention x is valid and 0 in any case)

$$P(y = j | X = x) = \frac{1}{K} \tag{2}$$

Classifier Section

In classifier section, we can analyze result performance of our model and other state-of-art models. Figure 6 dictates the performance results of the proposed KNN model. We can see that KNN model gives accuracy 100% and kappa statistic 1 and mean absolute error 0.0201. The detailed accuracy by class of each fish species is presented in the Fig. 6.

```
Correctly Classified Instances      88           100    %
Incorrectly Classified Instances     0             0    %
Kappa statistic                      1
Mean absolute error                  0.0201
Root mean squared error              0.0303
Relative absolute error              9.1778 %
Root relative squared error          9.173  %
Total Number of Instances           88

=== Detailed Accuracy By Class ===

           TP Rate  FP Rate  Precision  Recall  F-Measure  MCC    ROC Area  PRC Area  Class
           1.000    0.000    1.000      1.000   1.000      1.000  1.000     1.000     katla
           1.000    0.000    1.000      1.000   1.000      1.000  1.000     1.000     prawn
           1.000    0.000    1.000      1.000   1.000      1.000  1.000     1.000     rui
           1.000    0.000    1.000      1.000   1.000      1.000  1.000     1.000     koi
           1.000    0.000    1.000      1.000   1.000      1.000  1.000     1.000     sing
           1.000    0.000    1.000      1.000   1.000      1.000  1.000     1.000     pangas
           1.000    0.000    1.000      1.000   1.000      1.000  1.000     1.000     telepia
           1.000    0.000    1.000      1.000   1.000      1.000  1.000     1.000     silverCup
Weighted Avg.  1.000  0.000  1.000      1.000   1.000      1.000  1.000     1.000
```

Fig. 6. KNN model analysis

4 Flowchart and Cloud Server of Proposed IoT System

Figure 7 states the flow chart of our system.

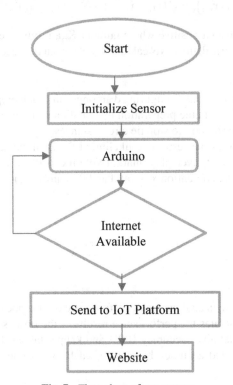

Fig. 7. Flow chart of our system

First of all, the proposed system can run whenever an Internet connection is available. If the system is not connected with the Internet, then the system will remain 'off'. Otherwise, when the system is connected with the Internet, the system will run and start to collect data from sensors which are joined with microcontroller by Ethernet shield. After that, these data will be saved automatically in the IoT cloud database. This IoT server is connected with a website by Rest-API [16]. Finally, those data are analyzed and classified by our proposed algorithm.

We create a channel for data parsing in the cloud server namely ThingSpeak cloud server. It is an Internet of Things (IoT) platform that lets us analyze and visualize the data in MATLAB without buying a license from Mathworks. IT allows us to collect and store sensor data in the cloud and develop IoT applications. It works with Arduino, Particle Photon and Electron, ESP8266 Wifi Module, BeagleBone Black, Raspberry Pi, Mobile and web apps, Twitter, Twilio, and MATLAB to end the sensor data to ThingSpeak. The ThingSpeak is mostly focused on sensor logging, location tracking, triggers and alerts, and analysis. The export things data with Rest-API is shown Fig. 8.

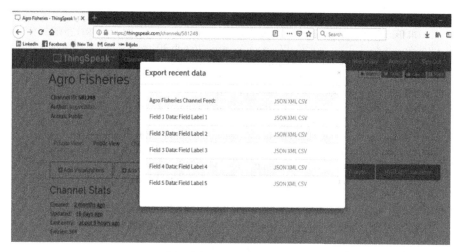

Fig. 8. Exported CSV data

In Fig. 8, we name our channel as Agro fisheries. End user of the proposed system can download JSON, XML or CSV formatted file.

5 Conclusion

We completed this research to establish an IoT system for selection fish species in an aquatic environment using Arduino and KNN machine learning model. The proposed KNN model is included for prediction fish species using analyzing aquatic data set. The paper talked about physical estimates, for example, temperature, water level, PH worth, CO and turbidity utilizing the A/D signal handling, by means of Ethernet shield move to the IoT cloud network. The information messages are logically prepared, shipped off the worker data set and showed on a cloud station. The framework additionally has an observing capacity. Framework control is introduced in fish ranch terminal gear to permit chairmen to screen the fish ranch status. We also make a dynamic website for end user. It is linked with cloud server. So farmers can easily monitor the system and take the necessary steps to rear up fishes. Also, it will lessen the time of the farmers to do the manual testing of each figure as they are getting it very effectively through this system. Our system is predictable, reliable, cost-efficient and time saving for the Aqua farmers.

References

1. FAO, The State of World Fisheries and Aquaculture 2020. https://www.fao.org/3/ca9231en/CA9231EN.pdf, last accessed 2020/1/21
2. Atzori, L., Iera, A., Morabito, G.: Understanding the Internet of Things: Definition, potentials, and societal role of a fast evolving paradigm. Ad Hoc Netw. **56**, 122–140 (2017)
3. Crab, R., Defoirdt, T., Bossier, P., Verstraete, W.: Biofloc technology in aquaculture: beneficial effects and future challenges. Aquaculture, {\bf 356–357)(1) (2012)

4. Sontakke, R., Haridas, H.: Economic viability of biofloc based system for the nursery rearing of Milkfish (Chanos chanos). Int. J. Current Microbiol. Appl. Sci. {\bf 7}(08) (2018)
5. https://www.cs.waikato.ac.nz/~ml/weka. Accessed 26 Feb 2020
6. https://thingspeak.com. Accessed 1 Mar 2020
7. Islam, M.M., Kashem, M.A., Jui, F.: Aqua fishing monitoring system using IoT devices. Int. J. Innov. Sci. Eng. Technol. 6(11), 109–114 (2019)
8. Ramya, A., Rohini, R., Ravi, S.: IoT based smart monitoring system for fish farming. Int. J. Eng. Adv. Technol. {\bf 8}(6S) (2019)
9. Sung, W., Chen, J., Wang, H.: Remote fish aquaculture monitoring system based on wireless transmission technology. In: International Conference on Information Science, Electronics and Electrical Engineering, Sapporo, pp. 540–544 (2014)
10. Prangchumpol, D.: A model of mobile application for automatic fish feeder aquariums system. Int. J. Model. Optim. 8, 277–280 (2018)
11. Joseph, T., et al.: Aquaculture monitoring and feedback system. In: IEEE International Symposium on Smart Electronic Systems (iSES), Rourkela, India, pp. 326–330 (2019)
12. Ullah, I., Kim, D.: An optimization scheme for water pump control in smart fish farm with efficient energy consumption. Processes 6, 65 (2018)
13. Preetham, K., Mallikarjun, B.C., Umesha, K., Mahesh, F.M., Neethan, S.: Aquaculture monitoring and control system: an IoT based approach. Int. J. Adv. Res. Ideas Innov. Technol. 1167–1170 (2019)
14. Zhang, W., Chen, X., Liu, Y., Xi, Q.: A distributed storage and computation k-Nearest neighbor algorithm based cloud-edge computing for cyber-physical-Social systems. IEEE Access 8, 50118–50130 (2020)
15. Wang, L., Zhou, W., Wang, H., Parmar, M., Han, X.: A novel density peaks clustering halo node assignment method based on k-nearest Neighbor Theory. IEEE Access 7, 174380–174390 (2019)
16. Neumann, A., Laranjeiro, N., Bernardino, J.: An Analysis of Public REST Web Service APIs. IEEE Trans. Serv. Comput. 1 (2018)
17. Rahman, F., Ritun, I.J., Ahmed Biplob, M.R., Farhin, N., Uddin, J.: Automated aeroponics system for indoor farming using arduino. In: Joint 7th International Conference on Informatics, Electronics & Vision and 2nd International Conference on Imaging, Vision & Pattern Recognition, Kitakyushu, Japan, pp. 137–141 (2018)
18. Muntasir Rahman, A.M., Hossain, M.R., Mehdi, M.Q., Alam Nirob, E., Uddin, J.: An automated zebra crossing using Arduino-UNO. In: International Conference on Computer, Communication, Chemical, Material and Electronic Engineering, Rajshahi, pp. 1–4 (2018)

A Fast and Secure Uniform Handover Authentication Scheme for 5G HetNets

Alican Ozhelvaci and Maodess Ma[✉]

School of Electrical and Electronic Engineering, Nanyang Technological University,
Singapore 639798, Singapore
emdma@ntu.edu.sg

Abstract. As the deployment of 5G has gained momentum, challenging cases to become more apparent such as handover authentication and the delay caused by frequent handovers. Therefore, to achieve handovers in heterogeneous networks which includes the 5G and other access networks is a challenging task that requires full access authentication procedures for different handover scenarios as the third Generation Partnership Project (3GPP) has suggested. So, this requirement not only increases the number of exchanged messages but also the system complexity in the network. Most importantly, the existing handover authentication mechanisms for different mobile networks are not suitable for handover cases in the 5G mobile network due to the different architecture of the 5G would make those schemes vulnerable to the various attacks. Therefore, in this paper, we propose a handover authentication scheme that is secure and efficient for every handover scenario in 5G Heterogeneous Networks. Compared with existed handover schemes and the current 3GPP standard, the proposed scheme can achieve seamless handover authentication with a simple process and robust security features against various attacks. The result of formal verification by using AVISPA and performance analysis shows that the scheme has robust security and efficiency for interconnectivity in the 5G mobile networks.

Keywords: Handover authentication · Key agreement · Security · Mobility · 5G

1 Introduction

The fifth-generation mobile network, the 5G, has been started deploying worldwide since 2019 and it is expected to support more than 1.7 billion subscribers worldwide by 2025 according to GSMA [1]. In order to support a massive number of subscribers and supply the promised network speed, bandwidth, coverage, efficiency with low latency and lower energy consumption, the 5G will encompass multiple technologies that consist of network virtualization function (NFV), software-defined networking (SDN), cloud computing and more according to an emerging consensus [2]. Therefore, in the 5G, a new radio access technology which is called 5G NR has been proposed [3] and embraces enhancement to LTE and Evolved Packet Core (EPC) to support mobile services not only over 5G NR but also other legacy networks and non-3GPP access networks such

© Springer Nature Switzerland AG 2021
M. Singh et al. (Eds.): IHCI 2020, LNCS 12616, pp. 119–131, 2021.
https://doi.org/10.1007/978-3-030-68452-5_12

as WLAN and WiMAX. And there are two types of non-3GPP access networks which are classified as trusted and untrusted networks [4].

The requirements have been specified to provide secure and seamless handover continuity for the user equipment (UE) when it moves between different access networks and to achieve seamless handovers between the 5G and non-3GPP access networks a few approaches have been proposed by the 3GPP organization [5]. But the problem is that a full authentication procedure has to be implemented between UE and the target network before handover can be happened for the UE to access a new network. Therefore, this requirement introduces a much longer handover delay because of the number of messages are exchanged with access and mobility management function (AMF) or proxy AMF when the UE moves from one network to another. Moreover, for each different handover scenarios need different handover authentication procedures according to the 3GPP specification, for example, if the UE moves from a non-3GPP access network to the 5G network or from 5G network to a non-3GPP access network, the handover procedures differ from one another depending on which network it comes from. The variety of handover authentication procedures would bring more computational complexity to the 5G system. Consequently, the next-generation network is considered to be much more heterogeneous and due to the densified small cell deployment, the user will face more frequent handoffs, thus, the handover authentication should be fast and efficient enough to ensure low latency.

In such a heterogeneous environment, a uniform handover mechanism is a crucial point for interconnectivity in heterogeneous networks to grand access to the UE which is registered and genuine. So, this will enable to use of the best distinctive features of available networks in the system. Moreover, while ensuring the interconnectivity for the UE which moves from one network to another, security measures have to be provided to protect against various attacks as well as providing data integrity, confidentiality with mutual authentication. It is inevitable that a uniform handover authentication scheme that is fast and secure is needed for the 5G heterogeneous networks.

2   Related Work

Designing a uniform handover authentication mechanism to allow the UE to have access in heterogeneous networks is a tough task. That's why there are only a few numbers of authentication schemes have been proposed for efficient and secure handover authentication for the 5G HetNets. These authentication schemes have been proposed to secure different scenarios and networks such as re-authentication [6], physical layer [7], LTE networks [8–10]. Han et al. [6] propose a scheme to achieve secure authentication and key exchange on 5G networks by utilizing extensible authentication protocol-authentication and key agreement (EAP-AKA) protocol. The scheme demonstrates a collation to implement a standardized security architectural model for 5G networks. Mobile Edge Computing (MEC) servers are implemented to the architecture but introducing an extra module between the core network and base station can add an additional delay in the case of roaming scenarios. Ma et al. [7] propose an authentication scheme for the physical layer in SDN-enabled 5G HetNets. The scheme utilizes the nonparametric Kolmogorov-Smirnov (K-S) hypothesis test for handover authentication. The scheme is more efficient in terms of computational and storage cost when compared with General Likelihood Ratio Test (GLRT) methods, however, in real-time interconnectivity scenarios the scheme can introduce new vulnerabilities in the system.

Haddad et al. [8] introduces a scheme that Home Subscriber Server is not involved during handover authentication to reduce the latency during handover. However, the scheme has three phases in order to utilize the proposed method. The first phase is authentication of base stations to the home network, the second phase is authentication and providing keys of the UE, finally, the last phase is the handover mechanism. Although the scheme is secure thanks to its three-phase authentication mechanism, it may cause an additional delay in 5G HetNets due to its small cell deployment. Cao et al. [9] designed a handover procedure between HeNBs and eNBs in the LTE networks to provide secure and seamless handover. Schnorr proxy signature provides authentication to the nodes without involving the HSS in the scheme. The scheme guarantees security for perfect forward secrecy (PFS) and master key forward secrecy (MKFS) with user anonymity. However, the connection between HeNBs and MMEs is through the internet which is an insecure environment that can cause new vulnerabilities to the LTE networks. In order to solve this problem, Cao et al. [10] propose new group-based handover authentication within E-UTRAN in LTE-A networks by utilizing a signing delegation technique, therefore, the authentication of the nodes has been secured in the system. Also, this group authentication is only in the same network which may not be suitable to the heterogeneous environment of 5G.

3 Motivation

It can be understood that most recent studies do not emphases on the heterogeneous environment of 5G, therefore, there is a need for a uniform handover authentication mechanism in 5G HetNets. The other reason is that the development of 5G has not been finished or finalized, therefore, there would be new use cases and application scenarios that can introduce new challenges and security issues in the system. Hence, we propose a fast and secure uniform handover scheme for heterogeneous networks (FaSUHA) using identity-based encryption (IBE). With this protocol, the user equipment can move from one network to another securely and with minimum delay. When the UE moves to a new base station or access point, authentication and the key agreement will be done between the UE and base station by using their long-term keys generated by Access and Mobility Management Functions (AMF) in the 5G core network to drive the shared session key with 3-handshake without any other entities involved.

Our major contribution in the paper can be summarized as follows. First, we propose a FaSUHA scheme that is lightweight, robust against various attacks while preserving privacy and efficiency to have a simple authentication process. The second, the proposed scheme is designed to support and ensure the security architecture in 5G Hetnets specified in technical standard release 15 of the 3GPP [11]. So that, the non-3GPP access network system is also supported for the handover authentication. Third, the proposed scheme has been formally verified by Automated Validation of Internet Security Protocols and Applications (AVISPA) in writing its language HLPSL to show that it is safe against various attacks.

The rest of the paper is organized as follows. Section II explains the system model of the 5G HetNets and attack model. Section III presents the proposed scheme with a thorough description. The security evaluation and formal verification are shown in Section IV and followed by performance evaluation in Section V. Finally, a conclusion is drowned in Section VI.

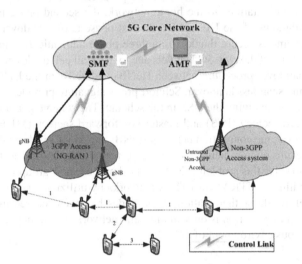

Fig. 1. A simplified model of 5G Heterogeneous Networks

4   System Model

4.1   System Model

First, we will introduce the system model of 5G HetNets as shown in Fig. 1. The UE can access the core network through the new generation 3GPP access network or non-3GPP access network in 5G HetNets [12]. In the system, basically, there are 4 types of entities that why we call it as a simplified system model of the 5G HetNets and these are the user equipment (UE), session management function (SMF), access and mobility management function (AMF), and the next generation node B (gNB). The detailed explanation of each entity is as follows.

The User Equipment: In the 5G system the devices that connect to the networks are called as the user equipment. The aim of the system is to authenticate and allow access to the user equipment when handover phenomena happen. It is assumed that the user equipment is within the coverage of the next-generation nodes which are connected to the 5G core system.

Access and Mobility Management Function: Because the Mobility Management Entity (MME) has been separated into AMF and SMF in the 5G core network, the AMF module has become responsible for the connection, registration, and mobility management and forwarding all related messages to the SMF module. Also, due to the heterogeneous environment of the 5G system, the UE will face frequent handovers, therefore, the AMF module is responsible to ensure seamless and efficient handovers between new gNBs and the UE.

gNB: In the 5G system the connection points for the UE are called as the new generation base station. It a bridge between the UE and the access system infrastructure and core network. It coordinates UEs in time, space, and frequency to provide resources of the network efficiently and reasonably. Moreover, it is responsible for the handover when the UE moves from one gNB to another in our proposed scheme.

4.2 Security and Threat Model

In our proposed protocol in order to keep the generalization simple enough the non-3GPP access system will be named as Foreign Network (FN) and the 3GPP access system will be named as Home Network (HN). Assume that the UE has registered to its Home Network (HN) and there is a signed roaming (inter-operator) agreement between the HN and the FN to have honesty and mutual authentication. Also, thanks to the signed roaming agreement, thus, both networks can authenticate each other and shared the public keys which can be used to establish a secure channel between the HN and the FN, and the UE knows the public keys of both networks.

Due to the nature of 5G HetNets, during the handover procedure, the UE's privacy can be exposed to many security threats such as eavesdropping, user privacy infringement, replay attack, and MitM attacks. Additionally, the UE can be resource-constraint that makes the UE vulnerable against various attacks. By using the Dolev-Yao intruder model which is one of the widely used models to explain the security threats for wireless communication, we can model security threats in order to improve the design of the security model of the proposed scheme. To countermeasure the above-mentioned potential attacks, the proposed protocol should have robust security and privacy features including identity information, key exchange protection, and backward and forward secrecy. Therefore, we aim to establish a secure communication channel between the proximity devices in 5G HetNets and achieve a universal and efficient privacy-preserving and key agreement protocol for D2D communication with robust resistance against various attacks.

5 The Proposed Scheme

In this section, the proposed scheme, FaSUHA, will be explained in two sections which are the handover initialization section and handover authentication section to address secure handover establishment in 5G HetNets. The handover authentication will be done by adopting identity-based encryption and the ECDS algorithm to secure connection during handover. The definition of the notations used in the protocol is defined in Table 1.

Table 1. The Definition of Notations in The Proposed Scheme.

Notation	Definition
p	a k-bit prime
F_p	a prime finite field
E/F_p	an elliptic curve E over F_p
G	$G = \{(x, y) : x, y \in F_p; (x, y) \in E/F_p\} \bigcup\{\Theta\}$
P	generator for the group G
$H_1()$	a secure hash function $H_1 : \{0, 1\}^* \times G \to Z_p^*$
$H_2()$	a secure hash function $H_2 : \{0, 1\}^* \times \{0, 1\}^* \times G \times G \times G \times G \to \{0, 1\}^k$
$H_3()$	a secure hash function $H_3 : \{0, 1\}^k \times G \to \{0, 1\}^k$
T_{exp}	expiration time
ID_X	identity of node X, which is $ID_X = (MAC\ address\ (or\ identifier) \parallel T_{exp})$
x/PK	private/public key of KGC, $x \in_R Z_p^*$ and $PK = xP$
(s_X, R_X)	X's private long-term key generated by KGC

5.1 Handover Initialization Section

In this subsection, the preparation for the handover initialization is explained. The AMF module contacts with both gNBs and the UE to provide their identifiers and get their secret keys. Only authenticated entities can get their secret keys from the AMF. The pseudonym of the UE which needs to be provided in order to protect the UE's privacy and the private key which is provided by AMF is valid within the expiration time T_{exp}. This section consists of the following steps: Initialization and Key Distribution. The KGC is the module that works within the AMF module and it has secure channels to communicate with SMF in the core network.

Initialization: The AMF module signs inter-operator (roaming) agreement and generates system parameters. The process is as follows:

(1) Chooses a security parameter 1^k, selecting a k-bit system prime p with determining a system elliptic curve ε over F_q with a cyclic additive points group G and a generator P.
(2) Chooses randomly a system secret master key and computes the system public key PK.
(3) Chooses three cryptographic hash functions *H1, H2, H3*.
(4) Publishes the system parameters par = {$Fp, E/F_p, G, P, PK, H1, H2, H3$} [9] and keep the master key as secret.

Key Distribution: The ID-based long-term key will be generated only when the UE register to the network for the first time or when its secret key expires by the KGC module. After the UE register with full access authentication to the network, the UE sends a request including with its identifier ID_{UE} with its pk_{UE} the private key to the KGC via the AMF by using the secret key negotiated between the UE and the AMF module. When the KGC receives the message, it starts generating the following for the UE.

(1) Chooses a random number $r \epsilon_R Z_p^*$, computes $h = H_1(ID_{UE} \| R_{UE})$.
(2) Computes $s_{UE} = r + hmpk$

The UE's long-term key pair (s_{UE}, R_{UE}) is sent to the AMF then the AMF module sends to the UE secretly by using the secret key. Upon receiving the long-term key by the UE, it validates the key by using Eq. 1.

$$s_{UE}P = R_{UE} + H_1(ID_{UE} \| R_{UE})PK \tag{1}$$

The same procedure above will be executed by each base station and each of them will be authenticated by 5G-AKA or EAP-5G authentication protocol with the AMF or SMF module. Again, each base station or access point sends a request to receive their long-term key pairs (s_{BS}, R_{BS}) and this process will be done in a secure communication establishment with the IPsec channel between base stations and AMF or SMF module.

5.2 Handover Authentication Section

In this subsection, the UE moves from the current base station to the new base station but before moving the mutual authentication is processed between the UE and the new gNB. The session key which is the Pairwise Transient Key (PTK) is generated directly and a symmetric key cryptosystem system is defined for encryption Enc_K and decryption Dec_K. So, the following is the process of the handover authentication step.

1. **Sample** $UE \rightarrow gNB_2$: **REQ** $(ID_{UE}, R_{UE}, T_{UE})$
    The UE randomly chooses $a \epsilon_R Z_p^*$ and computes $T_{UE} = aP$. Then it sends ID_{UE}, R_{UE}, and T_{UE} to the new gNB.
2. $gNB_2 \rightarrow UE$: **REP** $(ID_{gNB_2}, R_{gNB_2}, T_{gNB_2}, c, H_3(PTK \| T_{gNB_2}))$

After the gNB_2 receives the message from the UE, it randomly chooses $b \epsilon_R Z_p^*$ and computes $T_{gNB_2} = bP$. After that, it computes the UE's public key PK_{UE}, the shared secrets K_{gU}^1 and K_{gU}^2, and the session key PTK as shown in Eq. 2, Eq. 3, and Eq. 4 respectively, the random number b will be deleted from the memory. gNB_2 randomly chooses a temporary ID, TID for the UE and computes $c = Enc_{PTK}(TID)$ using the session key. Lastly, it sends the necessary parameters to the UE.

$$PK_{UE} = R_{UE} + H_1(ID_{UE} \| R_{UE})PK \tag{2}$$

$$K_{gU}^1 = s_{gNB_2} T_{gNB_2} + bPK_{gNB_2}, \quad K_{gU}^2 = bT_{UE} \tag{3}$$

$$PTK = H_2(ID_{UE} \| ID_{gNB_2} \| T_{UE} \| T_{gNB_2} \| K_{gU}^1 \| K_{gU}^2) \tag{4}$$

3. $UE \rightarrow gNB_2$: **ACK** $H_3(PTK \| T_{UE})$.

When the UE receives the message from the gNB$_2$, firstly it computes the shared secrets K_{Ug}^1 and K_{Ug}^2 as shown in the Eq. 5 and the session key PTK also is shown in Eq. 6, and a will be deleted from the memory. In the next step, if the UE verifies $H_3(PTK||T_{gNB_2})$ successfully, it decrypts c and stores TID. Lastly, the conformation of the PTK agreement is done as $H_3(PTK||T_{UE})$.

$$K_{Ug}^1 = s_{UE}T_{gNB_2} + a(R_{gNB_2}+H_1(ID_{gNB_2}||R_{gNB_2})PK)K_{Ug}^2 = aT_{gNB_2} \tag{5}$$

$$PTK = H_2(ID_{UE}||ID_{gNB_2}||T_{UE}||T_{gNB_2}||K_{Ug}^1||K_{uG}^2) \tag{6}$$

4. If gNB$_2$ validates $H_3(PTK||T_{UE})$, it keeps ID_{UE}, $PK_{UE,}$ and TID.

If the UE handovers to the same base station in the future, above-mentioned the handover authentication process will be repeated except that (ID_{UE}, R_{UE}) would be replaced with TID. This will be used to establish a new session key and a new temporary ID TID^i will be computed during the handover process. The new temporary ID TID^i replaces the old temporary ID TID which is done by gNB$_2$.

Our handover authentication protocol can be applied to all handover scenarios between 5GNR and non-3GPP access networks in the 5G heterogeneous networks including handovers scenarios from 5GNR to trusted non-3GPP access network, from 5GNR to untrusted non-3GPP access network, from the trusted non-3GPP access network to 5GNR, from the untrusted non-3GPP access network to 5GNR. Additionally, the proposed scheme is applicable to the handover scenarios where it happens from one network to another. This handover is called vertical handover (inter-domain) authentication. The only difference is that there should be an inter-operator (roaming) agreement which is signed by the 5G core network's AMF module between different networks. This agreement would help to authentication simpler without the complicated roaming authentication protocols and extra entities to connect.

6 Security Evaluations

In this section, the proposed protocol has been evaluated in terms of security analysis and formal verification by AVISPA which is one of the most used security verification tools. Hence, it is shown that the scheme has robust security features and not vulnerable to various attacks.

6.1 Security Analysis

In this subsection, the proposed protocol is investigated in terms of robustness and its security features meet the requirement of 5G HetNets.

User Identity Protection: The UE doesn't use their real identity to establish a connection with gNBs or Core Network. Instead of real identity, after the registration, the user gets its temporary ID from the AMF module to set up a session with gNB. Since all pseudonyms have an expiration time, the attacker cannot get the user's information from a temporary ID. Moreover, it is impossible to map real identities from different temporary IDs. Therefore, the protocol protects the user identity.

Mutual authentication: The protocol achieves mutual authentication between the UE and new gNB by utilizing IBC and the session key agreement is done by negotiation which utilizes the elliptic curve cryptosystem (ECC). Also, the hash values from the UE and gNBs are checked whether both parties agree on the shared secrets. Thus, both entities can obtain the session key PTK. Hence, only the registered and legitimate UE and gNB can compute the agreed session key PTK and valid hash values by held each side.

Impersonation attack: Our proposed protocol has resistance against impersonation attack and ensures non-repudiation. If an adversary impersonates the UE, it may only get the temporary ID and try to send its fake messages to the gNBs. However, since the temporary ID can be negotiated with legitimate entities and has an expiry time. Therefore, the attacker can't communicate with an expired temporary ID.

Replay Attack: If the adversary records any request messages during the handover authentication stage, it cannot access the gNB or the UE by replaying the captured messages because it cannot compute the valid request and acknowledgment without having the private key. The other reason is that the random parameters of the protocol are updated by gNBs and the UE for each handover authentication. Therefore, it can be said that the protocol can provide the freshness of parameters and session key. That's why the scheme is robust against a replay attack.

Perfect Forward Secrecy: PFS is one of the important security features. The proposed protocol leverages the identity-based cryptosystem to generate the user's public/private key. Therefore, if the long-term secret key is compromised, the adversary can only achieve a part of the private key. Even if the adversary compromises the master key of KGC, he cannot derive the session key without its secret values because the session key is computed by a random value of the UE and gNB.

Perfect Backward Secrecy: In the scenarios of compromised keys to support dynamic key management, the random values of the UE and gNBs are regenerated for each different handover authentication. If new the UE moves to new or existing networks, the UE and gNBs will re-compute the session key and gNBs will delete the old random values from memory. It implies that the adversary would not know the previous random values, therefore, it guarantees the freshness of the session key.

6.2 Formal Verification

After the investigation of the security features of the proposed protocol, the scheme needs to be further verified by AVISPA [13]. AVISPA is a formal verification tool that is used to do security analysis and validation of the proposed scheme. It is a push-button tool that provides a modular and expressive formal language which is called high-level protocol specification language (HLPSL) and applies various protocol analysis techniques; thus, it can be used to model and analyze security protocols. Therefore, this tool was chosen because it is one of the most common and sensitive tools among other security analysis tools.

To take advantage of the algebraic properties of the XOR operator two back-ends are chosen on the security analysis of FaSUHA which are On-the-Fly Model-Checker

```
% OFMC                                        SUMMARY
% Version of 2006/02/13                        SAFE
SUMMARY
  SAFE                                        DETAILS
DETAILS                                        BOUNDED_NUMBER_OF_SESSIONS
  BOUNDED_NUMBER_OF_SESSIONS                   TYPED_MODEL
PROTOCOL
  /home/span/span/testsuite/results/The_Proposed_Protocol.if  PROTOCOL
GOAL                                            /home/span/span/testsuite/results/The_Proposed_Protocol.if
  as_specified
BACKEND                                        GOAL
  OFMC                                          As Specified
COMMENTS
STATISTICS                                     BACKEND
  parseTime: 0.00s                              CL-AtSe
  searchTime: 0.06s                           STATISTICS
  visitedNodes: 7 nodes
  depth: 6 plies                                Analysed  : 2960 states
                                                Reachable : 1522 states
                                                Translation: 0.18 seconds
                                                Computation: 0.10 seconds
```

Fig. 2. The Output result of formal verification using OFMC and CL-AtSe back-ends in AVISPA

(OFMC), Constraint-Logic-Based Attack Searcher (Cl-AtSe). CL-AtSe converts the Intermediate format (IF) into a set of constraints on the intruder knowledge that can effectively detect attacks to the proposed scheme. The number of iterations can be bounded by using CL-AtSe to analyze the scheme. On the other hand, the OFMC backend uses symbolic techniques and a tree system to analyze the protocol with infinite state spaces.

The proposed scheme is investigated by these two back-ends and written in HLPSL. Assuming that the intruder knowledge is defined which includes the user's pseudonym, the hash function, the common system parameters, and all exchange messages over the channel. The output of the OFMC and CL-AtSe back-ends show that the proposed scheme is safe as shown in Fig. 2. Also, the goals are achieved such as mutual authentication, no vulnerable points, etc.

7   Performance Evaluation

Since the proposed protocol can be applied to different handover scenarios such as vertical and horizontal handovers, it can be said that the proposed scheme is uniform. In this section, the proposed scheme is evaluated in terms of communication and computation overhead by comparing it with other schemes.

7.1   Communication Cost

The comparison of the number of messages exchanged between two entities in the proposed scheme and other schemes are shown in Table 2. From the table, the proposed protocol has fewer exchanged messages than existed protocols. The proposed scheme exchanges two messages during handover initialization and three messages.

during handover authentication and the other schemes send more messages between the UE and base station which can introduce more delay during handover which makes them not suitable for the heterogeneous networks.

Table 2. The number of exchanged messages between entities in 5G HetNets

Protocols	UE-BS	BS-BS	BS-AMF	AMF-AMF	Total
[8]	3	0	4	2	9
[9]	3	0	7	4	14
[10]	8	0	4	4	16
FuSUHA	3	0	2	0	5

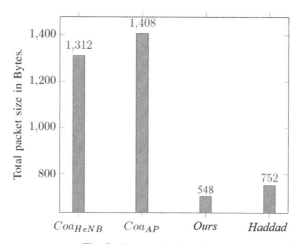

Fig. 3. Communication Cost

According to the 5G service requirements specified by 3GPP TS 22.261 [5], in general, city-wide application scenarios, the experienced data rate of downlink and uplink are 50 Mbps and 25 Mbps respectively. The parameters used to calculate communication costs are assigned as two bytes for each entity, 160 bits for q, 20 bytes for each elliptic curve point, 20 bytes for signatures, and 5 bytes for time stamps. The typical certificate size is 54 bytes which includes the public key, an identifier, the identifier of the issuer, the issuing date, the expiry date, and a signature. The comparison of communication cost with other schemes is shown In Fig. 3 and it can be said that our scheme requires less communication cost.

7.2 Computational Cost

In this subsection, the proposed scheme is evaluated on a laptop run 64-bit Windows 10 operating system with Intel Core I7-8550u (8 MB cache up to 4.0 GHz) and 8 GB RAM and pairing-based cryptography (PBC) library is used for algebraic operations [14]. The measured time for bilinear pairing operation is 3.79 ms, ECC-based scalar multiplication is 0.65 ms, an ECC point addition is 0.486 ms, and a hashing operation is 0.0036 ms/Byte.

Table 3. Computational Cost of Each Entity in the 5G System

Protocols	UE	gNB	AMF
[8]	0.531	7.65	7.6
[9]	5.43	8.11	5.34
[10]	5.43	8.11	5.34
FuSUHA	4.85	6.77	2.25

Table 3 shows the comparison of the computational cost of each protocol with using these measurements. According to the comparison, the computational performance of the proposed scheme is much better than other schemes. Therefore, it is safe to say that our protocol brings significant improvement to be suitable for all handover scenarios in 5G HetNets.

8   Conclusion

In this paper, a fast and secure uniform handover authentication scheme for 5G HetNets is proposed to achieve seamless inter-domain and intra-domain handover authentication for the UE in the 5G system. The crucial point of having seamless handovers is to avoid any connection termination and improve the quality of service.

The reason behind why we need such a fast and efficient uniform handover authentication lays under the deployment of small cells that cause much higher handovers in the 5G system. The proposed scheme needs a few numbers of messages to be exchanged between the UE and gNB for the handover procedure. And the security and performance analysis has demonstrated that the scheme is safe against various attacks without having any vulnerable points and has enough efficiency to be suitable for all handover scenarios in the 5G Heterogeneous Networks.

Acknowledgment. This work is supported by the MOE ACRF Tier 1 funding for the project RG 26/18 by the Ministry of Education, Singapore.

References

1. Alcaraz-Calero, J., et al.: Leading innovations towards 5G: Europe's perspective in 5G infrastructure public-private partnership (5G-PPP). In: 2017 IEEE 28th Annual International Symposium on Personal, Indoor, and Mobile Radio Communications (PIMRC), pp. 1–5. IEEE (2017)
2. Agiwal, M., Roy, A., Saxena, N.: Next generation 5G wireless networks: a comprehensive survey. IEEE Commun. Surv. Tutorials **18**(3), 1617–1655 (2016)
3. 5G; Study on New Radio (NR) access technology (3GPP TR 38.912 version 15.0.0 Release 15), 3GPP TR 38.912 V15.0.0 (2018)
4. 3rd Generation Partnership Project; Technical Specification Group Core Network and Terminals; Access to the 3GPP 5G Core Network (5GCN) via Non-3GPP Access Networks (N3AN); Stage 3(Rel 15), 3GPP TS 24.502 V15.2.0, December 2018

5. Technical specification group services and system aspects; service requirements for the 5g system; (release 16), 3GPP TS22.261 (2019)

6. Han, K., Ma, M., Li, X., Feng, Z., Hao, J.: An efficient handover authentication mechanism for 5G wireless network. In: 2019 IEEE Wireless Communications and Networking Conference (WCNC), pp. 1–8. IEEE (2019)

7. Ma, T., Hu, F., Ma, M.: Fast and efficient physical layer authentication for 5G HetNet handover. In: 2017 27th International Telecommunication Networks and Applications Conference (ITNAC), pp. 1–3. IEEE (2017)

8. Haddad, Z., Mahmoud, M., Saroit, I.A., Taha, S.: Secure and efficient uniform handover scheme for LTE-A networks. In: 2016 IEEE Wireless Communications and Networking Conference, pp. 1–6. IEEE (2016)

9. Cao, J., Ma, M., Li, H.: An uniform handover authentication between E-UTRAN and non-3GPP access networks. IEEE Trans. Wireless Commun. 11(10), 3644–3650 (2012)

10. Cao, J., Li, H., Ma, M., Li, F.: UGHA: uniform group-based handover authentication for MTC within E-UTRAN in LTE-A networks. In: 2015 IEEE International Conference on Communications (ICC), pp. 7246–7251. IEEE (2015)

11. 3rd Generation Partnership Project; Technical Specification Group Services and System Aspects; Security architecture and procedures for 5G system (Rel 15), 3GPP TS 33.501 V15.3.1, December 2018

12. Technical specification group services and system aspects; system architecture for the 5g system; (release 16), 3GPP TR23.501 (2019)

13. von Oheimb, D.: The high-level protocol specification language HLPSL developed in the EU project AVISPA. In: Proceedings of APPSEM 2005 workshop, p. 17 (2005)

14. IEEE standard for identity-based cryptographic techniques using pairings. IEEE P1363.3/D9, p. 136 (2013)

Assistive Living

An Automated Wheelchair for Physically Challenged People Using Hand Gesture and Mobile App

Md. Farhan Razy[1], Sabah Shahnoor Anis[1], Md. Touhiduzzaman Touhid[1], and Jia Uddin[2(✉)]

[1] Department of Computer Science and Engineering, BRAC University, Mohakhali, Dhaka 1212, Bangladesh
farhanrazyanik@gmail.com, shahnoor.evana@gmail.com, touhidshuvosb@gmail.com
[2] Technology Studies Department, Endicott College, Woosong University, Daejeon, South Korea
jia.uddin@wsu.ac.kr

Abstract. Traditional wheelchairs which operate manully are difficult to use without the assistance of physical strength. This paper presents a model of an automated wheelchair for disabled people. The proposed wheelchair is battery powered and uses an Arduino microcontroller to operate. The whole system consists of two parts, one of them is the sending end and another part is the receiving end. The sender end records instructions from the user's different hand gestures which are collected using an MPU-6050 gyro accelerometer sensor. The instructions are continuously recorded and sent to the receiver end using an RF transmitter- nRF24101 through an Arduino microcontroller. At the receiving end, an nRF module receives the signal for mobilizing the wheelchair on a plain surface. Additionally, the device contains a Bluetooth module which can be used as a secondary control system. Hence our chair can be operated by a smartphone application too. Experimental results demonstrate that the proposed model is functional under normal circumstances with a fast response time which can be very useful for disable people to commute with.

Keywords: Gesture Controlled Wheelchair · Gyro accelerometer · Arduino microcontroller · RF transmitter · Bluetooth module · Mobile application

1 Introduction

Wheelchairs are needed for physically challenged people of different age groups. Out of the total number of Wheelchair users, 1.825 million users are more than 65 years old [1]. Besides ageing, other factors are resulting in increased usage of wheelchairs which include genetic disorders, accidents etc. It is evident from a survey that shows, between the years 2009 and 2014 the revenue sector of wheelchair industry had annual growth rate of 2.5% in average [1]. Besides, the survey estimated that, almost 3.3 million

© Springer Nature Switzerland AG 2021
M. Singh et al. (Eds.): IHCI 2020, LNCS 12616, pp. 135–144, 2021.
https://doi.org/10.1007/978-3-030-68452-5_13

wheelchairs are being operated in the USA alone and about 2,000,000 new wheelchair users are adding to that number every year. Also among the working age wheelchair users, approximately 17.4% of them have jobs and they need to move outside of home regularly [1]. Previously, the traditional hand-rowed wheelchairs only had two wheels attached to a chair that had to be controlled manually either by the passenger himself or a third person. It is not an undisclosed fact that a lot of wonderful work has already been done in the domain of wheelchair technology. Nonetheless, there is still room for development in this sector. Some patients cannot run the wheelchair with their arms, because they cannot move most of the body parts. This is where the demand for automated wheelchairs arises. In the evolution of intelligent wheelchair system, one of the earliest move was a system which was proposed by Connell and Viola, in the year 1990 where a chair was made to locomote by mounting it on top of a robot [2]. The user could control the chair with the help of a joystick fitted across the arm of the chair connected to the robot. Nipanikar (2013) et al. implemented voice recognition system to control an automated wheelchair and ultrasonic and infrared sensor system to avoid obstacles [3]. However, the use of voice recognition and voice command to handle the wheelchair may sound innovative but is not so efficient and cannot be relied on. Because it can be risky for outdoors as the noise can manipulate and can even change the voice command of the user. Besides this, it also had the provision of an accelerometer sensor and joystick for disabled people who can easily move their hands. Thereafter, in 2014, Vishal et al. structured a wheelchair controlled by hand gesture recognition [4, 21]. This is designed using accelerometer sensor that made it a controllable wheelchair which is able to move easily in every direction. Furthermore, a wheelchair which was controllable by the head movement was designed by Rishi et al. [5]. Their system used two head orientation detection units (Ultrasonic Sensors). The system could detect the head position via the ultrasonic sensors and could control the motor rotation connected to the two wheels. This idea is suitable for those patients who can only move and use their heads. In (2017), Balsaraf et al. designed a wheelchair controlled by an android device like a mobile phone [6]. The Android Mobile was used as an input which operated the wheelchair to desired directions.

In this era of wireless communication, fixed joystick control is not efficient enough. As using this system, users cannot operate the wheelchair remotely in need. The projects where wireless controlling was used, were based on either gesture or phone application but not on both. So, the users did not have any freedom of option for controlling those wheelchairs according to their comfort.

To overcome all these limitations related with the existing automated wheelchairs, in this paper, a model is proposed for aiding people to handle the wheelchair remotely from a distance via Bluetooth features on a mobile device [7]. Additionally, it can be handled via hand gestures using gyroscope sensors for techno-phobic people who are not comfortable with phone functionalities, to reduce physical labor and aid in an advanced and progressive life [8].

The rest of the paper is presented as follows: Section 2 describes the proposed model with hardware implementation and Sect. 3 deals with experimental setup along with result analysis.

In the end, Sect. 4 concludes the paper.

2 Proposed Model

In our designed wheelchair, the wheels are powered by DC motors [9] to enable the user for using it automatically. It is much easier than operating manual wheelchairs. The DC motors take energy from rechargeable batteries which can be used for 2–3 h when fully charged.

We designed a smart automated wheelchair that can be useful for aged and handicapped patients. This wheelchair is primarily divided into two systems which are gesture control system and Bluetooth control system using mobile application (Bluetooth RC controller). In our system, the wheelchair can be controlled by hand gestures with the help of a gesture recognition system [10]. It is the accelerometer sensor on which gesture recognition system relies [10]. It costs very low and it is capable of providing the direction of the hand, hence it assists to recognize the gestures. This provides an easily controllable wheelchair for the disabled people. They can drive it by themselves effortlessly.

To understand the framework or procedure of the Hand Gesture Controlled system, let us divide the project into three isolated parts.

The first step is, the Arduino [11, 22] gets data from the MPU6050 Accelerometer Gyro Sensor [10]. The Arduino continuously receives data from the MPU6050 and depending on the predefined parameters, it sends data to the RF Transmitter [12].

Wireless Communication between the RF modules is the next part of the project. The RF Transmitter transmits received data from Arduino to the RF Receiver, through the RF Communication [12]. Also our wheelchair can be controlled by a mobile application which uses Bluetooth communication system.

Finally, the third part of the project is-the Data received by the RF Receiver has to be decoded. After that it should send the appropriate signals to the Motor Driver IC [13]. Thus the Wheel Motors of the wheelchair are activated.

2.1 Block Diagram of the Proposed Model

Figure 1 shows the block diagram of transmitter end, where the hand gesture is recognized by the MPU6050 gyro sensor [10] and processed. It sends commands to the microcontroller [11] accordingly and hence to the system. When the user tilts his/her hand, accelerometer sensor finds the change of position of the hand. It sends analog signal to microcontroller and convert it into appropriate digital signal for the motors of wheelchair. Then the digital signal is sent via nRF24L01 transceiver module [12].

Fig. 1. Sample block diagram of Transmitter

Figure 2 shows the block diagram of the receiver end. Here the data sent by transmitter end is received by nRF24L01 transceiver module [12] and converted it into appropriate signal. The signal is sent to the microcontroller [14] accordingly. Thus the motors are controlled to run the wheels. Microcontroller controls the movements of the wheelchair by sending appropriate instructions to the LN29D motor driver IC [13].

Fig. 2. Sample block diagram of Receiver

2.2 Materials Used

In this model, we have designed a simple Hand Gesture Controlled Robot which also can be used by simple mobile application. This Hand Gesture Controlled Robot is based on Arduino Nano [11], Arduino Uno [14], MPU6050 [10], (nRF24l01) RF Transmitter-Receiver Pair (for gesture control) [12], HC-05 Bluetooth module (for connect with mobile application) [15] and L293D Motor Driver [13].

The title of our project says it as a Hand Gesture Controlled wheelchair, but to be specific, this wheelchair will be operated by the tilt of the hand. However, Electronics Hub (2017) states that, the MPU6050 is one of the most popularly used Sensor Modules by hobbyists and enthusiasts. Accelerometer and Gyroscope on the same IC provides 6 degrees of free movement which are 3-axis of Accelerometer and 3-axis of Gyroscope.

2.3 Hardware Implementation of the Proposed Mode

Figure 3 shows how the components are connected in the transmitter end. The MPU6050 Accelerometer and Gyroscope Module's [10] SCL and SDA connected with A5, A4 pin of the Arduino NANO [11] as it shows in the diagram. For connection with the RF modules, NRF24L01's [12] MISO is connected to pin 12, MOSI is connected to pin 11 green wire, SCK is connected to pin 13, CE is connected to pin 9, CSN is to pin 10 of the NANO. The earth wire is connected to the ground pin and 5 V connects with VCC pin of both the sensors. Eventually the whole system is powered by a 9 V battery.

Figure 4 shows how the components are connected in the receiver end. With Arduino Uno's [14] pin 5, 6,9,10 firstly connected with DC M Driver L298's [13] INT1, INT2, INT3 and INT 4 accordingly. Moreover, DCMDriverL298 OUT1, OUT2, OUT3, OUT4 are connected to DC Motor Coil2, Coil1 Coil2, and Coil1 accordingly and RF module [12] connects with UNO similarly as shown in Fig. 3. Lastly the whole system uses 2, 3.7 V rechargeable batteries to power up the whole system.

Figure 5 shows the connection of Bluetooth module HC-05 [12] with Arduino UNO [14]. The module is powered simply with +5 V and the Rx pin of the module is connected to the Tx of MCU. Tx pin of module is connected to Rx of MCU, and the other connection are same as Fig. 4.

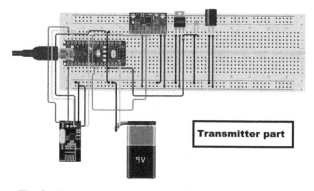

Fig. 3. Hardware implementation of the gesture control unit

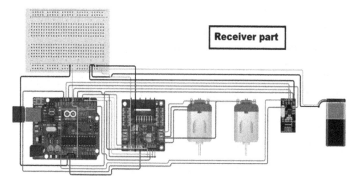

Fig. 4. Hardware implementation of the wheelchair (with nRF24101)

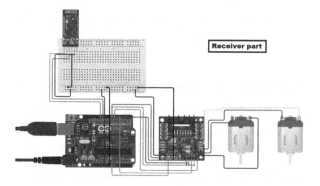

Fig. 5. Hardware implementation of the wheelchair (with Bluetooth module)

2.4   Program Pseudocode

The Pseudo code for gesture control (receiver part) is given below.

```
Begin
void setup ()
{
  Define the four input pins: forward, backward, left
and right;
  Define the four output pins: MotA1, MotA2, MotB1,
MotB2;
}
void loop ()
{
  Read four inputs;
  if(backward==1   &&   forward==0   &&   left==0   &&
right==0)
  backward();
  else if(backward==0  &&  forward==1  &&  left==0  &&
right==0)
  forward();
  else if(backward==0  &&  forward==0  &&  left==1  &&
right==0)
  left();
  else if(backward==0  &&  forward==0  &&  left==0  &&
right==1)
  right();
  else
  Stop();
}
End
```

Firstly, four input and four output pins are defined. The input pins take four inputs from the data sent by the MPU-6050 sensor. The four output pins send signals to the motor driver for its control. Afterwards, an infinite loop is run which keeps reading the inputs and checks them with the given conditions to determine the movement of the wheelchair.

3   Experimental Setup and Result Analysis

To begin the experiment, firstly both the prototype model wheel-chair and the signal transmitter device were connected to their power supplies. We tested the transmitter for its functionality and found that it was transmitting data properly. The data collected from the movement of the hand by the MPU-6050 Gyro Accelerometer sensor was encoded by the encoder and sent through the RF transmitter. There were five types of signals or gestures that the device will recognize: Stop, forward, backward, left and right. Afterwards, we tested the receiver end for its functionality. The RF receiver was collecting the data sent by the transmitter device and it was decided by the decoder. Then the decoder sent the data to the Arduino and the wheel-chair followed the instructions according to the signal that was sent.

The five different hand gestures are shown in Fig. 6, where the different movement recorded by the MPU-6050 sensor is sent to the wheelchair. Figure 6.(a) shows the gesture for the stop command, (b) shows the gesture for the forward command and (c), (d) and (e) show the gestures for backward, left and right command respectively.

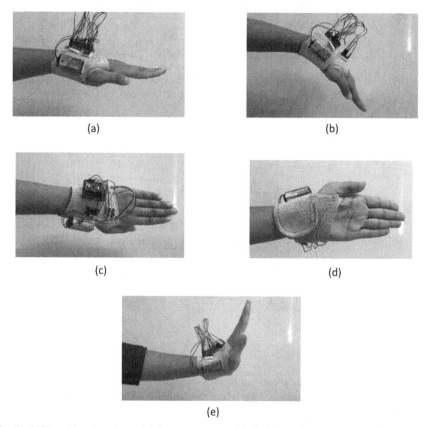

Fig. 6. Different hand gestures, (a) the stop gesture, (b) the forward gesture, (c) the right gesture, (d) the left gesture and (e) the backward gesture

After finishing the experimental procedure, we got the expected results. The wheelchair was moving according to the hand gesture given by the user. Figure 7 shows the model body of the wheelchair, the receiving end of the device.

The set of input received by the wheelchair and the movement it makes is shown in Table 1. This device will be very useful for disabled people as they only need to use intuitive hand gestures to control the chair. Moreover, the device will stop at once if the signal is lost. The user can easily reach the device from their bed so that there is minimum assistance required from someone else. This device will be very cost efficient so that people of all financial capability can afford it. Hence, the affordability and performance of this chair is satisfactory.

(a)   (b)

Fig. 7. The receiver end of the device, (a) back view and (b) side view

Table 1. The input for Arduino and their corresponding direction of movement

Movement of hand	Input for Arduino from gesture				
Side	D3	D2	D1	D0	Direction
Stable	0	0	0	0	Stop
Tilt right	0	0	0	1	Turn right
Tilt left	0	0	1	0	Turn left
Tilt back	1	0	0	0	Backward
Tilt front	0	1	0	0	Forward

4   Conclusion

This paper is to demonstrate solutions to real-world problems and one of those real-world problems is the growing need for advanced wheelchair machines. There has been a drastic increase in the rate of road-accident in comparison to the last few decades. More accidents, more injuries. The injured people and people who have physical disability deserve an equal opportunity at having a healthy and prosperous life just as the regular people. So to ease their lifestyle a proposal has been made for a wheelchair that can be comfortable, reliable, advanced, sustainable and most importantly affordable to the common people. It will also be efficient and economical and most importantly, eco-friendly as it will be rechargeable. Besides, upgraded features such as head motion, sonar detection, sensation and GPS location can be added in the future which will truly transform the way traditional hand rowed wheelchairs provide services. All in all, this system can be made highly efficient, durable and effective if hard and fast environmental conditions are maintained.

References

1. https://kdsmartchair.com/blogs/news/18706123-wheelchair-facts-numbersand-figures-inf ographic?fbclid=IwAR1104Ca-lsHn4w7AxTG7q-6Cqqwo8PgF6sb8D1AfpYSiO3DExA-IUZaAWI 12 Feb. 2020
2. https://www.cs.unc.edu/~welch/class/mobility/papers/yanco.pdf 1 Dec. 2019
3. Nipanikar, R.S., Gaikwad, V., Choudhari, C., Gosavi, R., Harne, V.: Automatic wheelchair for physically disabled persons. Int. J. Adv. Res. Electron. Commun. Eng. **2**(4), 466–478 (2013)
4. Pande, V.V., Ubale, N.S., Masurkar, D.P., Ingole, N.R., Mane, P.P.: Hand gesture based wheelchair movement control for disabled person using MEMS. Int. J. Eng. Res. Appl. **4** (2014)
5. Sajeevan, R.P., Yousaf, F., James, B., Kuruvilla, J.: Head orientation controlled wheelchair. Int. Res. J. Eng. Technol. **4**(4) (2017)
6. Balsaraf, M.D., Takate, V.S., Siddhant, B.: Android based wheelchair control using bluetooth. Int. J. Adv. Sci. Res. Eng. **3**(4) (2017)
7. Rai, N., Rasaily, D., Wangchuk, T.R., Gurung, M., Khawas, R.K.: Bluetooth remote controlled car using Arduino. Int. J. Eng. Trends Technol, **33**, 381–384. (2016)
8. Al-Neami, A.Q.H., Ahmed, S.M.: Controlled wheelchair system based on gyroscope sensor for disabled patients. Biosci. Biotechnol. Res. Asia, **15**(4), 921–927 (2018)
9. Electric DC Motors - Direct Current Motor Basics, Types and Application. (2016, March 28). Retrieved December 1, 2019, from https://www.elprocus.com/dc-motor-basics-types-applic ation/
10. MPU6050 Pinout, Configuration, Features, Arduino Interfacing & Datasheet. (2018, March 17). Retrieved December 1, 2019, from https://components101.com/sensors/mpu6050-module
11. Arduino Nano Board: Features, Pinout, Differences and Its Applications. (2019, June 24). Retrieved December 1, 2019, from https://www.elprocus.com/an-overview-of-arduino-nano-board/
12. nRF24L01 Pinout, Features, Circuit and Datasheet. (2018, April 30). Retrieved December 1, 2019, from https://components101.com/wireless/nrf24l01-pinout-features-datasheet
13. Instructables. (2017, October 5). How to Use the L298 Motor Driver Module -Arduino Tutorial. Retrieved December 1, 2019, from https://www.instructables.com/id/How-to-use-the-L298-Motor-Driver-Module-Arduino-Tu/
14. Arduino Uno Pin Diagram, Specifications, Pin Configuration & Programming. (2018, February 28). Retrieved December 1, 2019, from https://components101.com/microcontrollers/ard uino-uno
15. HC-05 Bluetooth Module Pinout, Specifications, Default Settings, Replacements & Datasheet. (2018, March 10). Retrieved December 1, 2019, https://components101.com/wir eless/hc-05-bluetooth-module
16. Jadhav, A., Pawar, D., Pathar, K., Sale, P., Thakare, R.: Hand Gesture Controlled Robot Using Arduino. In: International Journal for Research in Applied Science & Engineering Technology, vol. 6 (2018)
17. https://web.eecs.umich.edu/~kuipers/papers/Gulati-icra-08.pdf 1 Dec. 2019
18. Wada, M., Kameda, F.: A joystick type car drive interface for wheelchair users. In: RO-MAN 2009 - The 18th IEEE International Symposium on Robot and Human Interactive Communication (2009)
19. https://create.arduino.cc/projecthub/mayooghgirish/hand-gesture-controlled-robot-4d7587 1 Dec. 2019]
20. Karim, K.E., Nahiyan, H.A., Ahmed, H.: Design and Simulation of an Automated Wheelchair with Vertically Adjustable Seat. In: International Conference on Mechanical, Industrial and Energy Engineering, Bangladesh (2014)

21. Hassan, M.A., Shadman, Q., Chowdhury, M.H., Hasan, S.A., Uddin, J.: User Authentication using Password and Hand Gesture with Leap Motion Sensor. In: International Conference on Computational Intelligence, Data Science and Cloud Computing, Institute of Engineering and Management, Kolkata, India (2020)
22. Rahman, A.M., Mehdi, Q., Hossain, R., Shwon, M.R., Uddin, J.: An automated zebra crossing using Arduino-UNO. In: 2018 International Conference on Computer, Communication, Chemical, Material and Electronic Engineering (IC4ME2), Rajshahi, pp. 1–4 (2018)

Gait Analysis Using Video for Disabled People in Marginalized Communities

Achilles Vairis[1] , Johnathan Boyak[2] , Suzana Brown[2](✉) , Maurice Bess[2],
Kyu Hyun Bae[2] , and Markos Petousis[1]

[1] Hellenic Mediterranean University, Iraklio, Greece
[2] State University of NY Korea, Incheon, Korea
suzana.brown@sunykorea.ac.kr

Abstract. The goal of this project is to remotely analyze the gait of people walking with crutches. To that objective, the use of video analysis based on the open-source software OpenPose is compared with the data collected from a sensor mounted on a human subject. The results show that the average value of acceleration between the video analysis and the sensor differs by 0.05%. All steps are clearly identified and synchronized. As a consequence, it is possible to validate non-contact acceleration data from video analysis with an inexpensive setup described in this paper. The results show a promise that this non-contact method can be used to assess the gait of disabled people with assistive devices in remote locations.

Keywords: Gait analysis · Disability · Video analysis · OpenPose

1 Introduction

The current study aims to design better-suited mobility aids for people affected in insecure areas. A particular site for testing a design is a refugee camp in Bangladesh, specifically erected for the Rohingya people immigrating from Myanmar. Approximately 909,000 Rohingya refugees live in 34 highly congested camps. Since the humanitarian crises hit vulnerable groups hardest, such as people with disabilities, the goal was to improve the mobility of people with disabilities in navigating environmental obstacles in these limited-resource settings.

One challenge for this project was during the baseline study to remotely assess issues in the gait of the disabled population in a refugee camp in Bangladesh, so the team investigated the use of video analysis. The video analysis is based on the open-source software, OpenPose, which was developed in 2017 by the Carnegie Mellon University Perceptual Computing Lab. The software is based on Convolutional Neural Networks (CNN), and it can precisely identify 135 joint points on a human body. OpenPose can be used to capture either individual or multiple persons on a single image.

This study presents initial results in the lab while the team was assessing the suitability of the approach. This paper is also evaluating the use of the human computer interface (HCI) approach in low-resource settings, termed HCI4D or HCI for development, and the usefulness of OpenPose for remote gait assessment.

© Springer Nature Switzerland AG 2021
M. Singh et al. (Eds.): IHCI 2020, LNCS 12616, pp. 145–153, 2021.
https://doi.org/10.1007/978-3-030-68452-5_14

The main contribution of this paper that, to our knowledge, it is the first attempt to use video analyses software such as OpenPose to perform gait assessment.

2 Related Work

2.1 HCI4D

HCI4D research has been focused on development, low-resource settings, and/or marginalized groups [1]. In addition, some researchers viewed HCI4D as research defined by the location and certain infrastructural constraints, as well as motivated by regards for social justice [1]. It has been argued that user centered design and evaluation techniques do not easily translate from the Global North to the Global South, therefore they should be adapted to new cultural contexts and settings [2]. In addition, one of the ways to broaden the agenda in HCI is supporting marginalized communities [3]. For example, designing technologies to support low-income rural women in Bangladesh [4], which expands the HCI community's understanding of technology design within deeply patriarchal societies. The current study is aiming to support the marginalized population of disabled users in need of mobility support such as crutches. The intended users are Rohingya refugees but the diagnostic tool is useable for any distance gait evaluation by using a video tool.

2.2 Gait Studies

Gait analysis is the evaluation of the manner or style of walking usually done by observing the human as they walk in a straight line. This evaluation is commonly performed in two different ways. The first is by empirical analysis, while the second is based on sophisticated instrumentation measuring body movements, body mechanics and activity of the muscles. Various studies have been carried out using different tools such as force platform, optical markers and 3D-cameras. These motion capture systems are expensive and must be installed in appropriate rooms containing cumbersome expensive equipment and operated by trained personnel [5].

A simple way to analyze the motion is to use accelerometers. In motion analysis, the main interests are activity recognition [6] and the detection of specific gait events (e.g. falls). Most research on fall detection uses linear acceleration and gyroscopes. They typically detect falls by applying thresholds to accelerations, velocities and angles [7]. However, the performance of these systems depend strongly on the position of the sensors (e.g. the wrist, waist, ankle) [7, 8].

A popular current research challenge has been on the authentication of users based on gait recognition [9]. The research focused on the identification of a person based on the way that they walk using dynamic time warping [10]. However, during a natural walk, a person could change the way that they walk making it difficult for identification. A study most comparable to the current study uses video for gait analysis by detecting walking behavior based on the motion transfer by the user on the walker [11].

In recent years, OpenPose software has been widely used for gait analysis. OpenPose is an open-source software that is a real-time multi-person system designed to detect human body, hand, facial, and foot key points on images. It has been shown that OpenPose algorithm can learn to associate body parts with individuals in the image [12], that is reliable and valid in tracking bilateral squats [13], and an assessment system for a gross motor action recognition of preschool children [14]. To our knowledge no other study used OpenPose to access gait dynamically.

2.3 Theory

The current approach is leveraging an activity theory, a conceptual framework originally developed by Aleksei Leontiev. The essential concept of this theory is human activity, which is considered to be purposeful, mediated, and transformative interaction between human beings and the world. Along with some other frameworks, such as phenomenology and distributed cognition, activity theory established itself as a leading post-cognitivist approach in HCI [15]. HCI researchers use activity theory as a theoretical framework for empirical analysis to formulate specific questions for their studies for technology use in controlled experimental settings. One example closest to the present approach is a study that determines if hand posture can be used to determine the types of interactions in a desk/office environment [16].

3 Methodology and Data

Two iPhone 6s with 8-megapixel cameras were set up on tripods and placed at a distance that allowed for 30 frames per second video capture of several steps when using a swing-through gait[1]. The cameras were arranged to provide a stereoscopic view of the run, with one recording the side view of the walk and the second camera, at 90° to the first camera, recording the subject walking towards the camera. The two cameras were calibrated using a large checkerboard pattern to obtain the stereoscopic transformation matrix, which was used to obtain the 3D position data of an object in the covered area. A healthy male subject used axillary crutches and was instrumented with an accelerometer sensor firmly attached to his walking ankle while the video was recorded. The sensor setup attached to the leg of the subject (see Fig. 1) was the GY-521 module, which includes a gyroscope packaged together with a three-axis accelerometer connected to an Arduino Uno microcontroller together with a micro SD card for data collection. The protocol employed started with an initiation step (foot tap) and return to the initial position to signal the start of data collection. A short pause followed this initial step and steps with crutches were recorded using the stationary cameras.

The raw velocity data obtained from the sensors mounted on the healthy subject is presented in Fig. 2. From the velocities in three global coordinates (X, Y, and Z), we can identify individual steps. For comparison purpose, the horizontal axes in the plots are in time counts measured by the sensor and not in time units such as seconds, while the vertical axes units are in relative units, measured by the sensor. The velocity on the X axis (which is along the direction of movement) shows three steps that are in phase

---

[1] The swing-through gait requires the user to place both crutches ahead of themselves and swing both feet past the crutches to the next position. When performing a swing-through gait, the user's full body weight becomes temporarily supported by the crutches.

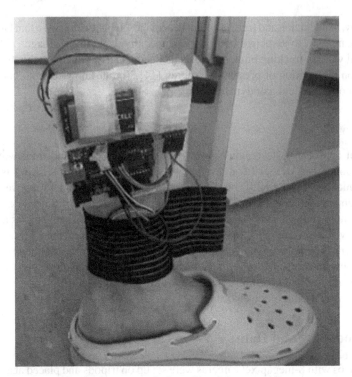

Fig. 1. Sensor attached to a leg of the test subject.

with the three steps shown in the Y axis (which is perpendicular to the floor). This is predictable as the foot moves in those two directions during each step. The same is true for the relative value of velocities between those two axes, as the forward movement of the foot results in the X velocity components being much larger than the ones in the Y direction. The third velocity component in the Z axis (which is along the floor, as the X components, but perpendicular to the other two components) shows depressions matching the steps of the other components, but at a much lower value than the X axis velocity. This results from the fact that the healthy subject's walk does not include a complex foot movement in space. In contrast, a disabled person may have a complex or even unstable foot movement, in which case there would be a much stronger Z velocity component. All traces in the three directions have practically zero value during the stance stage of the walk.

Subsequently, the raw data from the sensors are combined using the square root of the sum of the squares of the accelerations in the primary directions to produce the total acceleration experienced by the left ankle shown in Fig. 3. This equation is

$$Acc_{Total} = \sqrt{Acc_x^2 + Acc_y^2 + Acc_z^2}.$$

In addition, because of the complex movement of the foot in all three axes, the combined signal was further processed using the moving averages which smooths out the calculated acceleration while retaining the key characteristics. After the processing,

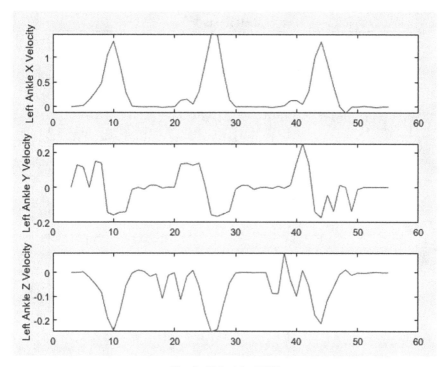

Fig. 2. Velocities XYZ

the individual steps are clearly identifiable while the stance period is stable and close to the acceleration of gravity (g). This operation allows for the qualitative identification of individual steps as it removes the signal fluctuations. One possible source of errors in the acceleration measurements may come from drift [18] in relative movement between the mounted sensor box and the body. However, measurements with the sensors mounted on the subject allow identification of the gait. As a consequence, it is possible to validate non-contact acceleration data from video analysis with this inexpensive setup.

With that objective, Fig. 4 combines on the same graph the raw data from both the accelerometer and the video analysis. Each step, both the initial and the walking steps, can be identified as a complex waveform. The step waveform for each source, the accelerometer and the video analysis, are different because of the nature and accuracy of the medium.

Figure 5 shows the processed acceleration recorded by the sensor and one calculated based on the video. The acceleration calculated from the video was passed through a low-pass filter with a 6 Hz cut off [17]. The data from the sensors was similarly processed to remove spurious very high acceleration values by cropping excessive values and then calculating the moving average of the signal. The average values of the processed acceleration between the video analysis and the sensor differ by 0.05%.

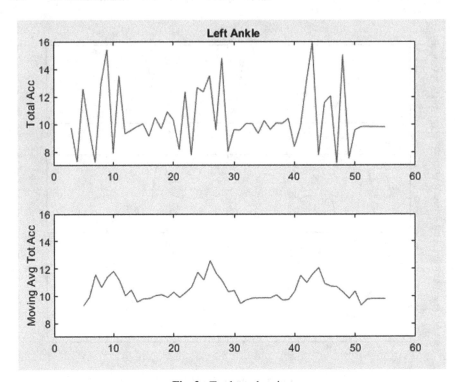

Fig. 3. Total acceleration

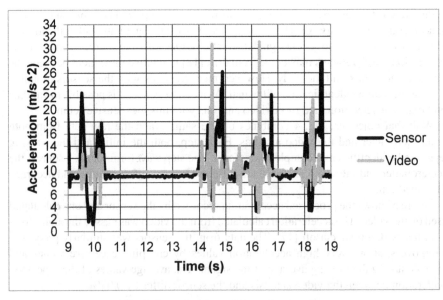

Fig. 4. Raw data from video and sensor data

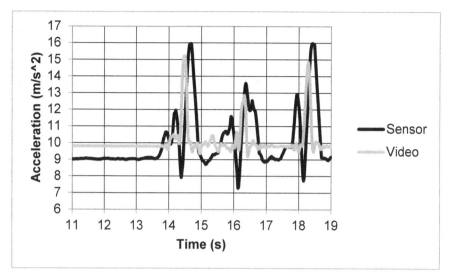

Fig. 5. Processed acceleration of video and sensor data.

4   Discussion and Conclusion

Comparing data from two different methods, video and sensors, it can be seen that all steps are clearly identified and synchronized. The idle stage before walking, at times between 11 and 13.6 s, and the intermediate stages between steps, for example, between 16.8 and 17.8 s, can be identified. Distinctive features of the acceleration from the sensor data are observed. Each step of the specific gait is represented by two maxima with a minimum value between them. These features cannot be identified in the video signal because the video is less sensitive to minute and fast changes in the acceleration and it identifies steps with single maxima. These differences between two completely different methods are expected because of their modes of operation. The sensors were strapped to the subject's ankle accurately following the movement but they have a limited relative movement to the actual bone, while the acceleration depends on the orientation of the sensor during walking [7, 8].

The acceleration calculated from the video analysis is calculated from the ankle movement using OpenPose. The software interprets each video frame separately identifying the various limbs of the person, and the time sequence of the spatial data allows the calculation of acceleration. As the software works on individual frames, minute differences in the pictures do not affect the identification of limbs but they affect the precise position of an ankle and limit the accuracy of accelerations. After the data was processed, the differences between the signals were minimal but the level of detail that the sensor provided was greater than the video signal. The main advantage of using a non-contact method, such as video analysis, to extract mechanics quantities, is that it is a fast and inexpensive method available in many real-life settings.

The results with OpenPose are promising. However, there are other multi-person pose estimation systems such as Alpha Pose[2], which is capable of following the same person across frames, that could provide more accurate measurements. Future work will consider such alternatives to verify video analysis for gait assessment. The goal of the current study is to contribute to the cost reduction of gait assessment. One benefit will be to reliably evaluate the gait of disabled people in remote locations. Future work is needed to verify that this non-contact method of video analysis can be used to assess the gait of disabled people with assistive devices in remote locations.

Acknowledgments. This study was supported in part by the MSIT (Ministry of Science and ICT), Korea, under the ICT Consilience Creative program (IITP-2020-2011-1-00783) supervised by the IITP (Institute for Information & Communications Technology Planning & Evaluation).

This study has been supported in part by Creating Hope in Conflict: A Humanitarian Grand Challenge; a partnership of USAID, The UK Government, and the Ministry of Foreign Affairs of the Netherlands, with support from Grand Challenges Canada.

References

1. Dell, N., Kumar, N.: The ins and outs of HCI for development. In: Proceedings of the 2016 CHI Conference on Human Factors in Computing Systems, pp. 2220–2232, May 2016
2. Chetty, M., Grinter, R.E.: HCI4D: HCI challenges in the global south. In: CHI 2007 extended abstracts on Human Factors in Computing Systems, pp. 2327–2332, April 2007
3. Bates, O., Thomas, V., Remy, C.: Doing good in HCI: can we broaden our agenda? Interactions **24**(5), 80–82 (2017)
4. Sultana, S., Guimbretière, F., Sengers, P., Dell, N.: Design within a patriarchal society: opportunities and challenges in designing for rural women in Bangladesh. In: Proceedings of the 2018 CHI Conference on Human Factors in Computing Systems, pp. 1–13, April 2018
5. Nwanna, O.: Validation of an Accelerometry Based Method of Human Gait Analysis (2014)
6. Satizábal, H.F., Rebetez, J., Pérez-Uribe, A.: Semi-supervised discovery of time-series templates for gesture spotting in activity recognition. In: ICPRAM, pp. 573–576 (2013)
7. Stroiescu, F., Daly, K., Kuris, B.: Event detection in an assisted living environment. In: 2011 Annual International Conference of the IEEE Engineering in Medicine and Biology Society, pp. 7581–7584. IEEE, January 2011
8. Gjoreski, H., Luštrek, M., Gams, M.: Context-based fall detection using inertial and location sensors. In: Paternò, F., de Ruyter, B., Markopoulos, P., Santoro, C., van Loenen, E., Luyten, K. (eds.) AmI 2012. LNCS, vol. 7683, pp. 1–16. Springer, Heidelberg (2012). https://doi.org/10.1007/978-3-642-34898-3_1
9. Baek, D., Musale, P., Ryoo, J.: Walk to show your identity: gait-based seamless user authentication framework using deep neural network. In: The 5th ACM Workshop on Wearable Systems and Applications, pp. 53–58, June 2019
10. Rong, L., Zhiguo, D., Jianzhong, Z., Ming, L.: Identification of individual walking patterns using gait acceleration. In: 2007 1st International Conference on Bioinformatics and Biomedical Engineering, pp. 543–546. IEEE, July 2007
11. Weiss, V., Bologna, G., Cloix, S., Hasler, D., Pun, T.: Walking behavior change detector for a "smart" walker. Proc. Comput. Sci. **39**, 43–50 (2014)

---

[2] https://www.mvig.org/research/alphapose.html.

12. Paulson, B., Cummings, D., Hammond, T.: Object interaction detection using hand posture cues in an office setting. Int. J. Hum Comput Stud. **69**(1–2), 19–29 (2011)
13. Ota, M., et al.: Verification of reliability and validity of motion analysis systems during bilateral squat using human pose tracking algorithm. Gait & Posture (2020)
14. Suzuki, S., Amemiya, Y., Sato, M.: Enhancement of gross-motor action recognition for children by CNN with OpenPose. In: IECON 2019–45th Annual Conference of the IEEE Industrial Electronics Society, vol. 1, pp. 5382–5387. IEEE, October 2019
15. Kaptelinin, V., Nardi, B.: Activity theory in HCI: fundamentals and reflections. Synt. Lect. Hum.-Centered Inf. **5**(1), 1–105 (2012)
16. Cao, Z., Simon, T., Wei, S.E., Sheikh, Y.: Real-time multi-person 2d pose estimation using part affinity fields. In: Proceedings of the IEEE Conference on Computer Vision and Pattern Recognition, pp. 7291–7299 (2017)
17. Lancini, M., Serpelloni, M., Pasinetti, S.: Instrumented crutches to measure the internal forces acting on upper limbs in powered exoskeleton users. In: 2015 6th International Workshop on Advances in Sensors and Interfaces (IWASI), pp. 175–180. IEEE, 18 June 2015
18. Neto, P., Pires, J.N., Moreira, A.P.: 3-D position estimation from inertial sensing: Minimizing the error from the process of double integration of accelerations. In: IECON 2013 – 39th Annual Conference of the IEEE Industrial Electronics Society, Vienna, pp. 4026–4031 (2013)

Implementation of CNN Model for Classification of User Sitting Posture Based on Pressure Distribution

Ji-Yun Seo[1], Ji-Su Lee[1], Sang-Joong Jung[1], Yun-Hong Noh[2], and Do-Un Jeong[1]($\boxtimes$)

[1] Dongseo University, 47 Jurye-ro, Sasang-gu, Busan 47011, Korea
{92sjy02,kkookkw3}@naver.com, {sjjung,
dujeong}@gdsu.dongseo.ac.kr
[2] Busan Digital University, 57 Jurye-ro, Sasang-gu, Busan 47011, Korea
yhnoh@bdu.ac.kr

Abstract. Musculoskeletal disease is often caused by sitting down for long period's time or by bad posture habits. In order to prevent musculoskeletal disease in daily life, it is the most important to correct the bad sitting posture to the right one through real-time monitoring. In this study, to detect the sitting information of user's without any constraints, we propose posture measurement system based on multi-channel pressure sensor and CNN model for classifying sitting posture types. The proposed CNN model can analyze 5 types of sitting postures based on sitting posture information. For the performance assessment of posture classification CNN model through field test, the accuracy, recall, precision, and F1 of the classification results were checked with 10 subjects. As the experiment results, 99.84% of accuracy, 99.6% of recall, 99.6% of precision, and 99.6% of F1 were verified.

Keywords: Musculoskeletal disease · Pressure distribution · CNN

1 Introduction

The incidence of musculoskeletal disease is increasing for modern people who spend most of their time sitting down on working and studying. The musculoskeletal disease is often caused by sitting down for a long time or bad sitting posture habits. Even though it is well known that bad posture strain a spine and can cause various diseases, most modern people do not take care of it. Studies on the determination and analysis of posture using acceleration sensor [1], pressure sensor [2], and camera image processing [3] were conducted to prevent musculoskeletal diseases through sitting posture monitoring. However, most of existing research requires measurement method with constraints, and expensive and high-performance measuring instruments. However, in case of the research using pressure sensor, posture's change rate can be measured by gravity center distribution and tilt posture in the front, back, left and right directions based on user's weight information without any constraints. Therefore, in this study, to detect the sitting information on a chair without any constraints, we propose posture measurement system

© Springer Nature Switzerland AG 2021
M. Singh et al. (Eds.): IHCI 2020, LNCS 12616, pp. 154–159, 2021.
https://doi.org/10.1007/978-3-030-68452-5_15

based on multi-channel pressure sensor and CNN model for classifying user's sitting posture types. The proposed posture measuring system is designed in cushion form, so it is easy to apply to existing products, and since it can measure without any constraints, it is easy to use in daily life. In addition, it is possible to analyze user's 5 types of postures according to pressure distribution based on sitting posture information.

2 Materials and Methods

2.1 Training Dataset for CNN Model

In this study, for composing dataset of CNN training model that can classify user's sitting posture, posture analysis system based on multi-pressure sensor was realized. The realized posture analysis system has 30 pressure sensors arrayed in 6 × 5 and measured the pressure distribution according to user's 5 types of sitting posture (right posture, left tilted, right tilted, front tilted, back tilted). Subsequently, the measured pressure distribution data were normalized from 0 to 100 and 2D-linear interpolation method is applied to create extended heat map image with the 600 × 500 matrix. After the images with interpolation went through a gray scale process, total of 10,000 sitting posture images (2,000 for each class) was used as a dataset for posture classification. Figure 1 shows an example of the training data preprocessing.

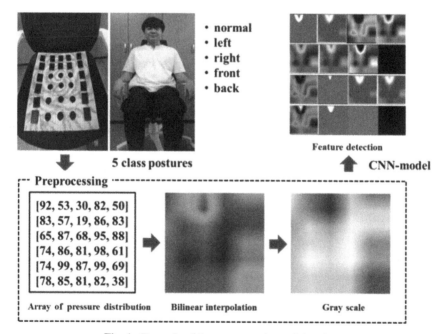

Fig. 1. Example of the preprocessing training data.

2.2 CNN Training Model for Classification of User Seating Posture

For posture classification based on pressure distribution according to user's sitting posture, 2D Convolutional neural network method was used [4]. Implemented neural network is consisting of 6 convolutional layers including convolution and pooling and 3 dense layers. The convolutional layer extracts input characteristics by generating a 3 × 3 feature map and used RELU as activation function. In addition, Drop-out was used to solve problems related to learning efficiencies, overfitting, and vanishing gradient. Dense layer serves as the connection and final output with the previous layer, and Soft-max was used as the output function. Configure of the implemented CNN model is shown in Fig. 2.

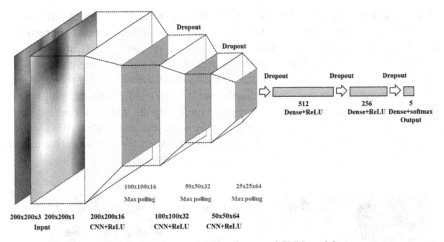

Fig. 2. Configure of the implemented CNN model.

The training environment was set to Adam for optimizer, 32 for batch size, 0.0001 for learning rate and early stop was added to prevent overfitting issue. Finally, for model evaluation, 10% of training data was set to validation data. The training result is shown in Fig. 3.

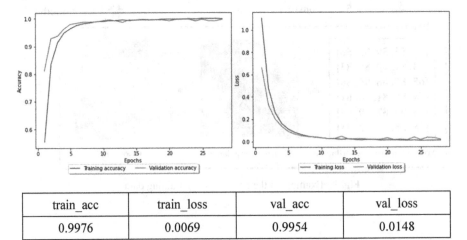

train_acc	train_loss	val_acc	val_loss
0.9976	0.0069	0.9954	0.0148

Fig. 3. Result of the training model.

3   Experiments and Results

For performance evaluation of the implemented posture classification CNN model, 500 data for evaluation was obtained for each class with 10 subjects. 5 types of sitting postures and pressure distributions are shown in Fig. 4. Then, with the obtained data set, the experiment to evaluate the posture classification performance of training model was conducted. The experiment calculated TP, TN, FP, and NP through confusion matrix regarding model's 5 type classification result and confirmed accuracy, recall, precision, and F1. Accuracy refers the number of truly predicted data in entire data, recall refers the number of data predicted as True in real True data, Precision refers the number of real True data in data predicted as True, and F1 refers the harmonic mean of Precision and Recall. As the result of the experiment, 99.84% of accuracy, 99.60% of recall, 99.60% of precision, 99.60% of F1 were confirmed. Classification confusion matrix of the implemented training model and experimental results are shown in Table 1.

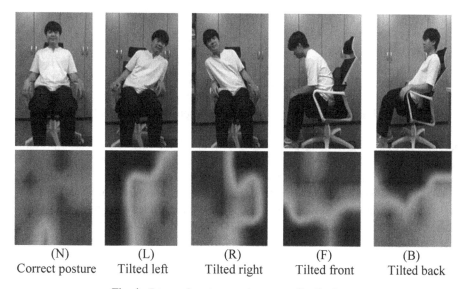

| (N) | (L) | (R) | (F) | (B) |
| Correct posture | Tilted left | Tilted right | Tilted front | Tilted back |

Fig. 4. 5 type of postures and pressure distributions.

Table 1. Result of the implemented training model classification with confusion matrix.

Confusion matrix		Predicted posture				
		N	L	R	F	B
True posture	N	0.99	0	0	0.01	0
	L	0	1	0	0	0
	R	0	0	1	0	0
	F	0.01	0	0	0.99	0
	B	0	0	0	0	1

Accuracy	Recall	Precision	F1
99.84	99.60	99.60	99.60

4   Conclusion

In this study, for the detect of sitting information on a chair without any constraints, multi-pressure sensor-based posture measuring system and CNN model for the classification of user's sitting posture are proposed. The proposed CNN model can analyze 5 posture types of users according to pressure distribution based on sitting posture information. For obtaining training data, the multi-channel pressure sensor-based posture analysis system was implemented, and the pressure distribution according to 5 sitting posture types were measured. 10,000 sitting posture images that performed the preprocessing of measured pressure distribution data were used for model training and 99.54% classification accuracy was confirmed. For performance evaluation of posture classification CNN training model through field-test, dataset for evaluation was obtained from 10 subjects and accuracy, recall, precision and F1 were identified. As a result, 99.84% of accuracy, 99.60% of recall, 99.60% of precision, and 99.60% of F1 were confirmed. In the future, we are to implement a high-performance neural network model that adds various posture classes and image processing methods for analyzing and monitoring the user's sitting posture by time zone and efficient posture correcting.

Acknowledgment. This research was supported by Basic Science Research Program through the National Research Foundation of Korea (NRF) funded by the Ministry of Education (No. 2018R1D1A1B07045337) and MSIT (Ministry of Science, ICT & Future Planning), Korean, under the National Program for Excellence in SW (2019-0-01817) supervised by the IITP (Institute of Information & communications Technology Planning & Evaluation).

References

1. Gao, L., Zhang, G., Yu, B., Qiao, Z., Wang, J.: Wearable human motion posture capture and medical health monitoring based on wireless sensor networks. Measurement, 108252 (2020)
2. Seo, J.Y., Noh, Y.H., Jeong, D.U.: Implementation of distracted estimation system based on sensor fusion through correlation analysis with concentration. Sensors **19**(9), 2053 (2019)
3. Byeon, Y.H., Lee, J.Y., Kim, D.H., Kwak, K.C.: Posture recognition using ensemble deep models under various home environments. Appl. Sci. **10**(4), 1287 (2020)
4. Kim, S.K., Huh, J.H.: Consistency of Medical Data Using Intelligent Neuron Faster R-CNN Algorithm for Smart Health Care Application

Implementation of Rehabilitation Exercise Posture Determination System Based on CNN Using EMG and Acceleration Sensors

Ji-Su Lee[1], Ji-Yun Seo[1], Sang-Joong Jung[1], Yun-Hong Noh[2], and Do-Un Jeong[1]($\boxtimes$)

[1] Dongseo University, 47 Jurye-ro, Sasang-gu, Busan 47011, Korea
{kkookkw3,92sjy02}@naver.com, {sjjung,
dujeong}@gdsu.dongseo.ac.kr
[2] Busan Digital University, 57 Jurye-ro, Sasang-gu, Busan 47011, Korea
yhnoh@bdu.ac.kr

Abstract. A number of musculoskeletal disorders occur worldwide in occupations that perform physically demanding tasks. In order to treat musculoskeletal disorders, rehabilitation must be performed, but it is not easy to correctly perform rehabilitation exercise by the patient at home where they spend a long time. Therefore, for effective rehabilitation exercise, a rehabilitation exercise posture determination system using body information of patients is required. In this paper, we implemented rehabilitation exercise posture determination system based on CNN using EMG and acceleration sensors. The implemented system measures data during rehabilitation exercise, performs pre-processing, and inputs it into the trained CNN model to determine the exercise posture. In order to evaluate performance of the implemented system, actual measurement data was input CNN model 50 times each to confirm the accuracy of posture determination. As a result of the experiment, the accuracy of 98.6% was confirmed.

Keywords: Rehabilitation exercise · Acceleration · EMG · CNN

1 Introduction

A number of musculoskeletal disorders occur worldwide in occupations that perform physically demanding tasks such as manufacturing workers and firefighters [1, 2]. In the early stages of musculoskeletal disease, non-surgical treatment is possible with simple medicine and rehabilitation exercises, but if the treatment time is missed, the condition deteriorated and surgical treatment is required. Rehabilitation exercises have a great rehabilitation effect only when repeatedly performed with a doctor and therapist [3]. However, it is not easy for patients to carry out rehabilitation exercises correctly at home where they spend a long time. Therefore, effective rehabilitation requires a system that can perform rehabilitation exercises based on physical information (muscular activation, movement, posture, etc.) of rehabilitation patients. In this paper, we implemented rehabilitation exercise posture determination system based on CNN using EMG and acceleration sensors to enable objective rehabilitation exercise based on user's body

© Springer Nature Switzerland AG 2021
M. Singh et al. (Eds.): IHCI 2020, LNCS 12616, pp. 160–166, 2021.
https://doi.org/10.1007/978-3-030-68452-5_16

information. CNN is deep learning, supervised learning neural network that analyzes the features of the input data to find the same pattern, and is used to classify various classes or predict values [4, 5]. Since bio-signals have unique patterns, it is easy to extract and classify features from signal patterns using CNN. The implemented rehabilitation exercise posture determination system uses EMG and acceleration sensors to measurement the user's body information during rehabilitation and classifies four classes (rest, lack of movement, lack of power, and exactly exercise) through the CNN.

2 Method and Materials

2.1 Rehabilitation Exercise Posture Determination System

In this paper, we implemented rehabilitation exercise posture determination system based on CNN using EMG and acceleration sensors for effective rehabilitation exercise. The implemented system uses an EMG sensor (PSL-iEMG2, PhysiLab Co.) and acceleration sensor (MPU-6050, InvenSense Co.) to measurement data according to rehabilitation exercises and perform a pre-processing using filter. After that, the pre-processed data is input to the CNN model to determine the posture of the rehabilitation exercise. Figure 1 shows the configuration of the rehabilitation exercise posture determination system.

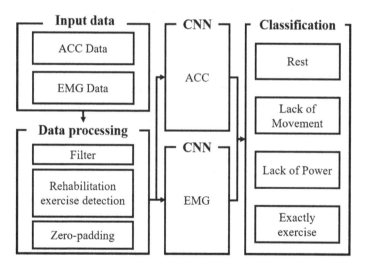

Fig. 1. Rehabilitation exercise posture determination system configure

2.2 Rehabilitation Exercise Data Set

Rehabilitation exercise data were measured by weight training (a) and squat (b), which are representative exercises that can be easily performed at home. (a) can recover weakened muscle strength by applying a constant load to the muscle by lifting and holding the object. (b) is an exercise that is performed when the pain is severe or the damaged area is fixed. Figure 2 shows the posture of weight training and squat.

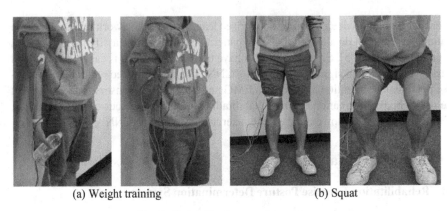

(a) Weight training   (b) Squat

Fig. 2. Rehabilitation exercise posture

The 2D CNN is determined by extracting the characteristics of the input image. However, it is difficult to distinguish the difference in waveform size in an image having the same waveform. Therefore, in order to detect the difference in the waveform size about the acceleration and EMG of the rehabilitation exercise, time series data with threshold values were used as training data. In order to construct rehabilitation exercise training data set were measured EMG and acceleration original data. The measured data were removed noise using a band pass (10–200 Hz) and a Kalman filter. In order to establish a relative threshold, all data converted to a value between 0 and 1 using min-max normalization. After that, one rehabilitation exercise was detected by extracting the minimum index on both sides of the acceleration waveform. Threshold values were set for the EMG and acceleration signals. When the threshold value was not exceeded, it was determined as Low, and when the threshold was exceeded, it was determined as High, and used as training data. Figure 3 shows the EMG and acceleration training data.

The rehabilitation exercise class was set up with four classes. When the two signals are Low, classified as Rest (R), when only the EMG is High, Lack of Movement (LM), when only the acceleration is High, Lack of Power (LP), and when two signals are High, Exactly exercise (E). The data set consisted of a total of 4,000 data, 1,000 for each class of EMG and acceleration training data. Table 1 shows the classification classes for rehabilitation exercises.

2.3 Implemented CNN Model for Posture Determination

In this paper, 1D CNN using time series data was used to determine rehabilitation exercise posture. The implemented CNN model consists of a total of 12 layers that convolution layer which extracts feature map that characteristics of training data using filters and kernel size, maxpooling layer which omits features except the maximum value in the feature map to prevent overfitting, dense layer which connects input and output. Figure 4 shows the configuration of the CNN model for rehabilitation exercise posture determination, and Table 2 shows the accuracy and loss of the CNN model.

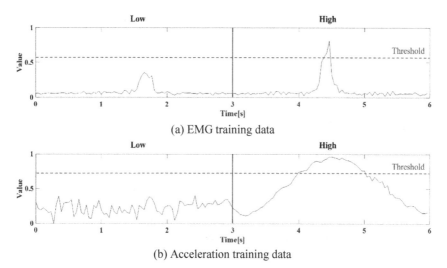

(a) EMG training data

(b) Acceleration training data

Fig. 3. EMG and acceleration training data

Table 1. Rehabilitation exercise classification class

Classification class	Acceleration	EMG
1. Rest (R)	Low	Low
2. Lack of Movement (LM)	Low	High
3. Lack of Power (LP)	High	Low
4. Exactly exercise (E)	High	High

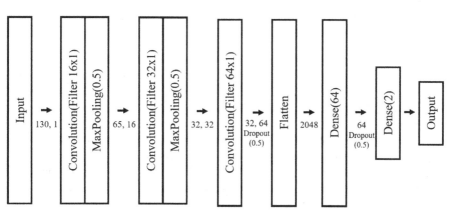

Fig. 4. CNN model configure

Table 2. Rehabilitation exercise posture determination CNN model accuracy and loss

Train accuracy	Train loss	Validation accuracy	Validation loss
1.0000	0.0012	0.9909	0.0141

3 Experimental Result

In this paper, we implemented rehabilitation exercise posture determination system based on CNN using EMG and acceleration sensors for correct rehabilitation exercise. The implemented system inputs the EMG and acceleration signals measured during the rehabilitation exercise into the CNN model to determine the rehabilitation exercise posture according to the exercise state. Figure 5 shows an example of experiment a posture determination system for rehabilitation exercises.

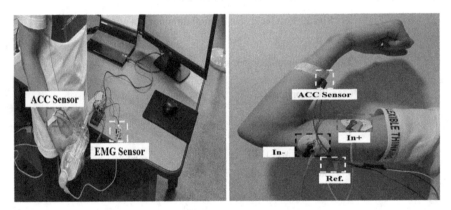

Fig. 5. Example of rehabilitation exercise experiment

To evaluate the performance of the implemented rehabilitation exercise posture determination system, a weight training experiment was conducted with 5 subjects. In the experiment, the EMG and acceleration signals were measured 50 times during rehabilitation exercise, and entered to the CNN model to determine the posture. In order to analyze the accuracy difference of the rehabilitation posture determination system according to the presence or absence of deep learning, a personalized rehabilitation monitoring system [6] and implemented system was compared the accuracy of the system's posture determination. The study in [6] is a personalized rehabilitation monitoring system that determines the posture of rehabilitation exercises according to changes in EMG and acceleration signals. The posture is determined by setting a specific threshold, and the proposed system uses CNN to determine the posture. As a result of the experiment, the accuracy of R was 97.6%, LM was 98%, LP was 99.6%, and E was 99.2% confirming a total accuracy of 98.6%. In addition, as a result of comparison with another thesis [6], the accuracy of the implemented rehabilitation exercise posture determination system using CNN was 98.6%, and the rehabilitation monitoring system according to the signal change was 94.9%, proposed system was 3.7% higher. Some errors in the implemented system are thought to have occurred by recognizing the slight shaking of the arm as an exercise when measuring weight training data in the R and LM classes with no change in acceleration. Table 3 shows the results of accuracy experiments of the proposed system (a) and the rehabilitation system using EMG and acceleration threshold value [6](b).

Table 3. Experiment results of rehabilitation exercise CNN model Accuracy

Sub.	Posture class							
	R		LM		LP		E	
	(a)	(b)	(a)	(b)	(a)	(b)	(a)	(b)
1	50	50	49	46	50	42	49	48
2	48	50	48	47	49	45	50	49
3	48	50	49	48	50	42	50	48
4	49	50	50	48	50	46	50	47
5	49	50	49	47	50	47	49	49
Acc. (%)	97.6	100	98	94.4	99.6	88.8	99.2	96.4
	(a) 98.6, (b) 94.9							

4 Conclusion

In this paper, we implemented rehabilitation exercise posture determination system based on CNN using EMG and acceleration sensors. The implemented system uses the data measured during the rehabilitation exercise to perform the pre-processing and then input it into the trained CNN model to determine the posture of the rehabilitation exercise. In order to evaluate the performance of the rehabilitation exercise posture determination system, actual measurement data were input to determine the posture, and an accuracy comparison experiment was conducted with the rehabilitation monitoring system using changes in EMG and acceleration signals. As a result of the experiment of the proposed system, the accuracy of 98.6% was confirmed, and the accuracy was 3.7% higher than rehabilitation monitoring system according to signal change. Some errors are thought to be occurred by slight arm shaking when measuring weight training data in the two classes with little change in acceleration. In future studies, we would like to conduct a more objective study of the rehabilitation posture determination system, such as measuring various posture data, expanding the classification class, and comparing the previous data with the current data.

Acknowledgement. This research was supported by Basic Science Research Program through the National Research Foundation of Korea (NRF) funded by the Ministry of Education (No. 2018R1D1A1B07045337) and MSIT (Ministry of Science, ICT & Future Planning), Korean, under the National Program for Excellence in SW (2019–0-01817) supervised by the IITP (Institute of Information & communications Technology Planning & Evaluation).

References

1. Carrillo-Castrillo, J.A., Perez-Mira, V., Pardo-Ferreira, M.D.C., Rubio-Romero, J.C.: Analysis of required investigations of work-related musculoskeletal disorders in Spain. Int. J. Environ. Res. Public Health **16**(10), 1682 (2019)
2. Kodom-Wiredu, J.K.: The relationship between firefighters' work demand and work-related musculoskeletal disorders: the moderating role of task characteristics. Saf. Health Work **10**(1), 61–66 (2019)
3. Hutting, N., Johnston, V., Staal, J.B., Heerkens, Y.F.: Promoting the use of self-management strategies for people with persistent musculoskeletal disorders: the role of physical therapists. J. Orthop. Sports Phys. Ther. **49**(4), 212–215 (2019)
4. Zhu, Z.A., Lu, Y.C., You, C.H., Chiang, C.K.: Deep learning for sensor-based rehabilitation exercise recognition and evaluation. Sensors **19**(4), 887 (2019)
5. Abo-Tabik, M., Costen, N., Darby, J., Benn, Y.: Towards a smart smoking cessation app: a 1D-CNN model predicting smoking events. Sensors **20**(4), 1099 (2020)
6. Jo, Y.H., Ha, J.Y., Seo, J.Y., Jeong, D.Y.: Implementation of customized rehabilitation monitoring system based on EMG and acceleration. In: Conference of Korea Institute of Convergence Signal Processing (2020)

Grabbing Pedestrian Attention with Interactive Signboard for Street Advertising

Heeyoon Jeong and Gerard Kim[(✉)]

Digital Experience Laboratory, Department of Computer Science and Engineering,
Korea University, Seoul, Korea
{nooyix,gjkim}@korea.ac.kr

Abstract. Digital interactive signboards are now very common. However, they are still mostly non-interactive and convey information only in one way. In this poster, we explore the idea of an advertisement that has a moving element according to that of the observer to grab and keep one's attention span and maximize the advertising effect. We have implemented a prototype of such an interactive signboard and conducted a pilot experiment comparing the gaze heat maps of observers of between our approach and traditional non-interactive signboard. Results have shown that the proposed approach was able to significantly increase the time the observer stayed with the advertisement while walking by it.

Keywords: Advertising · Attention · Interaction · Gaze tracking

1 Introduction

In today's digital era, traditional signboards are fast being replaced with digital signboards by the affordability of even large LCD displays. Among others, such signboards allow updating and switching of the contents (e.g. advertising, public announcements, exhibited art work) conveniently and quickly. For any signboards, it is most important that they capture the attention of the potential passer-bys and effectively convey the information at hand. In this poster, we explore an interactive advertisement in which the signboard advertisement is designed such that its main content object will "move" (within the display) along with the detected potential pedestrian observer as a way to grab one's attention. The hypothesis is that the observer might initially glance at the signage for its location and size, but to keep one's interest to keep looking, an additional conspicuous salient feature would be needed such as relative movement. Movement, in fact, is considered one of the very effective salient features that our low level vision system automatically perceives [1]. We validate such an effect by estimating the gaze direction of the observer and compute one's heat map and compare it to the case in which a traditional static and non-interactive signage is used.

© Springer Nature Switzerland AG 2021
M. Singh et al. (Eds.): IHCI 2020, LNCS 12616, pp. 167–171, 2021.
https://doi.org/10.1007/978-3-030-68452-5_17

2 Related Work

Digital signboards and such public displays are ubiquitous now. This is made possible in part by the increasingly lowering cost of even the large LCD displays and projectors. The digital signboards and kiosks are evolving into the interactive ones as well – but still mostly for the purpose of the user to directly make input, usually through touch or special buttons such as in public map/shop guidance systems [2]. Despite the projected advantages of the user interaction [3], advertisement with digital displays largely remains to be non-interactive – with the exceptions of special installation types – that is, no "general" framework for interactive advertisement stands out as yet. The main problem in realizing advertising to be more interactive has to do with serving a large mass of potential observers at once.

In this work, we leverage upon the visual saliency of relative movement as a way to grab the user's attention. Visual salience refers to is the state or quality by which it stands out from its neighbors. Saliency detection is considered to be a key attentional mechanism that facilitates learning and survival by enabling organisms to focus their limited perceptual and cognitive resources on the most pertinent subset of the available sensory data. Examples of visually salient features include contrasts in color, movement, size, and eccentricity (in visual location) [3].

3 System Implementation

The prototype of the proposed interactive advertisement was implemented as follows. A large display (43 in. monitor) is used as the signboard connected to a PC. The webcam is used to observe the area near the front of the advertising display. The image taken by the webcam is analyzed to detect the human figures (the passer-bys), more specifically their faces and orientations using the OpenPose [4] and OpenFace [5] libraries. Note that the gaze direction of the detected observer is only estimated by the front orientation of one's face. As there can be many potential observers detected, only one observer is picked at a time, e.g. the one who walks from the far left from the perspective of the advertisement webcam. The advertisement is designed such that the key element (arrow looking object in Fig. 2) in the content is initially located in the right end (or left from the perspective of the webcam), and assuming (for now) that the observer walks and passes by from the left to the right, it moves in the same pace toward the left until the passerby is out of sight from the camera. Then the key element is located back to its initial position and a new cycle of interactivity begins. The advertisement content was designed and implemented using the Unity. Figures 1 and 2 show the overall architecture of the system and snap shots of the workings of the interactive advertisement.

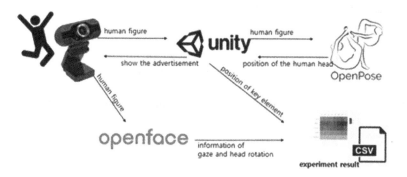

Fig. 1. The overall architecture of the proposed interactive advertisement.

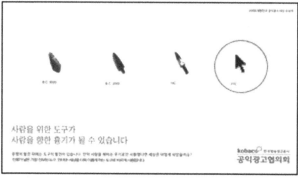

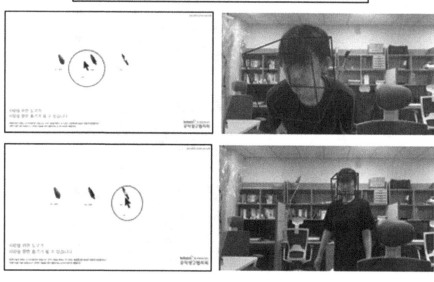

Fig. 2. Snap shots of the interactive advertisement whose key element moves along with the detected and designated observer.

4   Experiment and Result

To validate the effects of our proposed interactive advertising scheme, we ran a small pilot experiment comparing the heat maps and gaze stay time between the users of the proposed system and the non-interactive version. Note that the heat map or gaze time was recorded (and regarded as looking at the advertisement) only when the designated observer's head/gaze direction was within + or $-30°$ with respect to the viewing direction of the webcam. Five participants

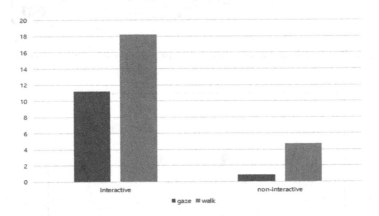

Fig. 3. The gaze time between the proposed interactive advertising system and the noninteractive one.

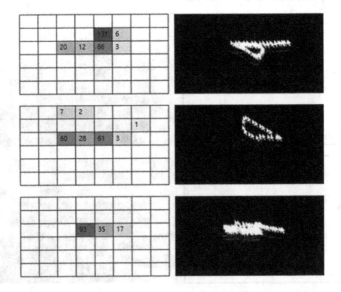

Fig. 4. Movement of the key element and the gaze heat maps of the users of the proposed interactive advertising system - follows the key element (left), and the non-interactive one (right)

(graduate students of age between 25 and 29) were recruited and instructed to walk by the two types of advertisement (in balanced order) and their heat map and gaze time recorded. Figure 3 shows that the users of the proposed interactive system exhibited a significantly higher gaze time toward and Fig. 4 shows representative heat maps that illustrate that the observers gaze followed that of the moving key element, contributing to this improved gaze time.

5   Conclusion

In this poster, we proposed for a general framework for interactive advertisement based on animating and moving the key element in the advertisement according to that of the observer. Our pilot experiment has shown that such an effect was indeed able to capture and maintain the one's attention and thereby increase the advertising effect. While the system only considers one observer at a time, we believe that the key element movement will also do the same for the other undetected/undesignated observers (however, more experiments and implementation would be required to show this formally). Other provisions will be needed such as how to deal with situations where the designated stops looking and having to switch to another observer. Perhaps, a method for grabbing collective attention (rather than for the chosen one) can be a future direction. It is not clear e.g. properly design such an advertisement, e.g. how to select the key element, how to strategically animate and move them.

References

1. LNCS. https://en.wikipedia.org/wiki/Digital_signage. Accessed 10 Oct 2020
2. Baltrusaitis, T., Zadeh, A., Lim, Y.C., Morency, L.P.: OpenFace 2.0: facial behavior analysis toolkit. In: 2018 13th IEEE International Conference on Automatic Face & Gesture Recognition (FG 2018), pp. 59–66. IEEE (2018)
3. Li, Z.: A saliency map in primary visual cortex. Trends Cogn. Sci. 6(1), 9–16 (2002)
4. Pavlou, P.A., Stewart, D.W.: Measuring the effects and effectiveness of interactive advertising: a research agenda. J. Interact. Advertising 1(1), 61–77 (2000)
5. Zhou, X., Wang, D., Krahenbuhl, P.: Objects as points. arXiv preprint arXiv:1904.07850 (2019)

Requirements for Upper-Limb Rehabilitation with FES and Exoskeleton

Woojin Kim[1](✉), Hyunwoo Joe[1], HyunSuk Kim[1], Seung-Jun Lee[1], Daesub Yoon[1],
Je Hyung Jung[2], Borja Bornail Acuña[2], Hooman Lee[3], Javier Fínez Raton[4],
Carlos Fernández Isoird[4], Iker Mariñelarena[4], Miguel Angel Aldudo Alonso[5],
Myung Jun Shin[6], and Tae Sung Park[6]

[1] ETRI, Daejeon, Republic of Korea
wjinkim@etri.re.kr
[2] TECNALIA, Donostia-San Sebastián, Spain
[3] EXOSYSTEMS, Seongnam-si, Gyeonggi-do, Republic of Korea
[4] Gogoa Mobility Robots, Abadiño, Bizkaia, Spain
[5] IHS WEIGLING S.L., Erandio, Bizkaia, Spain
[6] Pusan National University Hospital, Busan, Republic of Korea

Abstract. In the last work, we have presented the scope of our project, *i.e.* use cases of activities of daily living (ADL) for the on-going project a.k.a. iCARE. The project mainly handles the upper-limb rehabilitation in general, however, we have narrowed down the scope and focus on, in terms of the phase of the stroke recovery, the target body area of rehabilitation and the level of muscle function. In this paper, we have drawn the user and system requirements before design the specific functions of the targeted device. First, we have defined the stakeholders for the device and the rehabilitation service scenarios. Next, the user requirements are defined and finally the related system requirements are drawn.

Keywords: Rehabilitation · Upper limb · Requirement · Home-care

1 Introduction

The goal of this work is to facilitate the at-home rehabilitation in the automated manner, demanding smart and safe customization of rehabilitative systems to improve the stroke patient's status and recovery level by applying local and real-time intelligence while they are providing easy operation of the system.

The project iCARE has the following innovative research issues:

- In order to change the rehabilitation culture from at-clinic to at-home, the cost should be affordable and the rehabilitative training should be changed from therapy-centric to patient-centric.
- The rehabilitative training should be customized to the patient for the effective treatment.

M. Singh et al. (Eds.): IHCI 2020, LNCS 12616, pp. 172–177, 2021.
https://doi.org/10.1007/978-3-030-68452-5_18

Users' benefit has been our priority of iCARE project [2] while specifying the user and system requirements. These requirements are important and must be valid for patients, doctors, clinics and medical-device providers. Although the stakeholders' opinions are important, another main goal of this project is R&D for the next generation's technology and the researchers' opinions have also been included according to the research issues mentioned above.

During the project, important business-related factors, such as usability, utility and economics have be discussed in detail in accordance to the use cases developed in the previous work [1].

In this work, we develop user and system requirements for upper-limb rehabilitation with functional electric stimulators (FESs) and an exoskeleton. We have focused on the wrist and elbow rehabilitation and due to the different characteristics of two body parts, we have divided the requirements into two groups, i.e. requirements for the wrist rehabilitation and requirements for the elbow rehabilitation.

The rest of this paper is organized as follows: in Sect. 2, the stakeholders are defined. In Sect. 3, wrist and elbow rehabilitation service scenarios are described. In Sect. 4, the user requirements are introduced and the system requirements are specified in Sect. 5. Finally, the conclusions are followed in Sect. 6.

2   Major Interested Parties

Major interested parties, a.k.a. stakeholders, are parties that have an interest in the outcome and can either affect or be affected by the business. The basic stakeholders in the R&D project is the client (or funding agency), business operators and users.

2.1   Client

The client indicates the person or organization that finally confirms the completion of the research outcomes, or the person/organization that provides research funding. The iCARE project is co-funded by KIAT and CDTI from South Korea and Spain, relatively.

2.2   Business Operator

The business operator indicates the person or organization that conducts business using the research outcomes. Our project has two business partners, EXOSYSTEMS and Gogoa Mobility Robots from South Korea and Spain, relatively.

2.3   User

The user indicates the person or organization that uses the final research outcomes. The iCARE project aims to develop the rehabilitation devices, so the user can be defined as the patient.

2.4 Tester

Since our final outcome is the medical device, clinical test is mandatory. ETRI's quality assurance team will be in the charge of the technical testing and Pusan National University Hospital will be in the charge of the clinical testing.

It is important to note that the output of the "Sensing" is used as the input of the "Control". At the same time, the output of the "Control" is used as the input of the "Sensing", generally known as a closed-loop control system.

3 User Requirements

In order to develop a novel wrist/elbow rehabilitation device, we have drawn 37 user requirements (19 for wrist and 18 for elbow). The basic concept is that we are going to use the functional electrical stimulation (FES) for the wrist and both the exoskeletal robot and FES will be applied for the elbow rehabilitation. For example, user requirements specify the following items for the body actuation:

- User should be able to attach and detach the FES on the wrist.
- User should be able to attach and detach the exoskeleton robot and the FES on the elbow

Since this work aims to accomplish the closed-loop rehabilitation as mentioned in the previous section, sensors have been included in the devices. For example, user requirements specify the following items for the body movement sensing:

- User should be able to measure extension, flexion, radial and ulnar movement of the wrist via the device.
- User should be able to measure extension and flexion movement of the elbow via the device.

4 System Requirements

From the user requirements, we have developed the system requirements. First, we have define the system configuration. The targeted system aims to develop a system that can be used for rehabilitation training of the upper limbs (the wrist and the elbow) of patients in need of recovery of a loss of muscle functions caused by a stroke or an external injury.

For effective rehabilitation training of each part of the upper limbs, this system requires a high level of adaptability and elaboration and consists of the following three modules:

- A cognitive computing module for AI computation of sensors and actuators to achieve rehabilitation.
- An intelligent interaction software module for connection-oriented measurements and actuation solutions

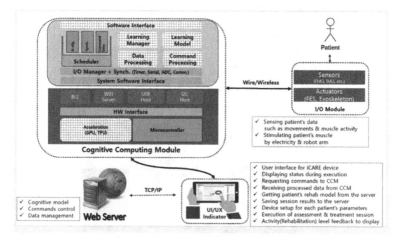

Fig. 1. Upper-limb rehabilitation architecture

- An auxiliary module that performs sensing, actuating and feed-backing for rehabilitation

In accordance with the architecture in Fig. 1, we have drawn 41 system functional requirements and 11 system non-functional requirements. Every item has the related user requirements and we have marked for tracking in the further development of the device. The followings are the system functional and non-functional requirements which are drawn from the user requirements mentioned in Sect. 4:

- The interacting part of the FES should be worn around the wrist.
- The weight of the worn part of the FES should be less than 500 g.
- The exoskeleton robot and the FES should be worn around the elbow.
- The exoskeleton robot and the FES should be made of cleanable materials.
- The device should measure the orientation of the wrist by means, *e.g.* inertia measurement unit (IMU).
- The device should measure activation of muscles linked to the wrist motion using multi-array electromyography (EMG) sensors.
- The exoskeleton robot should measure the angle of the elbow by a encoder (or similar means).
- The device should measure activation of muscles linked to the elbow motion using multi-array EMG sensors.

Figure 2 shows our current version of the rehabilitation system and its visualization software, which is implemented based on the system requirement specification aforementioned. Starting from the current version, we will collect data, revise the control algorithm and try the clinical test for the success of the project.

Fig. 2. Our prototype of the project including: (from top, left) the exoskeleton robot for elbow, the electric stimulator with printable electrodes and the charging dock, EMG and IMU sensors and their visualization software

5   Discussions

In this work, we have focused on drawing user and system requirements for the project iCARE. It is challenging that using both the robotic actuation and electrical stimulation at a time and we have drawn novel requirements for integrating those functions at one system.

6   Conclusions

In this paper, we have presented our system, named iCARE, which provides rehabilitation functionalities to the patients who need to recover upper limb muscle function. The goal functionality of our system is developing components enhancing AI-based computation, meeting real-time feedback, and actuation of the system.

In order to develop the novel rehabilitation device, we have considered various range of the stakeholders and interviewed for drawing fine user requirements. We have drawn 37 user requirements and 52 system requirements, consequently.

Based on the requirements, we have developed the proto-type and in the future, we will collect data, revise the control algorithm and try the clinical test with it.

Acknowledgment. This research was financially supported by the Ministry of Trade, Industry and Energy (MOTIE) and Korea Institute for Advancement of Technology (KIAT) through the International Cooperative R&D program (P0007114-InterConnected Intelligent Sensing and Actuation Solutions for At-home Rehabilitation (iCARE)(2020)).

References

1. Kim, W., et al.: Use case specification for upper-limb rehabilitation with FES and exoskeleton. In: 10th International Conference on ICT Convergence (ICTC 2019) (2019)
2. iCARE: project web page. https://www.icare-athome.com/

Development of Biosensor for Pressure Sores Diagnosis and Treatment Using Bio Photonics with Impedance

Eun-Bin Park and Jong-Ha Lee[✉]

Department of Biomedical Engineering, School of Medicine, Keimyung University,
Daegu, South Korea
segeberg@kmu.ac.kr

Abstract. Pressure Sores cause huge pain, financial burden and so on. Diagnosing Pressure Sores is a doctor's visual inspection that is the only available alternative. Because there is no quantitative diagnostic technique for Pressure Sores. The main problem of this methods is the difficulty of early diagnosis, which often makes fast care impossible. This study researched a bio-photonic sensor that can be attached to skin for diagnosing and treating Pressure Sores.

Keywords: Pressure sore · Pressure sore · Bedsore · Impedance · Laser irradiation · Bioimpedance

1 Introduction

Pressure Sores are a chronic skin disease that are caused by pressure. Pressure Sores cause huge pain, financial burden and so on. Diagnosing Pressure Sores is a doctor's visual inspection that is the only available alternative. Because there is no quantitative diagnostic technique for Pressure Sores. The main problem of this methods is the difficulty of early diagnosis, which often makes fast care impossible. This study researched a bio-photonic sensor that can be attached to skin for diagnosing and treating Pressure Sores.

1.1 Pressure Sore

A pressure sore is a state in which a pressure is constantly applied to a body part while sitting or lying down in a posture, or a circulatory disorder occurs in the part due to friction or the like to cause damage or soreation of the subcutaneous tissue (National Pressure Sore Advisory Panel 2009). These pressure sores are more common in patients who are hospitalized for short- and long-term periods compared to the general population, and they experience pain and discomfort at the time of the occurrence, lead to prolonged hospitalization and economic problems which are considered to be quite difficult to overcome in health care (NPUAP 2009). In Korea, 30–50% of inpatients in intensive care units are reported to have pressure sores, and 12–17% of foreign patients

© Springer Nature Switzerland AG 2021
M. Singh et al. (Eds.): IHCI 2020, LNCS 12616, pp. 178–187, 2021.
https://doi.org/10.1007/978-3-030-68452-5_19

have a relatively low incidence of pressure sores compared to Korean patients. The prevalence of pressure sores in patients with intensive care unit (ICU) was 5.2–43.3%. As shown in Fig. 1, critically ill patients have higher rates of prevalence and incidence of pressure sores than ordinary patients because they have a higher level of awareness and sensory perception, immobilization, edema, long-term treatment for respiratory as well as nutrition imbalance compared to general patients. The risk factors for the occurrence of pressure sores are reported to be related to the characteristics of diseases such as sensory perception, humidity, mobility, friction and shear force, long-term hospitalization, paralysis, edema, diabetes, and nutritional indicators. Preventing and early treatment of pressure sores is very important because patients have very low incidence of consciousness, limited activity and mobility, and bedsores occur very easily.

1.2 Impedance

Impedance is a value that interrupts the current flow when a voltage is applied in the circuit. The value of the impedance can be expressed by the ratio of the voltage and current of the AC circuit, and generally depends on the frequency of the AC voltage. Impedance in an AC circuit is an extension of the concept of resistance, which has a magnitude and phase value, unlike a resistor with only a magnitude value. In a DC circuit, the impedance and the resistance are the same, and the phase angle of the impedance is assumed to be zero. Impedance is often used when analyzing an AC circuit in an electrical network because the relationship between the voltage and the current in the form of a trigonometric function can be expressed in straight lines using impedance. Impedance is expressed as a complex number. In the international system, the unit is ohm (Ω). Usually, the method, which is shown in the complex plane, is most often used when analyzing a circuit. Unlike the electric resistance of a DC circuit, a large impedance can not be determined to have a higher resistance than a small one. When connecting two different circuits, the impedance of each circuit should be the same.

1.3 Low-Power Phototherapy Technology

Low-power phototherapy technology is a phototherapy technology that can show clinical effects by stimulating or suppressing the function of cells by irradiating light energy of appropriate level (1 mW–500 mW) with typical phototherapy. The low-power phototherapy technology uses a light source of visible light or near infrared light mainly in electromagnetic waves, or a combination of both. The main applications of medical care are pain and inflammation reduction, prevention of injury to the wounded tissue and promotion of tissue and nerve regeneration. Although the mechanism of general low-power phototherapy techniques is not accurately identified, the mechanisms assumed through some literatures are reviewed as follows. When the visible light or near-infrared light energy absorbed in the cells reaches the cell tissue to be irradiated, The mitochondrial respiratory chain cytochrome oxidase accepts the light as the photoacceptor. The resulting photons increase the synthesis of ROS (Reactive Oxygen Species) and ATP (Adenosine tipphosphate) and release Nitric Oxide. At an appropriate time and frequency, light of a selected wavelength leads to a continuous intracellular biochemical reaction, which is known to play an important role in cell regeneration and rehabilitation by actively performing the overall metabolic activity. Generally, light of a wavelength between 390 nm and 1,100 nm is continuously or pulsed. Relatively low fluence and power density were used in normal environments (AlGhamdi et al. 2012). Low-power phototherapy technology, known to be effective through a number of studies, was already widely used in clinical practice (Tables 1 and 2). The most clinically applicable field is dermatology, which is used for anti-aging and acne treatment. Acne bacteria produce potpyrin in the metabolic process, which acts as a photosensitizer that absorbs light energy and induces chemical reactions in cells or tissues. In other words, porphyrin absorbs light energy and causes a chemical reaction. In this process, free radicals are liberated and acne improves (Aziz-Jalali et al. 2012). Skin aging is associated with symptoms such as wrinkles, pigmentation, capillary dilatation, and loss of elasticity as the skin ages which is caused by tissue-based collagen reduction, reduced elasticity of elastic fibers, and shrinkage of skin blood vessels. Already, many studies have shown that bio-photonics research plays a role in helping cells regenerate and prevent aging of the skin. Microscopic structural changes in tissue were observed for 3 months using a light emitting diode (LED) with 830 nm wavelength (55 mW/cm^2, 66 J/cm^2) and 633 nm wavelength (105 mW/cm^2, 126 J/cm^2) And microscopic changes of tissue were observed for 3 months through skin elasticity, melanin photography, and profile method. As a result, increased the content of collagen and amount of elastic fibers were confirmed to reduce wrinkles (up to 36%) and increase skin elasticity (up to 19%) (Lee et al. 2007). In addition, subjects were exposed to light through 590 nm (0.1 J/cm^2) light emitting diodes for 8 consecutive times over 4 weeks, and then evaluated at 4, 8, 12, 18 weeks, 6 months and 12 months. Digital imaging data confirms that the surface of the skin is smooth, the surrounding glands are reduced, and the aging symptoms are reduced by 90% or more in subjects with erythema and pigmentation (Weiss et al. 2005). Low-power phototherapy technology is also known to be effective in wound healing and restoration. The low-power phototherapy technology also showed an effect in clinical trials of burn-harm therapy, which is known to be difficult to treat. Laser with 670 nm (soft laser, 400 mW) wavelength was randomly performed on 19 patients with burn scars for eight weeks. As a result, the control was decreased from 6.10 ± 2.86 to 5.88 ± 2.72 and from 7.10 ± 2.86 to 4.68 ± 2.05 after treatment in the VSS (Vancuber Scar Scale) evaluation and the experimental group respectively. As a result, 17 of the 19 scars identified positive effects (Gaida et al. 2004). Studies have shown that low-power phototherapy technology is also effective in promoting hair growth. Clinical trials were conducted on 300 patients suffering hair loss from the a low-level laser device, the US FDA-cleared HairMax Lasercomb®. The tests were conducted randomly by subjects receiving treatment for 26 weeks with three models and one of the fake packets. As a result, practitioners treated with actual low-power phototherapy techniques have had an effect on hair regrowth compared to patients treated with fake packets (Jimenez et al. 2014). In addition, many studies have demonstrated that low power phototherapy technology is effective for hair loss and hair regrowth. A number of studies have shown that chronic pain is effective in treating various types of musculoskeletal pain (Branco and Naeser 1999; Okuni et al. 2012; Ozkan et al. 2004). Nine men and 15 women with pain in their stomachs were studied to see if low-power phototherapy techniques worked for pain, ranging from elbow, wrist and finger. A semiconductor laser device with a wavelength of 1000 mW was evaluated twice a week for one month after treatment. The evaluation method was VAS. As a result, it was confirmed that the VAS score was significantly improved after treatment with the mean of VAS (visual analog scale) 59.2 ± 12.9 before laser irradiation and 33.1 ± 12.2 after irradiation. it was concluded that low-power phototherapy techniques at 1000 mW wavelengths were effective for chronic pain in the elbow, wrist and finger (Okuni et al. 2012). Currently, light sources such as lasers and light-emitting diodes are the most effective light sources in optogenetics based on phototherapy technology. The laser has a narrow emission pattern and can emit light intensively at a specific region. In addition, the coupling efficiency is high and uniformity of the same phase and wave is exhibited. However, since the laser means amplification of light by induced emission, a separate device for inducing light amplification is required, which may limit the miniaturization of the product (Cotler et al. 2015; Jin-Cheol 2016). On the other hand, a light emitting diode has a wider output spectrum than a laser light source and can be used for a wide range of irradiation. In addition, It also has the advantage of having strong but steady power and being able to control brightness and miniaturize it and extend its function to various types of wavelengths (Chung et al. 2012; Hashmi et al. 2010; Jin-Cheol 2016). Therefore, although laser is mainly used in the early stage of light treatment technology research, there have been many cases of using light emitting diodes recently. Over a long period of research, phototherapy has been used to treat various diseases in the human body. The effects of biochemical treatment are different depending on various variables such as specific wavelengths, energy densities, irradiation times, and pulses. In other words, it is most important to investigate the optimum light and dose values in a specific areas. Lower and higher doses may result lower treatment effectiveness and tissue damage respectively (Posten et al. 2005). Therefore, proper selection of light source and dose should be supported to minimize side effects and maximize therapeutic effect. Proper use and maintenance of the equipment is also required.

2   Material and Method

2.1   Impedance Measurement and Light Irradiation with Light Emitting Diode

In this study, impedance and electromyography were measured for the affected part of the pressure sore, and the degree of damage to the pressure sore area was rapidly made possible. A constant-voltage sinusoidal signal with a frequency of 100 Hz to 1 MHz was output using a E4980 Precision LCR Meter 20 Hz to 1 MHz at 100 mV for impedance magnitude and phase measurement. And Impedance and electromyography were measured for the affected part of the pressure sore, and the degree of damage to the pressure sore area was rapidly made possible. A constant-voltage sinusoidal signal with a frequency of 100 Hz to 1 MHz was output using a E4980 Precision LCR Meter 20 Hz to 1 MHz at 100 mV for impedance magnitude and phase measurement.

2.2   Experiments to Induce Pressure Sores Through Rat Model

The animals of this experiment were 12 male white rats (Sprague-Dawley) between 8 and 9 weeks of age, weighing 250–300 g and raised under the same conditions. After Rat was anesthetized with isofluorane mixed with oxygen, the back of Rat was shaved and the hair was removed using Veet hair cream removal, and then pressed with a forceps having a pressure of 140 mmHg to apply a constant pressure. A pressure of 140 mmHg was applied to the back of the rat for 3 h to induce a pressure sore of about two stages Fig. 1. The animals in this experiment were 12 male white rats (Sprage-Dawley) with a weight of between 250 and 300 g and 8 weeks to 9 weeks. After anesthetizing Rat with isofluoran mixed with oxygen, the back of Rat was shaved and the hair was removed using Veret deflated cream, and the part was taken with a tong with 140 mmHg of pressure to apply a constant pressure. A 140 mmHg pressure was applied to the back of Rat for 3 h, causing second level of pressure sore in Fig. 1.

2.3 Light Irradiation Using LED

As it shown in Fig. 2, The devices for irradiating light were a DC4104 (Thorlabs, USA) and a 4-Wavelength High-Power LED Head (LED4D231 model). The DC4104 has switch and brightness control through current setting, more safety than halogen lamp, and a long life that can be used for long time because it can be controlled by cooling fan. The operating temperature of LED4D231 is from 0 to 40 °C and the length, width, height, and weight are 164 mm, 150 mm, and 57 mm, and 1.6 kg respectively. For using 660 nm wavelength in the experiment, the minimum output power, the maximum current, the forward current. and the bandwidth are 210 mW, 1200 mA, 2.5 V, and 25 nm respectively. In this experiment, preliminary studies were carried out using 4 wavelengths (365 nm, 455 nm, 545 nm, 660 nm) to select 660 nm of the near–infrared region that induces cell proliferation without becoming toxic to the cell. Therefore, the wavelength of 660 nm was used in this study and the current was set to 500 mA.

2.4 Confirm Wound Healing Ability by Photonics

In this experiment, the mouse melanoma cell line, B16-F1 cells, 3×10^5 cells/ well was incubated in a 24-well plate (Becton Dickinson, San Jose, Calif.). When the cells were covered, the cells were scratched with 1000 μl tip and washed twice with PBS to remove the remaining cells.

1% FBS/DMEM medium was mixed with samples (rhVEGF, anti-human VEGF Ab, sFlt-1, SU5416) and it was added to the cells for 2 to 3 days and replaced with fresh medium every day. After the incubation, the extent of wound healing was observed with an optical microscope ($\times 100$).

2.5 Determination of Efficacy in Promoting Angioplasty by Photonics

In this experiment, HUVEC was replaced with medium M199 containing 1% FBS and starvation was performed for 6 h. 250 μl of ice-cold Matrigel (BD Bioscience) was added to the 24-well plate and the mixture was incubated at 37 °C for 30 min.

After completion of the starvation, the concentration of the cells was adjusted to 1×10^5 cells/well and Medium (M199/1% FBS) mixed with samples (rhVEGF, anti-human VEGF Ab, sFlt-1 and SU5416) was added. After culturing for 12 h, the degree of tube formation was observed with an optical microscope ($\times 100$).

3   Result and Discussion

3.1   Tissue Impedance Mapping of Multiple Electrode Arrays

As it shown in Fig. 3 we design and implement an electronic sensing device that measures spatially correlated complex impedance in vivo using a multi-electrode array. The device controls the electrode arrangement in contact with the skin and the hardware perform impedance spectroscopy in the array, and also developed 2 versions of the electrode array. In this paper, we developed a method to detect Pressure Sores by using impedance spectroscopy and at the same time, it was made flexible to use the method in vivo. Gold nano-particle ink has been printed in inkjet on thin (35 mm) polyethylene or phthalate substrates, which can improve the suitability of printed arrays and improve electrode-to-skin contact rates. It was essential to minimize process variability because several duplicated print arrangements were used during animal testing and the production process was found to be highly reproducible. Mechanical stability of printed lines was another major concern. Because the sensor array should be maintained in a worn state during data acquisition and the array is subjected to bending and twisting, the mechanical rigidity was tested by applying a torsional load to the array with twist angle $j = 30°$. Sampling the nearest neighbors of the array and configuring a map of the impedance measurements, as they circulate through the pair. As a result of confirming the wound healing ability by laser irradiation at 660 nm light irradiation, the result is similar to VEGF which is used as a positive control in laser irradiation group. The result is confirmed as compared with the non-treated control group. These results confirm that the laser irradiation at 660 nm light irradiation is effective for wound healing.

3.2   Relationship Between Impedance Spectrum and Tissue Health

Impedance spectra are associated with tissue health. 12 animals were used in this study, 11 animals and 9 from 12 were received 1 h and 3 h treatment respec-tively. fluorescence angiography was utilized to image real-time blood flow in the tissues, we found that the initial pressure relief as the blood returns to the affected tissue increases perfusion (reactive hypertension). When blood returns to Ischemic tissue, reactive oxygen species and free radicals, which can accelerate cell apoptosis. In our experimental model, 1 h and 3 h of pressure caused mild and severe reversible tissue damage respectively. After the Ischemic process, the wound was traced for at least 3 days using impedance spectroscopy. The system used a 100 mV RMS constant-voltage test signal to measure the impedance at frequencies between 100 Hz and 1000000 Hz for all the closest electrode pairs in the array. Impedance spectra of the sites showing Pressure Sores in all animals of the 3 h group which could be clearly distinguished by day 3. The tissue was less capacitive, good conductivity, and the integrity of the cell membrane was lost. In this paper, we have observed that the pole point for wound tissue in the impedance spectrum is lower frequency than the healthy tissue, and the impedance spectrum collected from the 14 wounds from 8 rats into a flexible electrode array was used to determine the specific threshold of impedance size and phase defining 'damaged tissue' (pressure sore). For each wound, the electrode is classified as 'pressure' or 'no pressure', depending on whether or not the tissue under pressure has been measured. Decrease in impedance amplitude and near zero phase angle for the three-hour pressure group was evident in the ensembled data. Rat pressure sore and photonics Impedance change by 660 nm light irradiation. We confirmed the possibility of diagnosis through the impedance in the Rat model of the pressure sore. As a result of checking the impedance values in various frequency ranges, the frequency range was confirmed which diagnosis of skin lesions or Pressure Sores was possible according to the difference of impedance values. The corresponding frequency range was confirmed to be 10 kHz to 100 kHz. In addition, it was confirmed that the impedance value of photonics treatment group is close to the normal range as a result of the animal experiment according to the laser irradiation at 660 nm irradiation. This shows that photonics treatment in the affected area is effective in treating of the pressure sore.

3.3 Confirmation of Wound Healing Ability by Laser Irradiation at 660 nm

As it shown in Fig. 4, as a result of confirming the wound healing ability by laser irradiation at 660 nm light irradiation, the result is similar to VEGF which is used as a positive control in laser irradiation group. The result is confirmed as compared with the non-treated control group. These results confirm that the laser irradiation at 660 nm light irradiation is effective for wound healing.

3.4 The Results of Confirming the Effect of Promoting Angiogenesis by Laser Irradiation at 660 nm

As it shown in Fig. 5, we confirmed that angiogenesis is induced more rapidly in the photonics-treated than the non-treated group through the tube formation which the angiogenic effect of Huvec Cell and promoting effect of the angiogenesis is confirmed by laser irradiation at 660 nm in this study.

4 Conclusions

The paper applied a skin patch biophotonic sensor for early diagnosis and treatment of Pressure Sores and verified the effectiveness of the proposed method. In this paper, we propose a method to diagnose pressure sore by impedance, and at the same time verify the treatment effect of pressure sore through laser irradiation at 660 nm. An impedance spectrometer was used to develop a noninvasive electrical detection method that detects pressure induced tissue damage (pressure sore) in the Rat model. The sensitivity of this detection method and calculated damage parameters are sufficient to detect mild physiological changes that are not evident by visual inspection. In this study, specific thresholds of tissue damage demonstrate the ability of impedance spectroscopy as a technique for the early detection of tissue wounds, but further studies will be needed to determine clinically relevant thresholds in patients in actual practice. Devices with this function can have a significant impact on the treatment of pressure sores. Disease prevention is a well-known approach in the medical field. Providing an early detection mechanism for the pressure sore allows the caregiver proactively prevent further damage to the tissue and closely monitor tissue status. In this paper, the antioxidant efficacy was verified by laser irradiation at 660 nm, and it was confirmed that the antioxidant activity was higher by photonics alone. The DPPH assay also inhibited the antioxidant effect and oxidative stress and confirmed the increase in cell viability. Laser irradiation at 660 nm also has a wound healing effect and confirmed the result of promoting angiogenesis.

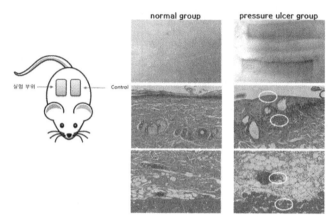

Fig. 1. Rat model to induce pressure sores

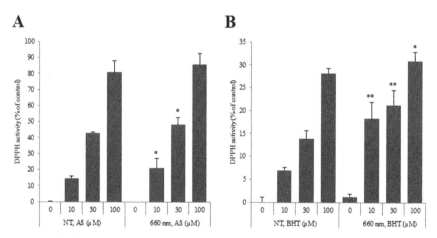

Fig. 2. Verification of antioxidant efficacy by laser irradiation at 660 nm

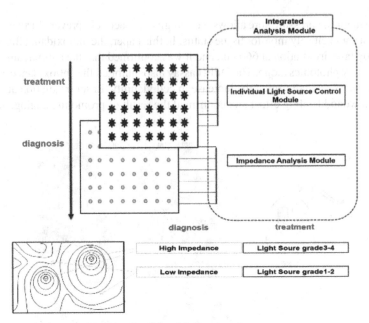

Fig. 3. The diagram of the proposed sores sensor

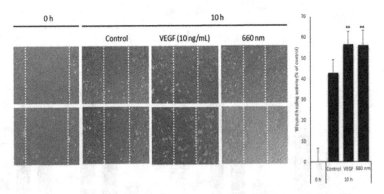

Fig. 4. Laser irradiation at 660 nm for wound healing ability

Control	660 nm

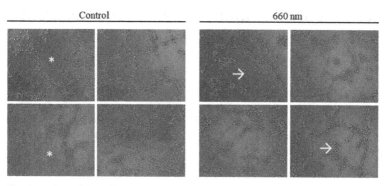

Fig. 5. Confirmation of the effect of laser irradiation at 660 nm on angiogenesis (*) where the blood vessels are not formed (→) where the blood vessels are formed

Acknoweldgment. This work is supported by the Foundation Assist Project of Future Advanced User Convenience Service" through the Ministry of Trade, Industry and Energy (MOTIE) (R0004840, 2020) and Basic Science Research Program through the National Research Foundation of Korea (NRF) funded by the Ministry of Education (NRF-2017R1D1A1B04031182).

References

Han, G., Ceilley, R.: Chronic wound healing: a review of current management and treatments. Adv. Ther. **34**(3), 599–610 (2017). https://doi.org/10.1007/s12325-017-0478-y

Sorg, H., Tilkorn, D.J., Hager, S., Hauser, J., Mirastschijski, U.: Skin wound healing: an update on the current knowledge and concepts. Eur. Surg. Res. **58**(1–2), 81–94 (2017)

Tatmatsu-Rocha, J.C., et al.: Mitochondrial dynamics (fission and fusion) and collagen production in a rat model of diabetic wound healing treated by photobiomodulation: comparison of 904 nm laser and 850 nm light-emitting diode (LED). J. Photochem. Photobiol. B Biol. **187**, 41–47 (2018)

Akhkand, S.S., Seidi, J., Ebadi, A., Gheshlagh, R.G.: Prevalence of pressure sore in Iran's intensive care units: a systematic review and a meta-analysis. Nurs. Pract. Today **7**(1), 21–29 (2020)

Lin, Y.A., Gupta, S., Pedtke, A., Loh, K.J.: Monitoring pressure distributions at human-socket prostheses interfaces using graphene-fabric sensors. In: Structural Health Monitoring 2019 (2019)

Lemmens, J.M., et al.: Tissue response to applied loading using different designs of penile compression clamps. Med. Dev. **12**, 235 (2019)

Wang, L., et al.: Preparation of engineered extracellular vesicles derived from human umbilical cord mesenchymal stem cells with ultrasonication for skin rejuvenation. ACS Omega **4**(27), 22638–22645 (2019)

Pal, A., Goswami, D., Cuellar, H.E., Castro, B., Kuang, S., Martinez, R.V.: Early detection and monitoring of chronic wounds using low-cost, omniphobic paper-based smart bandages. Biosens. Bioelectron. **117**, 696–705 (2018)

Ojarand, J., Min, M.: On the selection of excitation signals for the fast spectroscopy of electrical bioimpedance. J. Electr. Bioimpedance **9**(1), 133–141 (2018)

Sumarno, A.S.: Pressure Sores: the core, care and cure approach. Br. J. Community Nurs. **24**(Sup12), S38–S42 (2019)

Sustainably Stemming the Nursing Care Crisis in Germany

Thierry Edoh[1] and Madhusudan Singh[2]

[1] Department of Regulatory Drug Affairs, RFW-University of Bonn, Bonn, Germany
[2] School of Technology Studies, Endicott College of International Studies, Woosong University, Daejeon, South Korea
msingh@wsu.ac.kr

Abstract. The nursing care staff shortage is a worldwide phenomenon. The reasons for this are multiple. The older people population is the faster growing population in the western countries and their demand for accommodate nursing is also increasing day-in day-out. Though, at the same time the nursing profession is becoming less attractive for many reasons e.g. high burden on the caregivers, poor salary, poor quality of life, etc. All these ingredients lead to a severe nursing crisis worldwide.

This study aims at seeking alternative technology-supported solutions since all solutions in place, particularly in Germany, fail to stem the crisis.

The study evidence supports our theory that implementing smart home technology in the home care will increase family involvement in older people care at and thus decrease the burden on the inpatient-care-system. Consequently, this will mitigate the nursing shortage crisis.

Keywords: Nursing shortage · Home care · Smart home technology · Burden intensity · Older people quality of life · Family caregivers

1 Introduction

The share of older people is worldwide fast-growing. The high-income countries (HIC) are facing more rapid-growing older population than developing countries. The part of young people (up to 15 years) is continuously decreasing in HIC while developing and Low- and Middle-Income countries (LMIC) are bearing a high number of young people [1]. The demographic structure of the high-income countries (HIC) presents several health-, medical, and nursing challenges and issues for the health care systems in each given high-income country. The cultural means and societal structure in the HIC drive older people towards nursing homes.

Older people mostly suffer from cognitive impairments, such as dementia, Alzheimer Disease (AD), etc., and thus need more accommodate health, medical, and nursing care. Caring for elderlies is time manpower consuming. However, worldwide, the global health care system is facing a workforce shortage [2, 3] that leads to a care crisis in high-income countries and a brain drain issue in developing countries since Germany is presently struggling to recruit nurses from Mexico.

© Springer Nature Switzerland AG 2021
M. Singh et al. (Eds.): IHCI 2020, LNCS 12616, pp. 188–198, 2021.
https://doi.org/10.1007/978-3-030-68452-5_20

Faced with the workforce shortage issues, health care bodies are opting to recruit skilled nurses from abroad in order to stem the crisis. Unfortunately, this model does not contain the crisis until now. Beyond recruiting nurses from abroad, the nursing care system also adopts innovative technology to assist caregivers in their daily duties, to solve workforce shortage issues, and to provide a high level of quality of life to the older people [4]. However, the workforce still shortage remains and additionally many nursing care professionals face difficulties to use assistive technology and related devices.

Many researchers have addressed the term "workforce shortage" and proposed various models including electronic health/medical records systems (EHR/EMR), biosensors to collect physiological data, information and communication technology [3], and recently, assistive robots for older people to live independently are coming up, implemented and partially deployed in the USA and Japan. Technology in nursing care pursues the main objective of enabling well Aging-in-place, independent living [5] and assist caregivers. Many European countries, such as Germany, are facing a nursing care crisis. In Germany, the government is struggling day-by-day to address the crisis in reforming the nursing care training act, enacting the relaxation of the foreigner work-permit regulation, and motivate young people to embrace the profession of caregiver.

The remainder of the paper briefly presents in Sect. 2, research aims and objectives, in Sect. 3 research questions and hypotheses, in Sect. 4 Contextualized Model and Assumptions. To get insight into the crisis, a contextualized problem statement and analysis are conducted. A model is then developed with the objective to stem the crisis using IS including a multidimensional smart home automation technology. Assumptions are made to verify the contribution of the model. Section 5, Study Data and Methods, follows. It presents the study sample, study methodology, data collection, and proof-of-concept (model validation). The study's results are reported in Sect. 6, Study Results. The crisis leading causes are drawn, the readiness of the population to be involved in nursing care is presented. Though issues and challenges hindering their involvement are figured out. Finally, Sect. 7, Discussion, the authors discuss the study results and makes a recommendation for enacting paradigm and policy changes to support the proposed model and its adoption by the population and, thus, to contribute to stemming the crisis. The author pointed out the limitation of the study too. Section 8 concludes the paper.

2 Research Aims and Objectives

The nursing shortage crisis remains despite various technical and organizational solutions. Nursing care staff shortage increases in the COVID-19 period since most of existing workforce is concentrated on treating COVID-19-patient. The covid-19 has amplified and pointed out the extend of the nursing shortage.

The main aim of this study is to investigate the cause of the failures of these solutions and to explore another paradigm with a focus on 1) reduction of the intensity of the burden on caregivers, 2) improvement of the quality of life of the older and their care at home, 3) daily assistance to the older. 4) Fostering an environment conducive to home care, 5) medical education for caregivers, 6) autonomy of the older people.

Finally, the study aims at investigating to what extent a paradigm change that promotes home care could contribute to mitigate the nursing shortage. This aspect is high relevant in this COVID-19 period where many medical treatments are being postponed.

3 Questions and Hypotheses

Answering the main study question, "What are the main causes of the care crisis each country is facing?" (Q0), will lead to understand the main study question. This study is requested to answer the following question "Could an expansion of the home- and community-based care (aging-in-place) sustainably contribute to stemming the care crisis?" (Q1). Furthermore, to what extent aging-in-place can positively impact the quality of life of elderlies? (Q2). "Which modern information systems (IS) could assist to stem the care crisis and how?" (Q3).

Beyond the technical aspect of the approach, it is important to investigate the impact of existing paradigms and policies to get insight into their implications in the crisis and how to improve them. Though, the question arises to determine if "paradigm and policy changes are needed to (sustainably) stem the care crisis?" (Q4). The included reviews are enough exhaustive and comprehensive to build a solid fundament for the first hypothesis of the study:

H0: Launching a multidimensional smart home automation supporting home-based care will assist to increase the extent of home nursing in the German context, which in turn will contribute to stemming the care crisis. People with no serious illness but living today in residential nursing home could receive care at home and thus free up to 40% of the nursing resources since according to Krolop in [6] 40% of all beds in the nursing homes are occupied by "residents who do not require need much care".
H1: The family support is correlated with the patient's QoL
H2: The level of the intensity of burdens on family caregivers is correlate with using modern information technology in the home-based care system.

4 Contextualized Model and Assumptions

The model focuses on improving the quality of life in community and home-based care by including daily living assistance and the family involvement aspect, which commonly do not get consideration in existing home and community-based care solutions. Additionally, it is worth tearing down challenges like high burdens on caregivers in home-based care and the financial risks they are facing by caring for their parents.

4.1 Problem Statement Applied to the German Nursing Care

Traditionally, most German older, aged 65+ years and requiring constant nursing care, are living in a community or private nursing homes, a little number lives at home (informal care), or receive ambulant care [7]. At a nursing home, the older people are supposed to receive adequate nursing care and enjoy a better quality-of-life. However, the entire nursing care system is facing a series of scandals like reported in [8, 9]. These scandals seem to be caused by the extreme workforce shortage and the high burden on the caregivers. Rebecca Gale has pointed out that the burden on informal caregivers can increase when the caregivers have to make a decision to match a patient's wishes [10].

Today, 2.6 million people need nursing care. About 4.1 million by 2030 will need nursing care. Streit Matthias and Telgheder Maike have commented in a newspaper article in the column "Geriatric Reform" [6] a report on the German nursing home ranking [11]. The report paints a big picture of the nursing care crisis in Germany. It estimates the number of people to need nursing care to 4.1 million. The rapid growth of the older population implies a great need for nursing workforces since "Hospitals are increasingly sending patients straight to nursing homes and at the same time, the number of nursing home residents who do not require need much care increased sharply to around 40%, Mr. Krolop said. "Both developments are worrying" [6].

The German public health care and the nursing system is facing a severe workforce shortage [12] and is struggling to solve these issues by recruiting care staffs from abroad [13–15]. Recruiting care workers from abroad is becoming specific to the developed countries.

Most solution approaches to overcome the care crisis, are limited to recruit non-resident care staff, enroll more nursing trainees, regularly control the nursing home to prevent any abuse. However, the country still faces a care crisis.

4.2 Solution Approach

Based on the results of the qualitative surveys, literature reviews on the causes of the care crisis and the reasons why existing solutions do not work according to the design-science paradigm, served as guidelines in the search for an adequate approach to contain the care crisis.

The "Internet of health things" technology on the multidimensional and intelligent solution approach of home automation is proposed as a system to support older people and their caregivers at home with the aim of stemming the care crisis in Germany. The system covers the technical, medical, social, assisted-living, and independent living aspects. It allows the collection of medical and environmental data, monitoring of food and water, medication, assistance with indoor and outdoor activities in addition to the common features of smart homes such as energy, fire, safety and security, doors and windows, and room temperature management.

5 Nursing Data Analysis

We have analyzed the nursing data with study methods based on participant sampling, study data and analysis of data as shown in below Fig. 1 and explained at below sections.

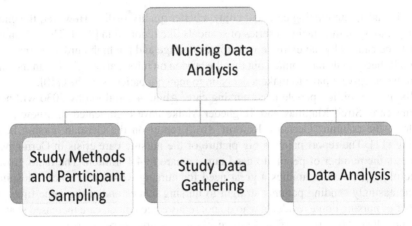

Fig. 1. Nursing data analysis

5.1 Study Methods and Participant Sampling

To get insight into the nursing care crisis, which implies to investigate the nursing policy, the preferences of older people to age-in-place, the willingness level of people to care for family members, as well as to evaluate existing approaches to tackle the care crisis, the authors conducted a qualitative study. People on the street, patient's relatives, and care staffs were in-depth interviewed.

Different probability and non-probability sampling methods. Outpatients (n = 32), inpatients (n = 32), their relatives (n = 40), and care workers (n = 120) were recruited using the cluster random sampling approach to select individuals from different cities. Informal and formal caregivers were thus sampled. Further, the snowball approach (non-probability approach) was used to select the outpatients and inpatients through their relatives and care staffs according to the author's defined eligibility criteria namely (i) "experience good or bad family support", (ii) "living in a nursing home or in a common community (aging in place)", "having cared or is caring for a family member". People on the street (n = 414 total, n = 397 adults, n = 17 youths) were selected using a probability sampling type, the simple random sampling.

5.2 Study Data Gathering

A survey was conducted. Outpatients and inpatients were interviewed. The survey was focused older people preference of living location, the importance of family support, and their quality of life in their current place of residence. Nurses were asked about the trend in admission and technical assistance for care of residents in the home. Relatives of the patient were asked about the difficulties and problems they encounter when caring for a sick family member, their professional and financial situation, and the impact of caring for family members on their work life and financial situation. Finally, the family caregivers were interviewed on modern technology impact on the intensity their burden. The burden on the caregivers was assessed. The Zarit burden interview is modified and questions to assess the impact of caring on the young caregivers' school performance were added to the questionnaire. The questions are then classified into eight (08) themes Four groups were built according to Zarit burden questions outcome classification (i) Very good: people with little or no burden, (ii) good: people facing mild to a moderate burden, (iii) passable: people facing moderate to a severe burden, and (iv) Bad: people facing severest burden level.

In addition to the surveys, a literature review was carried out to assess the state of the art in terms of assistive technology specific to home care. Research on "Ageing in Place" assistive technology was extracted from scientific databases such as Scopus, Web-of-Sciences and PubMed and was reviewed. Paper questionnaires, telephone and face-to-face interviews were conducted to assess the quality of life of outpatients, inpatients, and their relatives. The focus of the study was to determine the degree of quality of life of outpatients and their relatives in a home care system at different periods of the study by using a longitudinal survey to assess changes in the quality of life of outpatients and their relatives. The quality of life of family caregivers and elderly people were measured before the study began and after the deployment of the proposed solution to support long-term care at home.

5.3 Data Analysis

The causes of the crisis have been analyzed from an IT and information systems point of view in order to provide computerized and information systems-based solutions. The quantitative analysis was carried out using the ANOVA method, i.e., a comparison model where the difference between the quality of life of the groups before and after the use of the proposed solution was compared with a particular look at the degree of involvement of family members in home care.

6 Study Results

The study results have considered several components as follows shown in Fig. 2. Each Components are explained in below sections.

- **Care Crisis Leading Causes and Challenges:** The main leading causes driving into the care crisis is the workforce issues and the low number of trainees who enroll in each training session. Caregiving is challenging and physical work but also unattractive because of the low paid and shift work. The literature review reveals that the German nursing crisis is homemade, due to the saving politics applied over 22 years to high profits [16].
- **Workforces Issues and Employment Legislation towards Non-EU-Citizens:** The Germany employment legislation on foreign, especially non-EU citizens, is very complex and hinders the recruitment of skilled nurses from abroad. The Germany nursing system needs such workforces only because it cannot cover the demand with German skilled nurses. The direct cause is the low number of trainees. In Section "Discussion" the topic is deeply analyzed.
- **Low Paid: Unfortunately, the paid determines the motivation:** Caring for older people is challenging and requests enormous physical and psychological efforts from

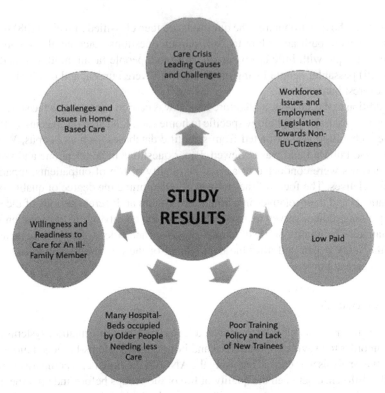

Fig. 2. Components of study results

the nurses. Though, many experts think that the nurses are underpaid and consequently make the nursing profession unattractive.

- **Poor Training:** Policy and Lack of New Trainees: As mentioned above, the present German government is struggling to enroll enough new trainees to overcome the care crisis. This struggle shows that the country lacks adequate plan ad policy in this field. It is obvious, that the lack of trainees is exacerbating the crisis.

- **Many Hospital-Beds occupied by Older People Needing less Care:** The country proceeds in 2008 and in 2017 with reform on the long-term care eligibility. Both reforms have consequently increased the number of people being considered as requiring nursing care. Though, between 2015 and 2017, the number of people "requiring" long-term care increases by 17.9% [17, 18]. An in-depth analysis of both reforms reveals that many people with no serious illness are being considered as nursing cases. The work-overload and the nursing workforce shortage are the direct and indirect consequences of these reforms, which in turn are the leading causes of the care crisis in Germany. A further consequence of these reforms is a decrease in places in nursing homes to accommodate people requiring nursing care. For example, in 2007, before the reform, it was an average of 37.7 places per 100 people in need of care against an average of 24 places per 100 people in need of care in 2015, after the reform in 2008 and before the reform in 2017 [17]. In conclusion, both reforms have caused workforce shortage and place lack issues, which in turn leading the crisis today.

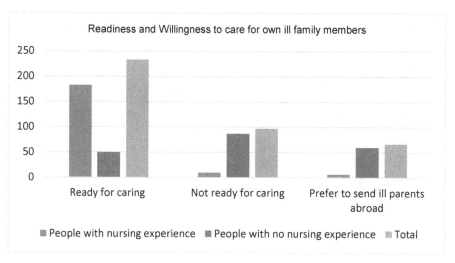

Fig. 3. Readiness to care for a relative

- **Willingness and Readiness to Care for An Ill-Family Member:** Despite that this study is limited in terms of the study population and testing period, it nevertheless confirms the finding of the survey conducted by Forsa and reported in [19]. 59.68% of the interviewed individuals on the street are willing to care for their ill-family members (see Fig. 3).
- **Challenges and Issues in Home-Based Care:** Despite many people willing to care for the parents at home, the share of home-based nursing care remains extremely low due to challenges and issues facing. The data analysis points out differences in the challenges regarding the age of the caregiver. Though, young, and adult caregivers are facing different challenges and issues when they are involved in family member care.

414 people (including n = 397 adults and n = 17 youth) were asked on the street about problems related to caring for older people from the point of view of family members at home. Both youth and adults face sleepless nights, social isolation and depression. Adults face financial stress, often unable to work due to the intensity of care, depression. All respondents (n = 414, 100%) face problems with night rest, social isolation (n = 290, 73.05%), high workload (n = 414, 100%), 24-h care (n = 397, 100%), poor quality of life (n = 414, 100%) and emergency problems (n = 375, 94.46%). In addition, adult caregivers face financial problems (n = 317, 79.84%). While youth caregivers are missive and have mental problems (n = 15 out of 17, 88.23%) and poor school performance (n = 10 out of 17, 59%), only a few adults have suffered from depression (n = 15, 3.78%) in the Fig. 4 and Fig. 5.

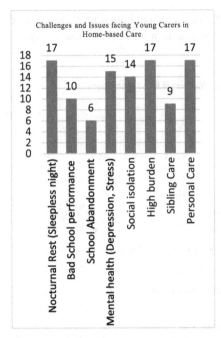

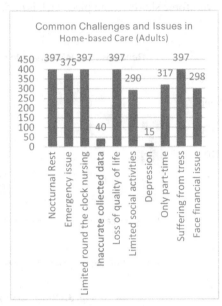

Fig. 4. Challenges facing youths in home   **Fig. 5.** Challenges facing adults in home care

7   Final Remarks

The principal goals of nursing care are not in the first line to treat the patient but provide palliative care and better quality-of-life to the older. Therefore, to ease the adoption of home-based by large-scale people, especially the caregivers, the system needs to assist to reduce burdens on caregivers. Additionally, its impacts should improve the patient's QoL. The system is, therefore, evaluated these aspects.

- **Impact of the proposed approach on the Quality-of-Life (Effect size):** The results revealed improvement in food and water intake against by using of the proposed solution. Improvement is also noticed in medication adherence, physical activities, and older people socialization. Cohen's determined differences strongly indicate that the smart home technology could increase older people quality of life in home care. The calculated Cohen's d shows a significant effect and the quality of life of 92% of all older people and their caregivers involved is improved. The obtained Cohen's differences have confirmed the hypothesis H1 with different effects.
- **Burden Effect Size (groups' Differences):** The obtained Cohen's differences have confirmed the hypothesis H2 with different effects. Cohen's determined differences show significant effects and thus a reduction in the burden. A significant improvement is seen in the group of caregivers caring for people aged 85+ years. The solution assists 92% of the caregivers. In addition, the solution shows potential to reduce the financial stress associated with caring for older people; caregivers can work part-time. Full-time or remotely (home office). The simulation showed that a patient could become

more independent and the caregiver would then have more free time to go to work. Unfortunately, this solution does not overcome privacy concerns. In this area, an improvement of the system is needed. Although a long-term evaluation is needed to support this outcome.

The young caregivers involved do not face financial stress. They face academic performance issues that have been improved by the solution.

8 Conclusions

This study has investigated the main causes of the care crisis in Germany. It, further, analyzed the different solutions the country is working out to tackle the crisis. Unfortunately, many proposed approaches do not solve the crisis. This study proposes a technical implication for solving the nursing care crisis. Evaluation evidence has shown promise for improvement. This study has shown that technology (smart home) implication can contribute to keeping at home old-aged people with no serious illness. The study mainly shows that unlike isolated IS solution, a multidimensional construct mostly contributes to stem the crisis. This contextualized model is a novelty since till today no such solution is proposed and tested. The contribution analysis of the solution confirms the contribution of this work to the literature too.

References

1. United Nations Department of Economic and Social Affairs/Population Division: World population prospects: the 2017 revision (2017). http://www.ncbi.nlm.nih.gov/pubmed/122 83219
2. Kamal, H., et al.: Policy changes key to promoting sustainability and growth of the specialty palliative care workforce. Health Aff. (Millwood) 38(6), 910–918 (2019). https://doi.org/10.1377/hlthaff.2019.00018
3. Krick, T., et al.: Digital technology and nursing care: a scoping review on acceptance, effectiveness and efficiency studies of informal and formal care technologies. BMC Health Serv. Res. 19(1), 1–15 (2019). https://doi.org/10.1186/s12913-019-4238-3
4. Isfort, M., Weidner, F., Rottländer, R., Gehlen, D., Hylla, J., Tucman, D.: Eine bundesweite Befragung von Leitungskräften zur. Dtsch. Inst. für Angew. Pflegeforsch. e.V, pp. 1–152 (2016)
5. Billings, J., Carretero, S., Kagialaris, G., Mastroyiannakis, T., Meriläinen-Porras, S.: The role of information technology in long-term care for older people. In: Leichsenring, K., Billings, J., Nies, H. (eds.) Long-Term Care in Europe, pp. 252–277. Palgrave Macmillan UK, London (2013). https://doi.org/10.1057/9781137032348_12
6. Streit, M., Telgheder, M.: In need of TLC nursing home care is a growing market in Germany, but it is hampered by staff shortages, a lack of capital and a lot of red tape. Handelsblatt GmbH, 23 November 2017. https://www.handelsblatt.com/today/companies/geriatric-reform-in-need-of-tlc/23573132.html. Accessed 18 Oct 2020

7. Prühl, S., Scmidt, W., Bätzing-Lichtenthäler, S., Müller, B., Jung, M., Hintzsche, B.: Pflege-politik in Deutschland. DEMO Impulse (2017). https://www.bundes-sgk.de/system/files/doc uments/impulse_8_2017.pdf. Accessed 15 Nov 2020

8. Hibbeler, B.: Pflegeheime Skandale statt Lösungen, Dtsch. Arztebl. **41**(102), 1 (2005). https://www.aerzteblatt.de/archiv/48694/Pflegeheime-Skandale-statt-Loesungen

9. Herz, A.: Pflegeskandal in Augsburger Heim: Erschreckende Zustände. Bayerischer Rund-funk (2019)

10. Gale, R.: Advance care planning with Alzheimer's: a tortuous path. Health Aff. **36**(7), 1170–1172 (2017). https://doi.org/10.1377/hlthaff.2017.0679

11. Augurzky, B., Krolop, S., Mennicken, R., Schmidt, H., Schmitz, H., Terkatzn, S.: Pflegeheim rating report 2007 Wachstum und Restrukturierung. Rheinisch-Westfälisch (2007)

12. Linsin, J., Zeeh, Schellenberg, J.: Care home market report 2016/2017 Germany'S care homes in the center of global attention (2017)

13. Bonin, H., Braeseke, G., Ganserer, A.: Internationale Fachkräfterekrutierung in der deutschen Pflegebranche Chancen und Hemmnisse aus Sicht der Einrichtungen, pp. 1–76 (2015)

14. Bonin, H., Ganserer, A., Braeseke, G.: The nursing-care industry in Germany is struggling with international recruitment efforts, pp. 1–4. Bretelsmann Stiftung (2015)

15. Knight, B.: Germany faces a care crisis Germany aims to revamp crisis-hit care industry amid worker shortage. Deutsche Welle, pp. 1–4 (2019)

16. Watanabe, K., Niemelä, M.: Aging and technology in Japan and Finland: comparative remarks. In: Toivonen, M., Saari, E. (eds.) Human-Centered Digitalization and Services. TSS, vol. 19, pp. 155–175. Springer, Singapore (2019). https://doi.org/10.1007/978-981-13-7725-9_9

17. Schulz, E.: The long-term care system in Germany (2010)

18. Rothgang, H., Müller, R.: Pflege-Report 2018. Springer, Heidelberg (2018)

19. Grießmeier, K., Peter, M.-A.: TK-Meinungspuls Pflege So steht Deutschland zur Pflege. Hamburg (2018). https://www.tk.de/resource/blob/2042934/1a33145a8bb2562010 3fcddd64316f75/studienband-meinungspuls-pflege-2018-data.pdf. Accessed 15 Nov 2020

First Eyetracking Results
of Dutch CoronaMelder Contact Tracing
and Notification App

Jan Willem Jaap Roderick van 't Klooster[1]([✉]), Peter Jan Hendrik Slijkhuis[1],
Joris van Gend[1], Britt Bente[2], and Lisette van Gemert-Pijnen[2]

[1] BMS LAB, Faculty of Behavioural, Management, and Social Sciences, University of Twente,
Enschede, The Netherlands
j.vantklooster@utwente.nl
[2] Department of Psychology, Health, and Technology, Faculty of Behavioural, Management,
and Social Sciences, University of Twente, Enschede, The Netherlands

Abstract. The coronavirus disease (COVID-19) has led to a global pandemic. Many countries are using contact tracing as one of the interventions to control virus transmission. In conventional contact tracing, index cases with positive test results are asked to provide contact information of close contacts who were at risk of acquiring infection from the index case, within a given time period before the positive test result. Mobile contact tracing apps can augment traditional contact tracing, as they exchange contact information wirelessly and semi-automatically, and allow for quicker warning of at-risk contacts.

However, little is known about user acceptance and perception of such contact tracing apps, or how people perceive the advice presented therein. In July 2020, the beta version of the official Dutch corona notification app was usability tested with participants from various age groups. This paper presents the results of the objective eye tracking measurements executed during these usability tests.

The results show that both health and process related information is easily overlooked in the app by participants. A closed-loop contact with the builders of the app allowed for a direct feedback channel and quick improvements during the development stage.

Keywords: Contact tracing app · COVID-19 · Eyetracking

1 Introduction

The coronavirus disease (COVID-19) has led to a global pandemic [1]. Many countries are using contact tracing as an intervention to control virus transmission. In conventional contact tracing, index cases with positive test results are asked to provide contact information of those close contacts, who were at risk of acquiring infection from the index case within a given time period before the positive test result. These contacts are then traced by health authorities and informed, warned, quarantined, and/or tested. Health authorities had to scale up their contact tracing staff extensively to execute this

M. Singh et al. (Eds.): IHCI 2020, LNCS 12616, pp. 199–207, 2021.
https://doi.org/10.1007/978-3-030-68452-5_21

labor-intensive task. However, index cases may not remember all their close contacts, or do not know them (e.g. in public transport). Mobile contact tracing apps can augment traditional contact tracing, as they exchange contact information wirelessly and allow for quicker warning of contacts.

Yet, so far only little is known about user acceptance and perception of such contact tracing apps [2] and how people perceive advice presented by such apps [3], or whether they understand the exposure logging mechanism. After it was decided that the Dutch government will develop such a contact tracing app, it was therefore usability tested with participants from various age groups. The hypothesis is that such tests allows the app to be improved in an early stage and better fit these heterogenous groups, increasing in the end the acceptance by the public.

This paper specifically presents the results of the objective eye tracking measurements executed during these usability tests. Our research questions were: what can we learn from using eyetracking in usability testing an early covid notification app, and how can the results be used to improve the notification ecosystem. This paper aims to present the results from the eyetracking usability test, and demonstrate the value of such methods in this area as an addition to qualitative, opinion based usability methods.

The rest of this paper is structured as follows. Section 2 briefly introduces the Dutch corona notification app and the working of this app. Section 3 describes the evaluation method. Section 4 discusses the eye-tracking results and Sect. 5 presents conclusions.

2   Corona Notification App

The Dutch 'CoronaMelder' corona notification app is a smartphone application built by the Dutch Ministry of Health. It is using Google and Apple's exposure notification framework (GAEN) [3]. This is a decentralized framework that allows modern smartphones to anonymously collect and log encounters with other smartphones at close range, and warn their users in case identifiers of other phones were later associated with cases having a viral infection as confirmed by a health authority. When turned on, the notification app tracks back encounters for 14 days for other app users at close range for 15 min. This is a good indication for exposure risk. In case the anonymous identifiers of positive cases available are matched in the local exposure log, the user gets an anonymous notification of risk of exposure, and advice on what to do: stay home, and get tested, if symptoms apply. If an app user is tested positive by the health authority themselves, they can notify other app users with the help of and confirmation by the local health authority.

Because of the Bluetooth low energy technology used for measuring distance between smartphones, sensitivity and specificity are around 70–75%. It is nevertheless believed that contact tracing technology is beneficial in fighting the pandemic [2, 4, 5] and their implementation is advised by multiple disease control expertise centra across the world [6]. Tracing apps, due to their nature, allow for much faster notification of possibly infected encounters as compared to traditional contact tracing methods, i.e., calling contacts by phone by the public health authorities. However, an internet connection and a relatively new smartphone is needed by the end users for tracing apps.

3   Methods

The present study consisted of scenario-based usability tests of CoronaMelder with think aloud method, interview questions and a questionnaire. The usability tests took place between June 29 and July 3, 2020. A beta version of the app was tested using smartphones with iOS (version 0.1, build 172) and Android (version 0.3.1, build 107). The study was conducted in the DesignLab of the University of Twente. The BMS Lab protocol for corona-safe human-related research that was followed, was approved by the Executive Board of the University of Twente. The study was approved by the University Twente's Ethical Committee under nr. BCE200953.

3.1   Participants

Participants were recruited via convenience sampling by the authors, with a focus on elderly and younger people.

The usability tests were conducted individually. Six participants wore Tobii eye tracker glasses for gaze analysis [7]. Prior to the test, the nature and purpose of the study was explained and permission for participation was obtained through an informed consent form that was signed by the participants.

The tests started with general questions about the impact of the pandemic on participants' lives and about what they have already heard about COVID-19 apps. Thereafter, each participant conducted 4 scenarios on the app, which represent actual use of the app while they were asked to think aloud. These scenarios are as follows:

1) **Introduction to the app:** In this scenario, the app was shown in the app store and additional information about the app could be read.
2) **Onboarding and activation of the app:** In this scenario, operation of the app is explained through onboarding steps in the app itself. Participants learn about the content of the app and confirm the right permissions to use the app (allow the app to use Bluetooth and to send notifications).
3) **Receive notification:** In this scenario, the participants receive a notification from the app about their increased risk of infection, because they have been in in close contact to an individual who had tested positively for the coronavirus.
4) **Sharing keys (telephone conversation with PHA):** In this scenario, participants were asked to imagine that they have been tested positive on coronavirus recently. During the scenario they receive a phone call from the PHA, in which participants are asked about their symptoms and receive help with uploading the authorisation key for the contact tracing and to warn the smart phones with whom they were in contact with.

After finishing the scenarios, final interview questions were asked about participants' attitude towards the app and their willingness to use the app.

Additionally, the UEQ-Dutch [8] was filled in. The interview and questionnaire aspects are outside the scope of this article. This paper focuses on the eye tracking study, as the other aspects will be reported in a separate article.

3.2 Eyetracking

Tobii AB Eye tracker Glasses 2 were used in this study. These glasses can provide objective data regarding gaze behavior in ambulant situations. Our goal in using them is to obtain insight into navigation behavior, search behavior, and the division of attention over text and visual aspects of the notification app [7].

The participants were measured during the execution of the abovementioned four scenarios. As such, the measurements record the first behavior of the participants that have no prior experience with the test object. The measurements were executed in 2 youngsters (secondary education), 2 youngsters (vocational education) and 2 adults.

4 Eye Tracking Results and Interpretation

Below results reveal both *heatmaps* and *gazeplots*. We present 2 different kinds of heatmaps. The first heatmap shows *how long* a person looks at a certain aspect. The second kind shows *how often* a test subject looks at a certain aspect.

The *gazeplot* shows the order in which the viewing behaviour occurs. In these plots, the size of the circle indicates the relative time duration of the viewing behaviour on that particular plave, and the numbers in the circles indicate the order of viewing. The color is indicative for the participant.

In the Fig. 1 gazeplots, the opening screen on the app is shown, which is visible after downloading and installing it. The opening screen of the app allows users to check that the GAEN mechanism is active ('de app is actief' translates to 'the app is active').

It allows users from top to bottom to select 4 menu item options to (i) get more information about the app; (ii) get information on what to do in case of an risk-of-exposure notification; (iii) get details on how to obtain a covid test; (iv) share an authorization key to notify other users in case the user had a positive lab-confirmed PCR test for covid-19.

The gazeplots show that participants pay significant attention to the image in the middle of the screen. The participants look at the image, and get back to that after scrolling and reading the menu items. Although the users have different behaviour, they look at both the image and the bold texts.

A relatively large part of the viewing time is devoted to looking at the image, in which an animation of a person on a bicycle attracts attention. This happens at the cost of reading the menu item explanations an understanding the structure. The app screens in Fig. 2 show that although there is attention for the text of the menu items, attention also goes to the visual components of the people composition. Before the fold attention goes to the visual aspects; after scrolling to the clickable menu options.

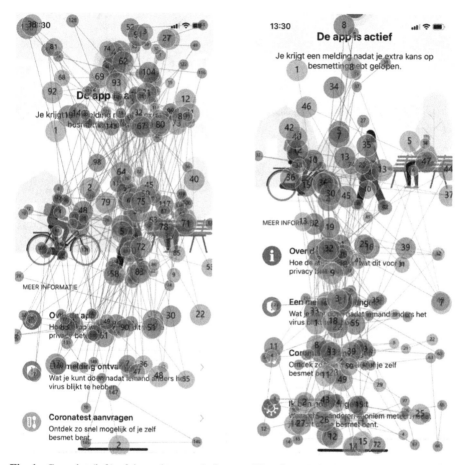

Fig. 1. Gazeplot (left) of the main menu before scrolling down (n = 4 out of 6). Gazeplot right) of the main menu (n = 4 uit 6) after scrolling down (under the 'fold' in the screen). Each color represents one participant. Standard attention filter, cut-off 100 %/s. Tobii Pro Lab 1.142 (tobii, 2020). n = 2 out of six participants received an inactive screen, which is further discussed in [10]. (Color figure online)

Thus, the questions arises what the main goal of the introductory screen of the app is. If promoting understandability of the navigation structure is important, this is an argument for less visual animations and more salient texts. A pitfall of too much animation is visual saturation.

Fig. 2. Heatmap (left) of the main menu (n = 4 of 6) before scrolling down. Heatmap (right) of the main menu after scrolling down. More intense red means viewed longer, more green means observed shorter. (Color figure online)

If the app onboarding is not completed successfully, the app becomes inactive ('niet actief', shown in Fig. 3). This was caused by a software bug in the beta version, but can become a reality in the released app too. We clearly see participants are able to find solution to active it nevertheless, which is important. They look long and often at the 'instellingen aanzetten' (activate settings) button. They look more at the icons and search for solutions all over the screen, looking for visual cues. The prominent image at the top is not used here, so a reason could also be that participants are not yet visually saturated.

When activating the app, a popup is used to enable exposure logging and notification. In Fig. 4, users are asked to accept turning on the GAEN framework to start logging encounters in the phone's logfile and to be able to receive exposure notifications. We observe that not all information is read, and that terminology ('COVID-19') receives more-than-average attention, both in time as well as in frequency. The title of this popup is barely read by these participants, yet terminology such as 'COVID-19' receives more-than-average attention. In addition, users also read texts that could have been blurred out because the popup is active.

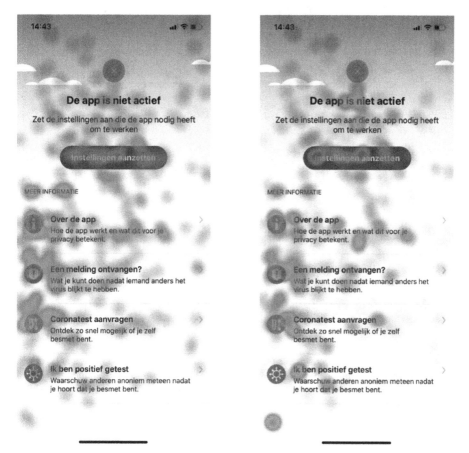

Fig. 3. Error in activating app, app becomes inactive and will not keep track of exposures (n = 2 out of 6, the 2 other test subjects). Left: Red means watched more frequently, green less often. Right: red means watched longer, green shorter time. (Color figure online)

It is not uncommon that users do not read all terms in a popup, neither so in health-related cases as compared to general use cases. The popups are often simply too long. In addition, prior screens already explain the to-be-expected behavior in this particular case. This is an intertwine between the app and the operating system. It is advised to show the most important information about rights in the first steps, then the pop-up notification text size can be reduced.

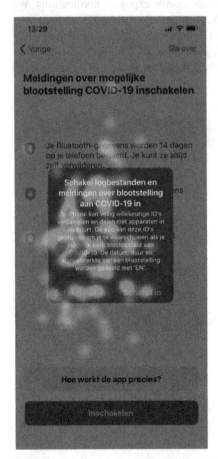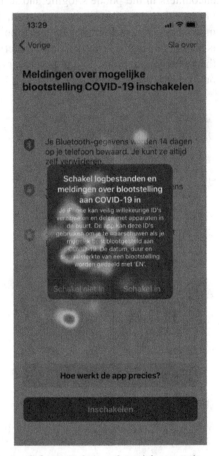

Fig. 4. Heatmap of the popup to enable GAEN Bluetooth framework (n = 3 participants; others could not be adequately mapped) before main menu is accessible. Left: Left: Red means watched more frequently, green less often. Right: red means watched longer, green shorter time. (Color figure online)

5   Conclusion

This article presented the first eye tracking study of the Dutch corona notification app.

Eye tracking has shown to provide insights into the way covid-specific information is being perceived in notification apps. Both health related and process related information is easily overlooked by the participants. The attention to visual information is relatively large compared to attention to explanatory text in the virus-related informative screens. Theses screens are crucial in understanding the app correctly.

Eyetracking also provides insight into the navigation and flow in using such an app, which in this case is especially relevant, given the covid- and notification mechanism related *explanatory content*. Moreover, eyetracking gives insight in the intensity and order of the users' perceptions. This showed that permission popup prompts in the app are important in validation steps in the usage process, yet were found to be improvable.

Improvement suggestions were shared with the builders of the app and the framework developers [9]. Importantly, a closed-loop contact with the builders allowed for a direct feedback channel and quick improvements during the app development.

A limitation of the study is that it had to be conducted under time pression because of the timeline for development set, and the need to come to countermeasures against the spread of the pandemic. This affected the number of participants, although for a usability study of a beta version, testing with few users is known to already provide important input [10]. Nevertheless, it is important to keep evaluating and comparing contact tracing apps as they are implemented, as our recommendation for further research.

References

1. WHO COVID-19 (2020). https://www.who.int/emergencies/diseases/novel-coronavirus-2019. Accessed 17 Feb 2020
2. Altmann, S., et al.: Acceptability of app-based contact tracing for COVID-19: cross-country survey study. JMIR Mhealth Uhealth **8**(8), e19857 (2020)
3. Exposure Notifications Frequently Asked Questions (2020). https://covid19-static.cdn-apple.com/applications/covid19/current/static/contact-tracing/pdf/ExposureNotification-FAQv1.2.pdf
4. Gleeson, M., Neville, K., Pope, A.: The design and development of a digital contact tracing application to better facilitate the tracing of passengers in the event of a biological threat/pandemic. J. Decis. Syst., 1–7 (2020)
5. ECDC: E.C.f.D.P.a.C.: Mobile applications in support of contact tracing for COVID-19 - a guidance for EU/EEA member states (2020)
6. Jacob, S., Lawarée, J.: The adoption of contact tracing applications of COVID-19 by European governments. Policy Des. Pract., 1–15 (2020)
7. Carter, B.T., Luke, S.G.: Best practices in eye tracking research. Int. J. Psychophysiol. **155**, 49–62 (2020)
8. Hinderks, A., et al.: Developing a UX KPI based on the user experience questionnaire. Comput. Stand. Interfaces **2019**(65), 38–44 (2019)
9. Rijksoverheid: Hoe is CoronaMelder getest? (2020). https://www.rijksoverheid.nl/onderwerpen/coronavirus-app/vraag-en-antwoord/hoe-wordt-de-corona-app-getest-voordat-deze-beschikbaar-wordt
10. Nielsen, J., Landauer, T.K.: A mathematical model of the finding of usability problems. In: Proceedings of ACM INTERCHI 1993 Conference, Amsterdam, The Netherlands, 24–29 April 1993, pp. 206–213. (1993)

Architecture of an IoT and Blockchain Based Medication Adherence Management System

Pravin Pawar[1,2], Colin K. Park[2], Injang Hwang[2], and Madhusudan Singh[3(✉)]

[1] Department of Computer Science, State University of New York, Korea Incheon, South Korea
`pravin.pawar@sunykorea.ac.kr`, `pravin.pawar@baswen.com`
[2] Baswen Medication, Seoul, South Korea
`{wpark,ijhwang}@baswen.com`
[3] School of Technology Studies, ECIS, Woosong University, Daejon, South Korea
`msingh@endicott.ac.kr`

Abstract. Ensuring optimal medication adherence is a complex problem and require involvement of multiple stakeholders in a healthcare system. Poor medication adherence leads to severe health consequences as well a serious increase in healthcare costs. To address the problem of medication nonadherence, technological solutions which use Medication Event Monitoring Systems (MEMS) combined with in-person interventions have been recommended by researchers. In this paper, we elaborate the architecture of an IoT and blockchain based medication adherence system that uses a newer generation MEMS device developed by Baswen Medication. The system consists of the functionalities for stakeholder registration, device configuration, medication adherence data management and reminder/notification support to increase degree of medication adherence. We justify the advantages of using blockchain technology and report on the progress of system development.

Keywords: Medication adherence · Medication Event Monitoring Systems (MEMS) · Internet of Things (IoT) · Blockchain technology

1 Introduction

The World Health Organization (WHO) defines medication adherence as the degree to which the person's behavior corresponds with the agreed recommendation from a health care provider [1]. A number of studies suggest that poor medication adherence leads to severe health consequences including increased risk of hospitalization, clinical complications, death as well as serious increase in healthcare costs [2, 3]. Adherence to proven and effective medication remains low [2]. Approximately, 50% of the patients do not take medication as prescribed [4, 5].

Approximately, a quarter of new prescriptions are never filled [6]. Moreover, in the patients suffering from chronic diseases, adherence decreases with time [2, 7]. Patients suffering from co-morbid conditions require to take multiple medications increasing the likelihood of medication nonadherence [2]. According to an estimate, medication

© Springer Nature Switzerland AG 2021
M. Singh et al. (Eds.): IHCI 2020, LNCS 12616, pp. 208–216, 2021.
https://doi.org/10.1007/978-3-030-68452-5_22

nonadherence results in approximately 125,000 deaths per year in the United States alone [6] and results in total costs of 100s of billion dollars every year [6, 8].

To address the problem of medication nonadherence, the solutions based on emerging technologies such as Internet of Things (IoT) and blockchain technology are being reported by the research community. In particular, Medication Event Monitoring System (MEMS) devices which keep track of pill-taking behavior provide effective solution for increasing medication adherence as compared to mechanisms such as pill count or patient report. In this paper, we present the architecture of an IoT and block-chain based system which collects medication dispensing data from a smart pill bottle to monitor medication consumption and provides context-aware interventions for improving medication adherence is proposed. Section 2 of this paper provides a brief overview of medication adherence field. Section 3 describes MEMS devices/solutions currently existing in the market. Section 4 briefly introduces blockchain technology and justifies why it is a lucrative choice for the management of health data and medication adherence data. Section 5 illustrates the architecture of proposed system that features modules for the configuration of MEMS devices, medication adherence data management and interventions. Section 5 also describes the current state of system development. Section 6 concludes the paper and provides directions for the future work.

2 A Brief Overview of Medication Adherence

The measures of medication fall into two categories: Subjective and objective. Subjective measures generally provide explanations for patient's nonadherence whereas objective measures contribute to a more precise record of patient's medication-taking behavior [5]. Subjective measures are patient self-report and healthcare professional assessments, while objective measures include electronic monitoring, biochemical measures, pill counts and secondary database analysis [5]. Patient self-reports are usually unreliable as it is observed that patients are reluctant to admit that they do not take medication as prescribed [9] to avoid disapproval from their healthcare providers [5].

The approaches prevalent for improving medication adherence could be categorized into educational/cognitive strategies (such as providing printed medication instructions, patient/family teaching and patient educational programs), counseling/behavioral interventions (such as counseling by a pharmacist, reminder systems, tailored medication regimen and using reports from electronic monitoring device), and psychological/affective strategies (such as providing telephonic assistance, peer-mentor programs, motivational interviewing and establishing partnership with patient) [10].

The use of Medication Event Monitoring System (MEMS) for measuring medication adherence was proposed in 1980's [11, 12]. The MEMS is a cap that fits on standard medicine bottles and records the time and date each time the bottle is opened and closed. Certain electronic systems also use blister packets with an invisible sensor grid [10]. Multiple researches [13, 14] have confirmed that the use of MEMS combined with other approaches such as pill count provides a reliable estimate of patient's medication adherence. Using reports from MEMS for medication adherence feedback is clearly effective, however, it is less frequently used in practice as these devices are rather expensive, may be complicated for some patients and time consuming for health care provider to administer [10]. Electronic monitoring is synonymous to a gold standard for assessing adherence, however, the patient may open pill bottle without taking medication or to take out more than prescribed resulting in medication overdose [14].

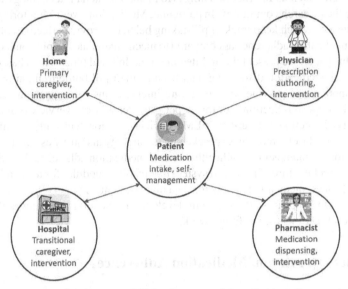

Fig. 1. Stakeholders and their responsibilities for successful medication adherence interventions (modified illustration based on example of Pharm Assist intervention [2])

A research reported by Granger et al. [2] presents the review of emerging technologies for medication adherence interventions. Among these, the information stored in pharmaceutical database allows identification and analysis of the proportion of days that patients have access to medication. A phone call reminder or electronic reminder on pagers, cell phones are programmed to notify the patients regarding the need for pill refill. The in-home telemonitoring systems which allow patients to generate and respond to their own data could also be used for medication-related self-management, However, it was observed that there is a risk of user error in connecting to electronic system and telemonitoring alone is ineffective to improve medication adherence. A combination of in-person intervention together with automated reminders and triggers has proved to be the best approach for improving medication adherence. The in-person interventions could be provided by a healthcare professional or a family member acting as caregiver at home or by a community-based pharmacist [2]. The stakeholders involved in medication adherence intervention are shown in Fig. 1.

3   Landscape of Medication Event Monitoring Systems (MEMS) Devices

The landscape of MEMS devices currently available in the market is shown in Fig. 2. These devices are of the following types:

Smart Pill with Digestible Sensor: Digestible sensor detects whether patients took medicine by reacting to gastric fluid producing micro electricity. It is required to attach a special patch to your body and swallow the pill containing an electronic chip. An electrical signal is sent to mobile to confirm that the medicine is consumed. This approach is the most accurate and confirms that the medicine is indeed consumed. However, the pill cost is very high. Proteus Discover is an example of pill containing an ingestible sensor and it's clinical evaluation is reported in [15].

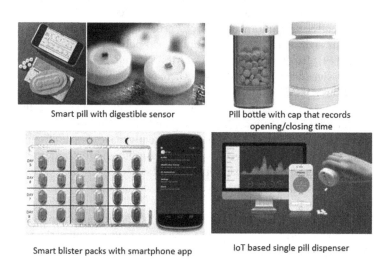

Smart pill with digestible sensor   Pill bottle with cap that records
                                          opening/closing time

Smart blister packs with smartphone app   IoT based single pill dispenser

Fig. 2. Landscape of Medication Event Monitoring System (MEMS) devices **Smart blister packs:** The sensor is attached to the wire to the back of the blister and it sends data to the mobile phone app whenever the patient takes the pill out of the package. This approach requires medicines to be pre-packed in a special pack equipped with blister sensors. Janssen iSTEP technology toolset employs smart blister pack and its clinical evaluation is being carried out [16].

Pill Bottle with MEMS Cap and Touch Sensor: These devices are equipped with several touch sensors inside the bottle, which provide an indication of remaining medicine weight. In addition, smart cap collects information about bottle opening and closing times. Data about cap opening/closing behavior and remaining medicine is collected and sent in real time to the mobile app/back-end server to determine when the medicine is taken and how many pills were taken. A possible limitation of such systems is that the data about remaining number of pills can be incorrect if the touch sensor does not detect accurate number of pills inside the container. SmartRxt Nomi sensor, Pillsy smart pill bottle, Adheretech pill bottle described in [17] are examples of such systems.

IoT Based Single Pill Dispenser: This type of device senses dispensing of and counts each pill dispensed from the bottle to record the medication data in real time, transfer and manage the data to the app and external management system. A spe- cially designed IoT cap has a weight lever based balancing mechanism that allows dispensing only one pill when the bottle is tilted down and prohibits dispensing further pills. The infra-red sensor attached to the opening accurately determines the dispensing time of the pill. This device is developed and patented by Baswen Medication [18] and this paper describes the medication adherence management system for this device.

4 Blockchain Technology in Medication Adherence

Recently, block-chain technology is being considered as a lucrative option for personal health information management due to its compelling features such as fault tolerant and distributed ledger, chronological and time-stamped data record, irreversible, auditable and cryptographically sealed information blocks, near real- time data updates, consensus-based transactions and ability of smart contracts and policy based access to facilitate data protection [19]. Blockchain is a chain of transactions built according to certain rules formed in a signed and validated block whereas each subsequent block contains a link to the previous one. Blockchains are cost-effective distributed ledgers which provide increased accessibility to information by connecting stakeholders directly without requirements for third-party brokers [20]. Compared to existing solutions, Blockchain technology provides the ability to safely and transparently make changes from an unlimited number of sources that can be geographically distributed. Moreover, use of block-chain technology ensures that each participant has a latest data version which is ensured using consensus algorithms, the system is decentralized and there is a confidence in the validity of each transaction [20].

Especially, blockchain technology is seen as a pathbreaking innovation and forerunner of fresh economic period giving rise to a new type of system called the Blockchain Economic System [21]. Major use cases of blockchain in healthcare are patient data management, medication adherence pharmaceutical research, supply chain management of medical goods, prescription management, billing claims management, analytics, and telemedicine [22, 23]. Companies such as ScriptDrop and ScalaMed use blockchain to the information about medication adherence [22]. The permanent chronological storage, traceability of medication dispensing and consumption events, ability of smart contracts to generate notifications for medication reminder events and ability to share medication adherence data in a privacy-sensitive manner make blockchain technology a suitable choice for the management of medication adherence.

5 The Architecture of Proposed Medication Adherence System

The architectural components of proposed system are shown in Fig. 3 and divided into following three parts:

Stakeholders registration and mobile app configuration: As shown in Fig. 1, the stakeholders involved in medication adherence domain are: patient, physician, pharmacist, caregiver and hospital. These stakeholders register their details with the proposed system using a web-based application. The physician prescribes a medi- cation, while pharmacist dispenses medication in a smart pill bottle. The patient downloads a mobile app on the smartphone, pairs the bottle and configures the med- ication details and timings in the mobile app with help of the caregiver.

Pill bottle usage and data management: In order to dispense medication, the pa- tient holds the bottle in tilted down position and a single pill dose is dispensed. In case the dose consists of multiple pills, the gesture is repeated desired number of times. The IoT cap attached to the pill bottle registers several events. These events and their significance are shown in Table 1:

Table 1. Events associated with a smart pill bottle and their significance

Event type	Event data	Significance
Pill dispensing	Dispensing success time	Medication is taken
Bottle cap	Cap open time	Bottle opened for refill
Bottle cap	Cap close time	Bottle is refilled
Smart phone connection	Device paired	Medication regime started
Smart phone connection	Device unpaired	Medication regime ended
Motain state change	Upward to downward	Bottle in inverted position
Motion state change	Downward to upward	Bottle in normal position

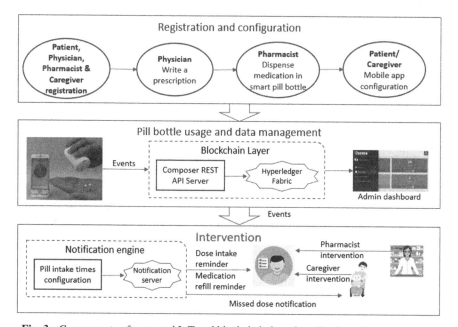

Fig. 3: Components of proposed IoT and blockchain based medication adherence system

The pill bottle is designed to store these events in the local memory of the IoT cap. When the bottle syncs with the mobile app, the app will read these events from the IoT cap using BLE connectivity. This data is sent to the backend on a periodic basis where blockchain layer has the responsibility of recording these events data and making this data available for medication adherence analysis and visualization. The blockchain layer provides functionality to implement and host blockchain database. While using blockchain for the storage of patient's medication intake in- formation, it is required to consider the choice of blockchain between public and private blockchain. A public blockchain network is completely open for joining and participation in the network, while private blockchains also called permissioned blockchain grant specific rights and restrictions to participants in the network. Hence private blockchains are more suitable for the medication adherence management.

Intervention: The proposed system sends reminder to the patient on patient's mo- bile device at the pre-configured time of medication intake. In case the mobile app does not detect the event of pill dispensing, another reminder is sent after certain interval (e.g. 15 min). If the pill dispensing event is still not detected within few minutes of the second reminder, a missed dosage notification is sent to the registered caregiver. It is then expected that the caregiver will intervene the patient either in person or by a phone call depending on the circumstances. Once all the medication in the pill bottle are consumed (as evident from remaining pill count), a notification is sent to the pharmacist, so that the medication refill can be arranged if necessary. This intervention strategy is in-line with the approach suggested by [2], where it is.observed that the combination of in-person intervention together with automated reminders and triggers has proved to be the best approach for improving medication adherence.

5.1 Current implementation status

Baswen Medication has developed a single pill dispenser bottle with IoT cap that records the events associated with medication dispensing and refilling. The backend system that supports registration of various stakeholders such as patient, caregiver, physician and hospital is completed. We are working on the development of pharmacist registration and prescription module. The mobile app under development is capable of connecting to one pill bottle, fetch pill bottle events and send these events to the backend. The patient reminder and caregiver notification functionalities are also completed. We are presently working on medication adherence data visualization on the mobile app as well as web-application. Also, we are experimenting with IBM Hyperledger as a blockchain platform for the storage of medication adherence data. Hyperledger fabric has been used as a blockchain platform in several healthcare applications [24–26]. Hyperledger Fabric is a permissioned blockchain framework implementation and one of the Hyperledger projects hosted by The Linux Foundation. Hyperledger Fabric leverages container technology to host smart contracts called 'chaincode' that comprise the application logic of the system.

6   Conclusion and Future Work

To address the problem of medication nonadherence, the research community proposes use of solutions based on emerging technologies such as Internet of Things (IoT) and blockchain technology in addition to prevailing methods such as pill count and patient diary. In particular, the Medication Event Monitoring System (MEMS) devices which keep track of pill-taking behavior are improving in capacity and functionality. In this paper, we presented the architecture of an IoT and blockchain based system which collects medication dispensing data from a newly developed smart pill bottle by Baswen Medication to monitor medication consumption and provides context-aware interventions for improving medication adherence. While the system components for stakeholders registration, mobile app configuration and reminder/notification functionalities are ready, we are experimenting with IBM Hyperledger as a platform for the storage of medication adherence data.

References

1. Kenreigh, C.A., Wagner, L.T. (2005). Medication Adherence: A Literature Review-Medscape
2. Granger, B.B., Bosworth, H.: Medication adherence: emerging use of technology. Curr. Opin. Cardiol. **26**(4), 279 (2011)
3. Roy, P.C., Abidi, S.R., Abidi, S.S.R.: Monitoring activities related to medication adherence in ambient assisted living environments. Stud. Health Technol. Informat. **235**, 28 (2017)
4. Brown, M.T., Bussell, J.K.: Medication adherence: WHO cares?. In Mayo Clinic Proceedings, vol. 86, no. 4, pp. 304–314). Elsevier, April 2011
5. Lam, W.Y., Fresco, P.: Medication adherence measures: an overview. BioMed research international, 2015 (2015)
6. Fischer, M.A., Choudhry, N.K., Brill, G., Avorn, J., Schneeweiss, S., Hutchins, D., Shrank, W.H.: Trouble getting started: Predictors of primary medication nonadherence. Am. J. Med. **124**(11), 1081-e9 (2011)
7. Maningat, P., Gordon, B.R., Breslow, J.L.: How do we improve patient compliance and adherence to long-term statin therapy? Curr. Atherosclerosis Reports **15**(1), 291 (2013)
8. Iuga, A.O., McGuire, M.J.: Adherence and health care costs. Risk Manag. Healthcare Policy **7**, 35 (2014)
9. Brown, M., Sinsky, C.A.: Medication Adherence - Improve Patient Outcomes and Reduce Costs, AMA STEPS Forward, June 2015
10. Berben, L., Dobbels, F., Kugler, C., Russell, C.L., De Geest, S.: Interventions used by health care professionals to enhance medication adherence in transplant patients: a survey of current clinical practice. Progress Transplantation **21**(4), 322–331 (2011)
11. Luc, M.Z.: Les Associations Fixes améliorent-elles l'observance? Une méta- analyse (Doctoral dissertation, Universite Claude Bernard-lyon 1) (1983)
12. Weber, E.: Compliance: a new monitoring method with the medication event monitoring system. Med. Monatsschr. Pharm. **11**(9), 308 (1988)
13. Lee, J.Y., et al.: Assessing medication adherence by pill count and electronic monitoring in the African American Study of Kidney Disease and Hypertension (AASK) Pilot Study. Am. J. Hypertens. **9**(8), 719–725 (1996)
14. Van Onzenoort, H.A., et al.: Assessing medication adherence simultaneously by electronic monitoring and pill count in patients with mild-to-moderate hypertension. Am. J. Hypertens. **23**(2), 149–154 (2010)
15. Osterberg, L., et al.: First clinical evaluation of a digital health offering to optimize treatment in patients with uncontrolled hypertension and type 2 diabetes. Journal of the American College of Cardiology, 67(13 Supplement) (2016)
16. Comstock, J., Janssen set to launch clinical trial system that uses smartphone app, smart blister packs, Mobihealthnews, October 2017. https://www.mobihealthnews.com/content/janssen-set-launch-clinical-trial-system-uses-smartphone-app-smart-blister-packs. Accessed on 15 March 2020

17. DeMeo, D., Morena, M.: Medication adherence using a smart pill bottle. In: 2014 11th International Conference and Expo on Emerging Technologies for a Smarter World (CEWIT), pp. 1–4. IEEE, October 2014

18. Park, H.: Discharge device of required dose and packing container having the discharge device, Patent no. KR20150106289A, Granted and published in October 2015 (2015)

19. Shafagh, H., et al.: Towards blockchain-based auditable storage and sharing of IoT data. In: Proceedings of the 2017 on Cloud Computing Security Workshop. ACM (2017)

20. Engelhardt, M.A.: Hitching healthcare to the chain: an introduction to blockchain technology in the healthcare sector, Technology Innovation Management Review, 7(10), 22–34 (2017)

21. Madhusudan, S., Kimb, S.: Blockchain technology for decentralized autonomous organizations. Role of Blockchain Technology in IoT Applications 115, 115 (2019)

22. Katuwal, G.J., Pandey, S., Hennessey, M., Lamichhane, B.: Applications of blockchain in healthcare: current landscape and challenges. arXiv preprint arXiv:1812.02776 (2018)

23. Whitaker, M.D., Pawar, P.: Commodity ecology: from smart cities to smart regions via a blockchain-based virtual community platform for ecological design in choosing all materials and wastes. In: Singh, D., Rajput, N.S. (eds.) Blockchain Technology for Smart Cities. BT, pp. 77–97. Springer, Singapore (2020). https://doi.org/10.1007/978-981-15-2205-5_4

24. Tanwar, S., Parekh, K., Evans, R.: Blockchain-based electronic healthcare record system for healthcare 4.0 applications. J. Inf. Secur. Appl. 50, 102407 (2020)

25. Attia, O., Khoufi, I., Laouiti, A., Adjih, C.: An IoT-Blockchain Architecture Based on Hyperledger Framework for Healthcare Monitoring Application. In: 2019 10th IFIP International Conference on New Technologies, Mobility and Security (NTMS), pp. 1–5. IEEE, June 2019

26. Ichikawa, D., Kashiyama, M., Ueno, T.: Tamper-resistant mobile health using blockchain technology. JMIR mHealth uHealth, 5(7) (2017)

Image Processing and Deep Learning

Face Anti-spoofing Based on Deep Neural Network Using Brightness Augmentation

Kun Ha Suh[1] and Eui Chul Lee[2(✉)]

[1] Department of Computer Science, Sangmyung University, Seoul, South Korea
[2] Department of Human Centered Artificial Intelligence, Sangmyung University, Seoul, South Korea
eclee@smu.ac.kr

Abstract. Recently, due to the risk of virus spread due to contact between humans, non-contact based biometrics are preferred, of which face recognition is one of the most promising non-contact based biometrics. However, since the face is exposed to the outside, it is easy to be the target of spoofing attacks. Many hand-crafted features have been devised to prevent spoofing attacks, but they are optimized for a particular environment and have the limitation that they do not work well in other recording conditions. To extract features with higher generalization performance, we used a deep learning approach in which the features are self-learned from the data itself. In this paper, the performance of four state-of-the-art CNN architectures is compared for face spoofing detection. In addition, in order to improve generalization performance and reduce overfitting problem, data augmentation technique is considered. In particular, in order to alleviate the difference in brightness between data obtained from different environments, we examined brightness augmentation. Experiments have shown that the ResNet-18 model has the great performances than other models in both intra-/cross-database scenarios. In addition, the application of brightness augmentation resulted in a decrease performance in the intra-database test, while increased in the cross-database test. Based on our analysis brightness can be a significant factor for improving the face spoofing detection performances, especially in cross-database protocol with different recording conditions.

Keywords: Convolutional neual networks · Deep learning · Face recognition · Face anti-spoofing

1 Introduction

Recently, concerns such as virus transmission have increased the need for contactless based authentication systems. Face recognition is one of the typical non-contact methods, and has the advantage of being more intuitive and convenient than other biometrics. However, with the increasing popularity of facial recognition technology, security threats are also growing. An attacker can attempt to fool the recognition system by using papers containing an authenticated human face, tablet displays, and so forth, without much effort.

© Springer Nature Switzerland AG 2021
M. Singh et al. (Eds.): IHCI 2020, LNCS 12616, pp. 219–228, 2021.
https://doi.org/10.1007/978-3-030-68452-5_23

There have been many attempts in solving the problems of face spoofing attacks in the last decades. An overall description of the existing face anti-spoofing method is summarized in the following literatures [1–4]. Most of the early face anti-spoofing approaches were hand-crafted feature-based methods such as local binary patterns (LBPs) and its variants, spectral methods, and color moments, and so on forth. These methods showed impressive results in the intra-database scenarios, but showed the limit of drastic drop in performance in the cross-database scenarios in which the training data and the test data were obtained in different conditions.

On the other hand, deep learning has achieved impressive results in various tasks and recent studies applied deep neural networks for face anti-spoofing. One approach has partially leveraged deep neural networks to improve the performance of human-designed hand-crafted features. In [5], a pulse energy based face representation using the remote photoplethysmography (rPPG) principle was designed and then fed into the deep convolutional neural network (CNN) to learn discriminative features for liveness detection. De Souza [6] proposed LBPnet, which extracts deep texture features from images by integrating the LBP descriptor into a CNN. Nguyen et al. [7] proposed a hybrid feature that combines the deep feature and the hand-crafted feature. The multi-level local binary pattern (MLBP) and the second fully connected layer on the fine-tuned VGG-19 network were used as a hand-crafted feature and a deep feature, respectively.

In another approach, the features were fully learned from the training data in an end-to-end manner without human knowledge intervention. To extract temporal features as well as spatial features, Xu et al. [8] proposed a deep neutral network structure that combines a CNN with long short-term memory (LSTM) units. Inspired by the idea that each convolutional kernel can be regarded as a part filter, Li et al. [9] investigated the extraction of deep partial features from the fine-tuned VGG-face model. In [10], an approach based on transfer learning was proposed that utilizes a pre-trained VGG-16 model using only static features to recognize photos, videos, or mask attacks.

In this paper, we investigated the face spoofing detection performance of four state-of-the-art networks, including unreported CNN architectures. In addition, data enhancement techniques were applied to prevent overfitting and improve generalized performance. In the experiment, not only spatial augmentation but also brightness augmentation was considered to examine the lighting change scenario in the face spoofing attacks.

Section 2 introduces the four CNN architectures used in the experiment and how to modify them to fit our binary decision task. In addition, data preparation and learning process for model training are explained. In Sect. 3, we provide the experimental results of the four trained models on intra-database and cross-database scenarios, and discussed in Sect. 4. Finally, we draw our conclusions in Sect. 5.

2 Method

2.1 Model Configuration

We reduced the number of parameters by modifying the networks in the following ways. First, we removed fully connected layers and instead directly output the spatial average of the feature map by applying a global average pooling layer to the output of the last convolutional layer in each network. Additionally, we adjusted the number of layers of the networks to have a similar number of parameters while keeping their main architecture. The last 1,000-unit output layer, originally designed to predict 1,000 classes, is replaced by a 2-unit output layer (i.e., real and fake). The approximate number of parameters for the four modified networks are summarized in Table 1. Further specifications and key architectures of each network are described follows.

Table 1. Number of parameters of each network (unit: millions).

Network	VGG-13	Xception-5	ResNet-18	DenseNet-121
# of parameters	9	11	11	7

VGGNet. The VGGNet take the 2nd place prize in the ILSVRC, 2014; it has since received much attention for its simplicity and high performance [11]. It is still one of the popular choices for various CNN-based applications. This network uses only 3×3 convolutional layers stacked on top of each other in increasing depth. The subsequent max-pooling layer reduces volume size. Stacking two 3×3 convolutional filters has the effect of using the same receptive field, as in a single 5×5 convolutional filter, but with fewer parameters. We used the configuration of the 13-layer version from the original paper, i.e., VGG-13, except for the fully connected layers.

ResNet. ResNet, the winner of the ILSVRC, 2015, uses skip connections (also called shortcut connections) [12]. It reported a top-5 error rate, under 3.6%, using an extremely deep CNN composed of 152 layers. The signal feeding into a layer is also added to the output of a layer located a bit higher up the stack. When training a neural network, ResNet is forced to model residual (i.e., difference) between the target and identity functions. It is thus called residual learning and speeds up training considerably. ResNet, which is much deeper than VGG and Xception, demonstrated that extremely deep networks can be trained through the use of residual modules, without the vanishing gradient problem. In our experiments, we used a version of 18 convolutional layers.

Xception. The Xception, stands for extreme version of inception [13], is a modified version of Inception-v3, which earned a 1st place prize in the ILSVRC, 2015 [14]. The goal of the inception module is to act as a multi-level feature extractor by computing different sizes of convolutional filters within the same module of the network. The output of these filters are then stacked along the channel dimension before being fed into the next layer in the network. Xception modified the Inception architecture by combining the use of residual modules and depth-wise separable convolutions. In the Xception architecture, the data first goes through the entry flow, then through the middle flow, and finally through the exit flow. In our work, to reduce the number of parameters, repetition of the middle flow was reduced from eight to five times. Thus we named it Xception-5.

DenseNet. The densely connected convolutional network, (DenseNet), connects each layer to every other layer in a feed-forward fashion [15]. For each layer, the feature-maps of all preceding layers are used as inputs, and its own feature-maps are used as inputs into all subsequent layers. The all preceding feature maps are concatenated before being fed into the next layer in the network. Additional techniques including composite functions, pooling layers, growth rates, bottleneck layers, and compression are applied. The DenseNet architecture alleviates the vanishing-gradient problem, strengthens feature propagation, encourages feature reuse, and substantially reduces the number of parameters. In our experiment, we utilized the configuration version DenseNet-121 (k = 32).

2.2 Data Preparation

In our approach, we used only the face region that did not include the background to determine if the input face was real or fake. The videos in the face anti-spoofing database have different face sizes, so they should be pre-processed for feeding to the CNN input layer. First, after detecting the position of the face region in each frame by the Dlib face detector [16], the face was cropped and resized to a 224 × 224-pixels and saved on the disk.

Furthermore, we used two types of data augmentation techniques on training phase. Firstly, we applied spatial augmentation of preserving pixel intensity such as random cropping, horizontal/vertical flipping. These type of augmentation prevent the risk of image distortion caused by interpolation process. In addition, the brightness or color of the face image can be biased by being affected by variations in the recording device and lighting conditions. But changing the color of the RGB channel can distort the context of the image, so we only adopted brightness augmentation.

In order to vary the brightness of the facial images, we considered two brightness augmentation approaches: brightness adjustment and gamma adjustment. The brightness adjustment method first converts the RGB image to the HSV color space and adjusts the V value at a random ratio to lower or increase the brightness of the image. In the gamma adjustment method, given a face image taken by a particular camera model with a gamma value of γ_1, the gamma adjustment to γ_2 was performed by the following equation:

$$I_{aug} = \left\lfloor \left(\left(\frac{I}{255} \right)^{\frac{\gamma_1}{\gamma_2}} \right) \right\rfloor * 255 \tag{1}$$

where I and I_{aug} are the original intensity and augmented intensity, respectively.

As a result, we have increased the data by ten times through two types of augmentation methods.

2.3 Model Selection and Training

In our experiments, we compared four state-of-the-art architectures that have shown great performance at ImageNet large scale visual recognition challenge (ILSVRC): VGG-13, Xception-5, ResNet-18, and DenseNet-121. Architecture. To fit our task, in each network we removed fully connected layers and instead directly output the spatial average of the feature map by applying a global average pooling layer to the output of the last convolutional layer. Also, the last 1,000-unit out-put layer, originally designed to predict 1,000 classes, was replaced by a 2-unit output layer (i.e., real and fake).

The face anti-spoofing dataset has a class imbalance problem, so we used weighted cross entropy as a loss function to alleviate it and optimized with the mini-batch stochastic gradient descent (SGD) optimizer. We used the learning rate of 10–4, with decay of 10–6, and set the mini-batch size to 16. The models were trained up to 100 epochs, with an early stopping technique to prevent overfitting.

3   Experimental Result

3.1   Database

In this paper, two public face spoofing databases were used to compare the face spoofing detection performance: Replay-Attack [17] and CASIA Face Anti-Spoofing Database (CASIA-FASD) [18]. These databases are also a pair mainly used for cross-database performance evaluation in face anti-spoofing papers (Fig. 1).

Fig. 1. Samples of Replay-Attack and CASIA-FASD databases.

First, the Replay-Attack database consists of 1,200 face videos with 50 subjects. The videos were captured by only the front-facing camera of an Apple 13-inch MacBook laptop with a resolution of 320 × 240 px at 25 fps, for 15 s each (a total of 375 frames). Two different stationary conditions were considered when taking the videos: (1) controlled, wherein the background of the scene is uniform and the light of a fluorescent lamp illuminates the scene; and (2) adverse, wherein the background of the scene is nonuniform and day-light illuminates the scene. For the spoofing medium, iPad 1 (with a size of 1024 × 768 px), iPhone 3GS (with a size of 480 × 320 px), and A4 printed paper are considered to display the face for spoofing purposes. Each attack video is captured for about 10 s in two different attack modes: (1) hand-based attacks, wherein the operator holds the attack media or device using their own hands; and (2) fixed-support attacks, wherein the operator sets the attack device on a fixed support such that they do not move during the spoof attempt.

CASIA-FASD has 600 face videos from a total of 50 subjects. Compared to the Replay-Attack database, the acquisition camera models of the CASIA database are more diverse. The quality levels of capturing devices range from low resolution, medium resolution (two different USB cameras with a resolution of 480 × 640 px) and high resolution (Sony NEX-5 camera with a resolution of 1280 × 720 px). The CASIA database has diverse attack types, including warping, cutting, and video replay attacks. Though the camera models and attacking types are more diverse, compared to the Replay-Attack database, the capturing background and the ethnicity of subjects (all Chinese) are limited.

3.2 Performance Measure

A face spoofing detection system has two types of errors: false acceptance, i.e. incorrect acceptance an access attempt by an unauthorized user, and false rejection, i.e. incorrect rejection and access attempt by an authorized user. Typically, when false acceptance increases, false rejection decreases, and vice-versa. To report on a detection system's performance, the equal error rate (EER) and half total error rate (HTER) are often measured. The EER is defined as the point at which the false acceptance rate (FAR) equals the false rejection rate (FRR). The HTER is defined as half of the sum of the FAR and FRR:

$$HTER(D_2) = \frac{FAR(\tau(D_1), D_2) + FRR(\tau(D_1), D_2)}{2} \tag{2}$$

where D_n is a database and $\tau(D_n)$ is a threshold estimated on D_n. The threshold $\tau(D_n)$ is determined as the point corresponding to the EER on the development set of the database D_n.

We followed the official overall test protocols of the two databases, thus allowing a fair comparison with other methods proposed in the literature. The Replay-Attack database provides a separate development set for tuning the model parameters. Thus, the results were given in terms of EER on the development set and the HTER on the testing set following the official test protocol. Since CASIA-FASD lacks a pre-defined development set, the models were trained and tuned using a subject-disjoint five-folds cross-validation on the training set, with the results reported in terms of EER and HTER on the testing set.

3.3 Intra-database Evaluation

We first analyzed the performance of the four CNN models in the intra-database protocol on Replay-Attack. As shown in Table 2, the ResNet-18 model showed the best performance in most cases, and the VGG-13 model showed the worst performance. In particular, when the brightness augmentation was applied, performances were significantly decreased in all cases except ResNet-13, which had a small performance drop. This may be hindered by the brightness augmentation when the model learns features from each real and fake sample with similar brightness.

Table 2. Intra-database evaluation on Replay-Attack (unit: %).

Network	W/o augmentation		With Gamma adjustment		With Brightness adjustment		With Gamma + Brightness adjustment	
	EER	HTER	EER	HTER	EER	HTER	EER	HTER
VGG-13	8.88	9.86	24.85	15.61	17.06	11.73	13.12	10.53
Xception-5	**0.04**	0.12	22.63	17.67	12.63	8.16	8.80	7.88
ResNet-18	0.06	**0.04**	**2.75**	**1.34**	**1.73**	**0.96**	**2.00**	**1.87**
DenseNet-121	0.05	0.08	23.81	16.67	15.11	8.44	7.89	7.24

3.4 Cross-Database Evaluation

Additionally, we conducted a cross-database evaluation to confirm the performance on the data from a different environment they had never seen on the training phase. The models trained on training set of Replay-Attack were evaluated on the test set of CASIA-FASD that was collected in different capture devices and conditions with Replay-Attack. The result of the cross-database protocol is shown in Table 3.

Table 3. Cross-database evaluation on CASIA-FASD, training on Replay-Attack (unit: %).

Network	w/o augmentation		with Gamma adjustment		with Brightness adjustment		with Gamma + Brightness adjustment	
	EER	HTER	EER	HTER	EER	HTER	EER	HTER
VGG-13	59.55	59.55	55.20	55.19	61.36	61.26	53.59	53.59
Xception-5	58.52	58.52	**41.76**	**41.78**	56.37	**53.39**	47.30	47.30
ResNet-18	59.65	59.65	55.45	55.47	**53.67**	53.69	**39.04**	**39.03**
DenseNet-121	**57.07**	**57.07**	51.23	51.25	54.86	54.88	47.90	47.91

From the results, it can be seen that the model trained on specific datasets shows a significant performance drop on the other datasets. It can be seen that the model has failed to generalize to the spoofed samples obtained in other environments by learning certain patterns in the training set. Brightness augmentation, on the other hand, has alleviated these performance gaps. Especially when performing data augmentation both gamma and brightness, it led to a returned error reduction of about 18%.

The brightness distributions of the images in each database are shown in Fig. 2. In terms of the overall brightness distribution range, the two databases have similar distributions. However, there is a slight difference in the distribution of the frequency for each bin. Thus, in the cross-database scenario, it is possible to obtain a mitigated result in which the generalization performance is slightly improved by considering the difference according to the distribution of these brightness of the images. The brightness augmentation in the training phase can be a way to do this.

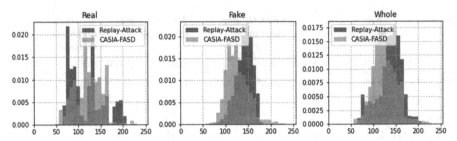

Fig. 2. Face image brightness distributions of Replay-Attack and CASIA-FASD. Each bins were quantized in 32-level for better visualization.

4   Discussion

In this paper, we compare face spoofing detection performances based on the models that were trained on four state-of-the-art CNN architectures using data augmentation. For fair comparison, we set the hyperparameters to be the same, even though the optimal parameters might be different for each neural network. It is possible to improve the models by using more elaborate deep learning architectures, hyperparameters, pre-processing techniques, and so forth. Therefore, these results are valid only in our experimental environment in which the hyperparameters were used the same, and they are not an absolute performance comparison that is always satisfied. Additionally, since the database of the face spoofing field is provided in the form of a video clip, performance may vary depending on how sampling is performed and on which face detector is used. Challengingly, comparison with other methods is difficult under exactly the same conditions.

On the other hand, as the data itself is the most important factor in deep learning, it is important to construct data that can represent the domain well. Our cross-database results using brightness augmentation emphasize the importance of collecting more diverse data. In the future investigations in the field of face anti-spoofing, databases should be built to include various factors analogous to real life scenarios.

The use of an appropriate data augmentation technique is still a good choice for face anti-spoofing field where has relatively insufficient data. Although data augmentation was applied, only a few data augmentation methods were used. Additionally, the lack in the amount and diversity of the training data itself presents a limitation. It is thus necessary to utilize more data augmentation methods and verify their effectiveness on various databases. If time and resources allow, an auto-augmentation technique should be further considered to better search augmentation combinations.

5 Conclusion

In this paper, we proposed a 2D video-based deep learning model for detecting face spoofing attacks. The ResNet-18 model among the four candidate CNN architectures showed the best classification performance. Additionally, data augmentation techniques were used to further improve generalization performance and reduce the risk of overfitting. According to the experimental results, brightness augmentation has resulted in poor performance in intra-database scenarios by hindering better feature learning, whereas in cross-database scenarios, it has improved performance by mitigating the brightness distribution gap of the between different recording conditions.

In future work, we will continue to explore suitable neural network architectures for face spoof detection purposes. In addition, we plan to analyze the performance of the face spoofing detection model in modified attack scenarios such as intentionally changing the brightness on a tablet display, and study a method to improve the performance by mitigating it.

Acknowledgement. This research was supported by the Bio & Medical Technology Development Program of the NRF funded by the Korean government, MSIT(NRF-2016M3A9E1915855).

References

1. Souza, L., Olivera, L., Pamplona, M., et al.: How far did we get in face spoofing detection? Eng. Appl. Artif. Intell. **72**, 368–381 (2019)
2. Galbally, J., Marcel, S., Fierrez, J.: Biometric antispoofing methods: a survey in face recognition. IEEE Access **2**, 1530–1552 (2014)
3. Edmunds, T.: Protection of 2D face identification systems against spoofing attacks. Doctoral dissertation (2017)
4. Anjos, A., Komulainen, J., Marcel, S., Hadid, A., Pietikäinen, M.: Face anti-spoofing: visual approach. In: Marcel, S., Nixon, M.S., Li, S.Z. (eds.) Handbook of Biometric Anti-Spoofing. ACVPR, pp. 65–82. Springer, London (2014). https://doi.org/10.1007/978-1-4471-6524-8_4
5. Lakshminarayana, N.N., Narayan N., Napp N., et al.: A discriminative spatio-temporal mapping of face for liveness detection. In: Proceedings of International Conference on Identity, Security and Behavior Analysis (ISBA), pp. 1–7. IEEE, New Delhi, India (2017)
6. De Souza G.B., da Silva da D.F., Pries R.G., et al.: Deep texture features for robust face spoofing detection. IEEE Trans. Circuits Syst. II: Express Briefs **64**(12), 1397–1401 (2017)
7. Nguyen, D.T., Pham, T.D., Baek, N.R., et al.: Combining deep and handcrafted image features for presentation attack detection in face recognition systems using visible-light camera sensors. Sensors **18**(3), 699 (2018)
8. Xu, Z., Li, S., Deng, W.: Learning temporal features using LSTM-CNN architecture for face anti-spoofing. In: Proceedings IAPR Asian Conference on Pattern Recognition (ACPR), pp. 141–145. Kuala Lumpur, Malaysia, IEEE (2015)
9. Li, L., Feng, X., Boulkenafet, Z., et al.: An original face anti-spoofing approach using partial convolutional neural network. In: Proceedings of International Conference on Image Processing Theory, Tools and Applications (IPTA), pp. 1–6. Oulu, Finland, IEEE (2016)
10. Lucena, O., Junior, A., Moia, V., Souza, R., Valle, E., Lotufo, R.: Transfer learning using convolutional neural networks for face anti-spoofing. In: Karray, F., Campilho, A., Cheriet, F. (eds.) ICIAR 2017. LNCS, vol. 10317, pp. 27–34. Springer, Cham (2017). https://doi.org/10.1007/978-3-319-59876-5_4

11. Simonyan, K., Zisserman, A.: Very deep convolutional networks for large-scale image recognition. arXiv preprint arXiv:1409.1556 (2014)
12. He, K., Zhang, X., Ren, S., Sun, J.: Deep residual learning for image recognition. In: Proceedings of the IEEE Conference on Computer Vision and Pattern Recognition (CVPR), Las Vegas, NV, USA, pp. 770–778. IEEE (2016). https://doi.org/10.1109/CVPR.2016.90
13. Chollet, F.: Xception: deep learning with depthwise separable convolutions. In: Proceedings of the IEEE Conference on Computer Vision and Pattern Recognition, Honolulu, HI, USA, pp. 1251–1258. IEEE (2017). https://doi.org/10.1109/CVPR.2017.195
14. Szegedy, C., Vanhoucke, V., Ioffe, S., Shlens, J., Wojna, Z.: Rethinking the inception architecture for computer vision. In: Proceedings of the IEEE Conference on Computer Vision and Pattern Recognition, Las Vegas, NV, USA, pp. 2818–2826. IEEE (2016). https://doi.org/10.1109/CVPR.2016.308
15. Huang, G., Liu, Z., Van Der Maaten, L., Weinberger, K.Q.: Densely connected convolutional networks. In: Proceedings of the IEEE Conference on Computer Vision and Pattern Recognition, Honolulu, HI, USA, pp. 4700–4708. IEEE (2017). https://doi.org/10.1109/CVPR.2017.243
16. King, D.E.: Dlib-ml: A Machine Learning Toolkit. J. Mach. Learn. Res. **10**, 1755–1758 (2009)
17. Chingovska, I., Anjos, A., Marcel, S.: On the effectiveness of local binary patterns in face anti-spoofing. In: Proceedings of International Conference of Biometrics Special Interest group (BIOSIG), pp. 1–7. Darmstadt, Germany, IEEE (2012)
18. Zhang, Z., Yan, J., Lei, S., Yi, D., Li, S.Z.: A Face Antispoofing Database with Diverse Attacks. In: Proceedings of IAPR International Conference on Biometrics (ICB' 12), New Delhi, India (2012)

Face Spoofing Detection Using DenseNet

Su-Gyeong Yu[1] ⓘ, So-Eui kim[1] ⓘ, Kun Ha Suh[2] ⓘ, and Eui Chul Lee[3(✉)] ⓘ

[1] Department of Artificial Intelligence and Informatics, Graduate School, Sangmyung University, Hongjimun 2-Gil 20, Jongno-Gu, Seoul 03016, Republic of Korea
[2] Department of Computer Science, Graduate School, Sangmyung University, Hongjimun 2-Gil 20, Jongno-Gu, Seoul 03016, Republic of Korea
[3] Department of Human-Centered Artificial Intelligence, Sangmyung University, Hongjimun 2-Gil 20, Jongno-Gu, Seoul 03016, Republic of Korea
eclee@smu.ac.kr

Abstract. Today, face recognition is the most widely used biometric recognition technology. However, face recognition is a biometric method that is vulnerable to spoofing. Types of spoofing attack include print, replay, and 3D mask. Methods based on hand-crafted features such as local binary patterns which use high-frequency features of images, are therefore vulnerable to blur caused by optical factors or motion. Recently, face spoofing detection methods based on learned features using the convolutional neural network series have been introduced. Among them, DenseNet-121 has a densely connected structure unlike the other structure, so it can widely reflect the characteristics of various frequency bands of an image. In this paper, we study face-spoofing detection using DenseNet-121. For the performance measurement, CASIA-FASD and a lab-made PR-FSAD were used. As a result of the experiment, it was confirmed that DenseNet showed good face spoofing detection performance in both DBs. This result can be analyzed because of the structural characteristics that DenseNet-121 well reflects the wide frequency characteristics of the image.

Keywords: Face recognition · Spoofing detection · Convolutional neural network · DenseNet · Densely connection

1 Introduction

Biometrics recognition refers to a technology that authenticates a user using physical or behavioral characteristics, which are unique characteristics of an individual. Face recognition is more convenient and low rejection methods among many biometric technologies. However, as face recognition technology is widely used, several security weaknesses have been reported [1]. Therefore, it is required to develop a safer and more accurate face recognition system.

Various studies have been conducted to detect face spoofing attacks. A spoofing attack refers to an attack that tricks a system using fake data. Common attack types include print, replay, and 3D mask. For that, Tian and Xiang extracted spatial information of each frame using LBP (Local Binary Patterns), and performed multiscale DCT (Discrete Cosine

© Springer Nature Switzerland AG 2021
M. Singh et al. (Eds.): IHCI 2020, LNCS 12616, pp. 229–238, 2021.
https://doi.org/10.1007/978-3-030-68452-5_24

Transform) along the vertical axis of the extracted features. The spatial and temporal information obtained through this was input to the SVM (Support Vector Machine) classifier to detect face spoofing attacks [2]. In addition, Benlamoudi et al. extracted features of each area of the image using LBP and input it to a nonlinear SVM classifier to detect face spoofing attacks [3]. This is a handcraft-based feature based method that requires human intervention and has relatively low classification accuracy. Also, it is based on the texture of the image, which is a high-frequency feature, and this information is easily lost depending on the resolution or focus of the image [4].

Recently, CNN (convolutional neural network) models showing high performance in image recognition and classification has been widely used in face spoofing detection. For example, studies using DPCNN (Deep Part Con-volution Neural Network) and LSTM-CNN have been investigated [5, 6]. In the DPCNN study, deep part features are extracted by inputting and training data into the preprocessed VGG model. At this time, in order to prevent overfitting during learning, the dimension of features was reduced using PCA. The extracted features were classified into real faces and fake faces using SVM. EER through this was 4.5%, showing better performance than the existing methods such as LBP and DOG. Next, in the LSTM-CNN study, a circulatory neural network was implemented using Long Short Term Memory units, and the temporal structure of video was learned through the LSTM-CNN architecture. Experimental results show that HTER is 5.93%, and this temporal feature improves the performance of face spoofing attack detection.

As such, CNN shows high performance in image recognition and classification. CNN automatically extracts image features from the convolution layer and it classifies in the Fully connected layer based on them [7]. At this time, the low-frequency feature is extracted from the low level layer, and as the layer progresses, the high-frequency feature is learned by combining the features obtained from the previous layer. However, the previous neural network model of such a structure has a disadvantage that the gradient vanishing problem occurs as the layer gets deeper. Even though several methods have been proposed to solve this problem, the same problem has been repeated again when the layer deepens. In contrast, DenseNet has a structure in which information of the previous layer is connected to the input of the next layer through concatenation operation. Through this structure, the feature of the initial layer is learned to be well preserved down to the relatively deep layer [8]. Therefore, in this paper, we studied a face spoofing attack detection technique using DenseNet-121 with these characteristics. The databases used are CASIA-FASD (Face Anti-Spoofing Database) [9], one of the public databases, and our lab made PR-FSAD (Pattern Recognition lab - Face Spoofing Advancement Database) [10].

2 Proposed Methodology

2.1 Database

The database used for learning is CASIA-FASD and PR-FSAD. First, CASIA-FASD is an open database released for scientific research purposes. This DB consists of real face image and fake face image and the types of attack are printed photos, blinking eyes, and video replay. Using three different cameras, three image qualities are included such as low, normal and high [7]. Examples of CASIA-FASD are as shown in Fig. 1. Next, PR-FSAD has real face images and fake face images with printed photos and replayed videos. These images have different characteristics from the existing databases such as unfixed background and lighting, three distances (near, halfway, distance), and three angles (bottom, middle, top). Face recognition in daily life takes place in various angles, backgrounds and lighting. Therefore, if data that does not reflect these characteristics are used for training, spoofing attack detection performance may be degraded. Therefore, PR-FSAD, which includes various environmental factors, shows excellent performance in face spoofing detection technology. [10]. An example of PR-FSAD can be seen in Fig. 2.

The construction process of PR-FSAD is as follows. Real and fake face data were obtained from 30 subjects and photographed with 2 tablets and 2 smartphones. Among the three angles, the top and bottom were obtained at about ± 30° based on the middle. In the case of distance information, there is a difference in body structure for each subject, so an image for each distance was obtained using the ratio of the relative face occupying the camera. In addition, the shooting was taken in an unfixed background and lighting environment. The fake face data was constructed by printing or replaying the obtained face data.

Fig. 1. Three examples of CASIA-FASD. (from the left, real, print, blinking eyes, replay)

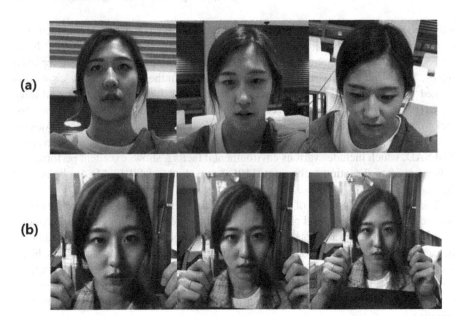

Fig. 2. Examples of PR-FSAD. (a): live cases with capture angle variation (from the left, bottom, middle, and top), (b): fake cases with distance variation (from the left, near, halfway, and distant)

2.2 Model

In this paper, DenseNet-121 was used as a neural network model. As explained in Sect. 1, DenseNet-121 has a different structure from the other CNN model. The previous CNN neural network model consists of a structure that uses the value output from the layer as an input to the next layer. As the layer gets deeper, the problem of gradient vanishing/exploding occurs, so learning may be difficult. DenseNet solved this problem through densely connectivity structure. Densely connectivity is a structure that concatenates input data to output data that comes out through a layer and uses the result as an input to the next layer. In a similar way, there is a shortcut connection of ResNet, which uses the added value of the output and input of the layer as the input of the next layer [11]. At this time, the biggest difference between ResNet and DenseNet is the calculation method of output and input. ResNet's add operation can be implemented without requiring additional operations or parameters and by preserving previously learned information and proceeding with learning, the aforementioned problems of the existing CNN have been solved. However, in the add operation, the reflection of the low-level information of the image may decrease as the layer becomes deeper. On the other hand, the concatenation operation of DenseNet preserves the low-level image information in the initial layer relatively well until the learning is completed as the feature maps are combined in the progression direction. In other words, this allows the characteristics of input data extracted from the previous Layer to be relatively well preserved and learned to the deep Layer. In addition, there are advantages such as mitigation of vanishing-gradient problem, recommendation of feature reuse, and reduction of computational amount and number of parameters. For this reason, DenseNet was selected as a learning model and an experiment was conducted. Figure 3 shows the basic structure of existing CNN, ResNet and DenseNet.

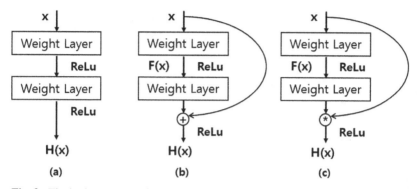

Fig. 3. The basic structure of the three models (a): CNN, (b): ResNet, (c): DenseNet.

DenseNet-121, which was not trained in advance, was used as a learning model, and a face image with a pixel size of 224 × 224 is used as the input data of the model. The learning model parameter setting is as follows. The loss function and the optimize function were respectively used with binary cross entropy and stochastic gradient descent (SGD), and learning rate = 0.001, stride = 2, and batch size were set to 8. In the case of PR-FSAD, epoch, which is the maximum number of learning repetitions, was trained as 100, and in case of CASIA-FASD, the optimal epoch obtained using the k-fold method was designated as 17. At this time, k-fold method was performed with k = 4. In addition, verification was performed every time one epoch was completed with a pre-divided validation set. In the learning process, images are randomly loaded to avoid learning by the features of the image sequence. The final output of this process is classified into two classes such as real and fake through the sigmoid function of the fully connected layer after global average pooling. Finally, in order to prevent overfitting in the learning process, learning was conducted using an early stopping technique that stops learning when performance is not improved more than 10 times. The DB configuration used in the experiment can be seen in detail in the following Table 1 and 2. Fig. 4 below shows the structure of the built DenseNet-121.

Table 1. Summary of the number of CASIA-FSAD data used for train and test

CASIA-FASD	Train	Test	Total
Real	6,007	9,962	15,969
Fake	21,894	30,427	52,321
Total	27,901	40,389	68,290

Table 2. Summary of the number of PR-FSAD data used for train and test

PR-FSAD	Train	Validation	Test	Total
Real	13,680	10,080	18,720	42,480
Fake	27,360	20,160	37,440	84,960
Total	41,040	30,240	56,160	127,440

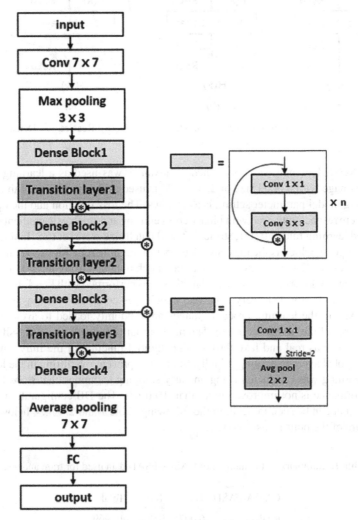

Fig. 4. Structure of DenseNet-121.

3 Experimental Results

In this section, we describe the performance results of face-spoofing detection using DenseNet-121 for two databases. The results were evaluated using Half Total Error Rate (HTER), Accuracy, and confusion matrix. HTER is a value using only the ratio of misclassification. The smaller the HTER, the better the performance. Confusion matrix is a table for comparing predicted values and actual values to measure prediction performance through transcription. The meanings of TP, FP, TN, FN, and GP, GN used are as follows. TP (True Positive) is a value that actually correctly predicts the reality, FP (False Positive) is a value that incorrectly predicts Fake as Real, TN (True Negative) is a value that correctly predicts Fake as Fake, and FN (False Negative) is It represents the value that was incorrectly predicted as Fake of Real. GP and GN used to represent the ratio of normal classification and misclassification mean the total number of actual real data and actual fake data, respectively.

First, when using PR-FSAD as the training data, HTER and accuracy were 2.34% and 97.66%, showing excellent performance. In the case of CASIA-FASD, HTER and accuracy were 2.35% and 97.65%, respectively, showing similar results to PR-FSAD. Table 3 shows the performance of DenseNet-121 and its comparison with other study. Next, the confusion matrix for the result is shown in Table 4 and Table 5 below. Through the matrix of CASIA-FASD, it can be seen that the rate of false classification of fakes as real is about 2.346% (714/30427). In the case of PR-FSAD, the rate of misclassifying Fake as Real is about 2.337% (875/37440), and the rate of misclassifying Real as Fake is about 2.339% (438/18720). As a result, DenseNet-121 exhibits better performance than the research results using the existing CNN model. In addition, not only PR-FSAD containing various deviation characteristics but also CASIA-FASD, one of the widely used public databases, shows excellent spoofing detection performance.

Table 3. Comparison of spoofing attack detection performance by neural network model

Method	DB	HTER (%)	ACCURACY (%)
CNN [6]	CASIA-FASD	7.34	——
LSTM-CNN [6]		5.93	——
DenseNet-121	CASIA-FASD	2.35	97.65
	PR-FSAD	**2.34**	**97.66**

Next, Receiver Operating Characteristic (ROC) curves were used to evaluate and visualize the performance of the results. The ROC curve is a graph using False Positive Rate (FPR) and True Positive Rate (TPR). The closer the graph is to the upper left and the wider the area under the curve mean the better classification performance. The ROCs of our experimental result are shown in Fig. 5. In here, the graphs are close to the upper left of both databases. In the other neural network model, the degree of

Table 4. CASIA-FASD confusion matrix

Confusion matrix	Predicted fake	Predicted real
Actual Fake	29713/30,427 (TN/GN)	714/30,427 (FP/GN)
Actual Real	234/9,962 (FN/GP)	9728/9,962 (TP/GP)

Table 5. PR-FSAD confusion matrix

Confusion matrix	Predicted fake	Predicted real
Actual Fake	36565/37440 (TN/GN)	875/37440 (FP/GN)
Actual Real	438/18720 (FN/GP)	18282/18720 (TP/GP)

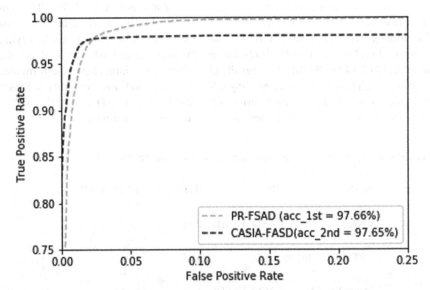

Fig. 5. ROC curve for spoofing attack detection performance evaluation of DenseNet-121

reflection of initial layer information decreases as learning progresses. However, due to the densely connection structure, DenseNet preserves the information from the initial layer relatively well up to the deep layer and reflects it in learning. This means that the low-level information of the image is well preserved and learned together with the high-level information extracted from the terminal layer. In other words, it can be said that the face spoofing attack was well detected by reflecting the characteristics of not only the high frequency region but also the low frequency region well.

4 Conclusion

In this paper, we studied face spoofing detection based on CNN model that automatically extracts and learns features of input data. As a neural network model, DenseNet-121 with densely connection structure was used. CASIA-FASD, one of the common DBs, and PR-FSAD, which contains various deviation information of angle and distance, were used. HTER and accuracy were used as outcome indicators. As a result, CASIA-FASD was 2.35% and 97.65%, respectively and PR-FSAD was 2.34% and 97.66%, respectively, showing good performance in both databases. As well, performance evaluation was performed by visualizing the results as a ROC curve. The differentiated dense connectivity structure of DenseNet has the characteristic of preserving low-frequency image characteristic from the initial layer to the deeper layer relatively well, indicating that face spoofing was well detected.

In future works, we will study how the maintenance of low-frequency feature of images affects performance in face spoofing detection through comparison with ResNet with a structure similar to DenseNet. In addition, we plan to analyze the shape and texture information of the face through detailed experiments on protocols for each angle and distance.

Acknowledgement. This research was supported by the Bio & Medical Technology Development Program of the NRF funded by the Korean government, MSIT(NRF-2016M3A9E1915855). Also, this work was financially supported by a Grant (2018000210004) from the Ministry of Environment, Republic of Korea.

References

1. Biggio, B., Akhtar, Z., Fumera, G., Marcialis, G.L., Roli, F.: Security evaluation of biometric authentication systems under real spoofing attacks. IET Biometrics 1(1), 11–24 (2012)
2. Tian, Y., Xiang, S.: Detection of video-based face spoofing using LBP and multiscale DCT. In: Shi, Y.Q., Kim, H.J., Perez-Gonzalez, F., Liu, F. (eds.) IWDW 2016. LNCS, vol. 10082, pp. 16–28. Springer, Cham (2017). https://doi.org/10.1007/978-3-319-53465-7_2
3. Benlamoudi, A., Samai, D., Ouafi, A., Taleb-Ahmed, A., Bekhouche, S.E., Hadid, A.: Face spoofing detection from single images using active shape models with stasm and lbp. In: Proceeding of the Troisime Conference Internationale sur la vision artificielle CVA, vol. 2015, p. 31 (2015)
4. Määttä, J., Hadid, A., Pietikäinen, M.: Face spoofing detection from single images using micro-texture analysis. In: 2011 International Joint Conference on Biometrics (IJCB), pp. 1–7. IEEE (2011).
5. Li, L., Feng, X., Boulkenafet, Z., Xia, Z., Li, M., Hadid, A.: An original face anti-spoofing approach using partial convolutional neural network. In: 2016 Sixth International Conference on Image Processing Theory, Tools and Applications (IPTA), pp. 1–6. IEEE (2016)
6. Xu, Z., Li, S., Deng, W.: Learning temporal features using LSTM-CNN architecture for face anti-spoofing. In: 2015 3rd IAPR Asian Conference on Pattern Recognition (ACPR), pp. 141–145 IEEE (2015)
7. LeCun, Y., Bengio, Y., Hinton, G.: Deep learning. Nature **521**(7553), 436–444 (2015)

8. Huang, G., Liu, Z., Van Der Maaten, L., Weinberger, K.Q.: Densely connected convolutional networks. In: Proceedings of the IEEE Conference on Computer Vision and Pattern Recognition, pp. 4700–4708 (2017)
9. Zhang, Z., Yan, J., Liu, S., Lei, Z., Yi, D., Li, S.Z.: A face antispoofing database with diverse attacks. In: 2012 5th IAPR International Conference on Biometrics (ICB), IEEE, pp. 26–31 (2012)
10. Bok, J.Y., Suh, K.H., Lee, E.C.: Verifying the effectiveness of new face spoofing DB with capture angle and distance. Electronics 9(4), 661 (2020)
11. He, K., Zhang, X., Ren, S., Sun, J.: Deep residual learning for image recognition. In: Proceedings of the IEEE Conference on Computer Vision and Pattern Recognition, pp. 770–778 (2016)

1-Stage Face Landmark Detection Using Deep Learning

Taehyung Kim[1] , Ji Won Mok[1] , and Eui Chul Lee[2]([✉])

[1] Department of Artificial Intelligence and Informatics, Graduate School, Sangmyung University, Seoul 03016, Republic of Korea
[2] Department of Human-Centered Artificial Intelligence, Sangmyung University, Seoul 03016, Republic of Korea
eclee@smu.ac.kr

Abstract. In this paper, we propose a new face landmark detection method. In previous methods, face detection was essential before a face landmark detection. The disadvantage of these methods is that they are greatly affected by the performance of the face detection model. In order to overcome this disadvantage, we proposed a method to simultaneously detect the face region and the face landmark. The basic idea came from 1-stage object detection. The structure of the Yolo v3 model, a representative 1-stage object detection model, was modified to find the landmark, and the loss function for training was modified to learn the coordinates of the landmark. In addition, MobileNet was used as the backbone network to increase the processing speed. In order to check the performance of the proposed model, the model was trained using the 300 W-LP database. It was then tested using Helen and LFPW databases, and the average normalized error was used as the evaluation metric. As a result of the evaluation, it was confirmed that the proposed model has improved performance over the previous methods.

Keywords: Face landmark detection · Face detection · 1-stage detection · Deep learning

1 Introduction

Facial landmark detection is a work of finding a face from an image and extracting a feature points of the face, which is a basic element of various face analysis task such as facial recognition [1], face verification [2], and face 3D modeling [3]. Therefore, it is important to accurately detect the face region and automatically select the feature points. However, in real-world, there are challenges such as variable pose, illumination, and facial expression, so it is difficult to detect with high accuracy.

There are three traditional methods of facial landmark detection algorithms that utilize facial appearance and facial shape information: Holistic methods, Constrained Local Model (CLM) methods, regression-based methods. These three methods differ in how to use facial appearance and shape information. Holistic methods use the global facial appearance and shape pattern information to detect facial landmarks, and explicitly

© Springer Nature Switzerland AG 2021
M. Singh et al. (Eds.): IHCI 2020, LNCS 12616, pp. 239–247, 2021.
https://doi.org/10.1007/978-3-030-68452-5_25

created a model this. Although it is a method that focuses on improving the fitting algorithms, but it was found that the overall performance was significantly decrease in cross database experiments [4]. CLM methods utilize the global facial shape model and build a local appearance model, inferring the location of landmarks based on independent local appearance information around each landmark and the overall facial shape pattern. However, CLM methods have an accuracy-robust tradeoff phenomenon in which the accuracy of landmark localization is reduced using large local appearance [5]. Regression-based methods implicitly capture facial appearance and shape information, then directly learn the mapping of landmark location in image appearance. Unlike Holistic methods and CLM methods, it does not build a global face shape model. In general, regression-based methods can be classified into direct regression methods, cascaded regression methods, and deep-learning based regression methods. However, these methods have the shortcoming that trained algorithms may not work properly if other face detectors are used to extract landmarks from face bounding box region [6]. The above methods affect the landmark detect performance according to the face area detection model. Landmarks are basic features for tasks such as facial recognition and facial expression recognition, so they must perform consistently in various environments. Therefore, face landmark detection model that is not affected by the face area detection model is required.

In this paper, we propose a 1-stage landmark detection model to solve the shortcomings of previous researches. The proposed model has the following strong points. (1) The proposed model is not affected by the performance of the face area detection model; (2) accurate and fast; (3) high generalization performance.

2 Proposed Method

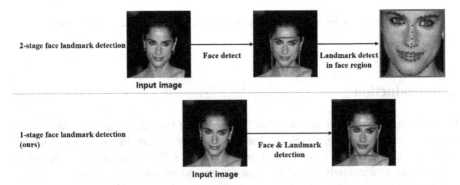

Fig. 1. Comparison between the previous face landmark detection and the proposed method.

As shown in Fig. 1, in the existing face landmark detection method, an input image is entered into a face region detection model and bounding box coordinates corresponding to the face area are obtained. Then, the face area is cropped through the bounding box coordinates and entered the face landmark detection model to obtain the landmark coordinates. Therefore, this method requires two steps of face area and landmark detection.

As mentioned above, it is affected by the face region detection model. Also, the speed of predict is not consistent because face landmark must be detected as many as the number of faces in the image. To solve these problems, we propose 1-stage landmark detection model that simultaneously detect face area and landmark without going through a face region detection step. The proposed method is not affected by the face region detection model, which is the shortcoming of the existing method, and guarantees the consistency of prediction speed.

2.1 Model Architecture

The proposed method was inspired by [7]. It is a method to obtain multiple regression values from a single feature map obtained by entering a 40×40 image into CNN. In [7], face landmark, glasses, gender, smile, and pose were predicted. In [7], face region detection is indispensable as preprocessing because the cropped image to fit the face region is entered. Therefore, the performance of the face region detection model can affect the landmark detection performance. To solve this weakness, we propose the landmark detection method based on the 1-Stage object detection model.

Typical 1-stage object detection models are You Only Look Once v3 (YOLOv3) and Single Shot Multibox Detector (SSD) [8, 9]. In terms of accuracy, the two models are observed to be similar, and in terms of speed, YOLOv3 is faster. SSD uses a Pyramidal feature hierarchy method that extracts feature maps of different scales from the input image and performs object detection by independently extracting features from each level. This is good for detecting small objects, but it doesn't use strong semantics of a small scale at a high level. Yolov3 uses a method similar to the Feature Pyramid Network, which adds a method of combining high level features with low level features using the skip connection concept in the Pyramidal feature hierarchy method. Each method can be seen in Fig. 2. So, features of various scales can be extracted well, so using Yolov3 can find the faces of various scales better. Therefore, the YOLOv3 model which can use strong semantics, was chosen as the base model. The difference from the original model is to predict the coordinates of the landmark instead of predicting the class of the object in the bounding box. This can be easily checked in Fig. 3.

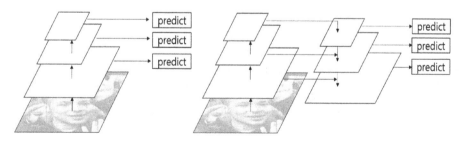

a) Pyramidal feature hierarchy   b) Feature Pyramid Network

Fig. 2. a) is the Pyramidal feature hierarchy method used in SSD, and b) is the Feature Pyramid Network method used in Yolov3.

2.2 Loss Function

The loss function for learning was also based on YOLO's one. The original method used a combination of localization loss, confidence loss, and classification loss. The classification loss is calculated as a squared error and can be expressed as Eq. (1) below. 1_{ij}^{obj} becomes 1 if an object exists in the corresponding cell, and 0 if it does not exist.

$$\sum_{i=0}^{S^2} 1_{ij}^{obj} \sum_{c \in classes} \left(p_i(c) - \hat{p}_i(c)\right)^2 \tag{1}$$

For 1-stage face landmark detection, the above Eq. (1) is removed and Sum Squared Error (SSE) is added as shown in the following Eq. (2). Likewise, if no object is detected, 1_{ij}^{obj} becomes 0 and the loss value is not calculated. λ_{coord} is a weight added to improve detection performance. S^2 is the size of the feature map (width x height), and B is the number of anchor boxes.

$$\lambda_{coord} \sum_{i=0}^{S^2} \sum_{i=0}^{S^2} 1_{ij}^{obj} \left[\left(x_i - \hat{x}_i\right)^2 + \left(y_i - \hat{y}_i\right)^2\right] \tag{2}$$

Consequently, the landmark loss (SSE) is added in the original loss function that the classification loss is removed.

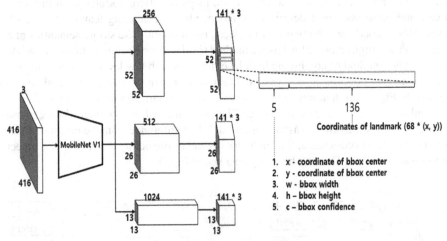

Fig. 3. The proposed model. Instead of predicting the probability value for each class, it was modified to predict the coordinate value of the landmark. If you input 416 × 416 × 3 input image, three scale feature maps are output (52 × 52 × 256, 26 × 26 × 512, 13 × 13 × 1024). Each feature map is matched to 141 × 3 channel to obtain face area information and face landmark information.

3   Experimental

3.1   Experimental Setup

Backbone Network

The performance and processing speed of the CNN model are greatly affected by the backbone network. In the original YOLOv3, Darknet-53 is used as the backbone network. This network is designed as a deep network to improve the accuracy of multiple class predictions. In this study, only one class (a human face) is used, so we used MobileNetv1 to increase the processing speed, although the accuracy is slightly reduced [10]. In Mobilenetv1, the Depthwise Separable Convolution operation, which combines Depthwise Convolution and Pointwise Convolution, is used. Depthwise Convolution is a convolution operation using each kernel by separating each channel, and the input channel and the output channel are always the same. Pointwise Convolution is a convolution operation that can change the output channel by convolution with 1 × 1 kernel. Through this, the amount of parameters is drastically reduced to improve speed, and performance doesn't decrease significantly. The Depthwise Separable Convolution method can be seen in detail through Fig. 4. Therefore, the MobileNetv1 was used for 1-stage detection.

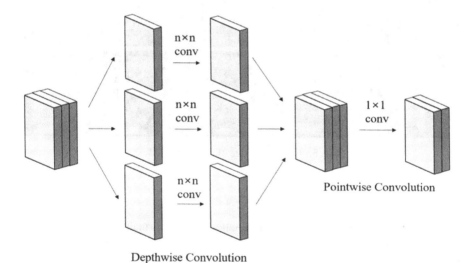

Fig. 4. The structure of Depthwise Separable Convolution of MobileNetv1. If you input 3 × 3 × 3 input image, the 3 channels are separated and convolutional, then recombined and convolution with 1 × 1 kernel.

Database

We used 300 W across Large Poses (300 W-LP) [11] during training the model. 300 W-LP is a database that applies rotation of 3D images to 300 W [6] for data augmentation. 300-W is a standardized database of 68 landmarks for HELEN [12], LFPW [13], and AFW [14]. The total number of data of 300 W-LP is 122,450. The 300 W-LP consists of processed images as if one image was taken from several angles, and the example image can be seen in Fig. 5. This data was divided into a training set (97,967) and a validation set (22,035) for training. And then, the remaining data (Helen: 1616, LFPW test set: 832) that were not used for the training were used for evaluation.

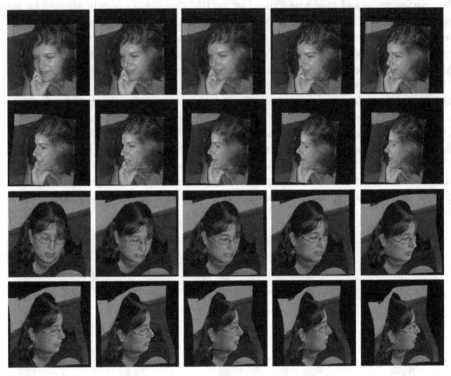

Fig. 5. Example image of 300 W-LP DB.

Evaluation Metric

We used average normalized error as an evaluation metric. Normalized error is generally used as an evaluation matrix because the face area in the image varies depending on the object, location and environment [15]. In the equation below, N means the number of landmarks, and d_i means the coordinates of the landmark predicted by $\{d_{x,i}, d_{y,i}\}$. And g_i is $\{g_{x,i}, g_{y,i}\}$, which is the coordinates of the ground truth landmark coordinates. g_{le} and g_{re} are criteria to prevent the error from changing due to changes in image and face size. In this paper, the left eye inner corner and the right eye inner corner are used.

$$\frac{1}{N} \sum_i^N \frac{\|d_i - g_i\|_2}{\|g_{le} - g_{re}\|_2} \tag{3}$$

3.2 Experimental Result

The training and testing were conducted in the NVIDIA GeForce GTX1080 Ti (Santa Clara, CA, USA) environment. The size of the input image is 416 × 416. As a framework for learning, PyTorch was used. Adam was used as the optimizer, and the learning rate was set to 0.001. The training was stopped early at training loss 0.0039 and validation

Helen   LFPW

Fig. 6. Result of the test of the 1-stage face region and landmark detection model. Blue dot: ground truth of face landmark, red dot: predicted face landmark points, red box: predicted face region. (Color figure online)

loss 0.0847. The landmark extraction result of the trained model can be confirmed in Fig. 6. The processing speed per image is 20 ms.

In order to confirm the accuracy of the proposed model quantitatively, we compared it with previous studies that measured the accuracy using LFPW and Helen. The number of landmark points extracted from the proposed model was 68, and we used the normalized error mentioned above as the evaluation metric. Since the purpose is to measure the distance difference between the ground truth and the predicted landmark, the general error rate is evaluated [15]. The performance of previous studies was obtained in [15], and the evaluation results can be seen in Table 1. The evaluation results for each DB, LFPW is 3.16, Helen is 3.93. The lowest error rate was confirmed in both DBs

Table 1. Performance comparison with previous methods using normalized error as evaluation matrix.

Helen [12]		LFPW [13]	
FPLL	8.16	FPLL	8.29
DRMF	6.70	DRMF	6.57
RCPR	5.93	RCPR	6.56
Gaussian-Newton DPM	5.69	Gaussian-Newton DPM	5.92
SDM	5.53	SDM	5.67
CFAN	5.50	CFAN	5.44
CFSS	4.63	CFSS	4.87
OURS	**3.93**	**OURS**	**3.16**

4   Conclusion

In this paper, we proposed the 1-stage face landmark detection model that performs face detection and face landmark detection at the same time. The proposed model does face detection and landmark detection at same time, so no additional face detection step is required. Therefore, constant detection performance can be expected. As a result of the evaluation, the proposed model performed better than previous studies tested using LFPW and Helen. Based on YOLOv3 but using Mobilenetv1 as a backbone network, the processing speed became faster. Therefore, it will be easier to apply additional processing such as gaze tracking, face recognition, facial expression recognition. In future work, we will implement a model that can be used in CPU and mobile environments. For this, we will use advanced networks such as MobileNetV3 and Efficientnet [16, 17].

Acknowledgement. This work was supported by the NRF(National Research Foundation) of Korea funded by the Korea government (Ministry of Science and ICT) (NRF-2019R1A2C4070681).

References

1. Ahonen, T., Hadid, A., Pietikainen, M.: Face description with local binary patterns: application to face recognition. IEEE Trans. Pattern Anal. Mach. Intell. **28**(12), 2037–2041 (2006)
2. Sun, Y., Wang, X., Tang, X.: Deep learning face representation from predicting 10,000 classes. In: Proceedings of the IEEE Conference on Computer Vision and Pattern Recognition. pp. 1891–1898 (2014)
3. Kemelmacher-Shlizerman, I., Basri, R.: 3D face reconstruction from a single image using a single reference face shape. IEEE Trans. Pattern Anal. Mach. Intell. **33**(2), 394–405 (2010)
4. Saragih, J., Göcke, R.: Learning AAM fitting through simulation. Pattern Recogn. **42**(11), 2628–2636 (2009)
5. Ren, S., Cao, X., Wei, Y., Sun, J.: Face alignment at 3000 fps via regressing local binary features. In: Proceedings of the IEEE Conference on Computer Vision and Pattern Recognition. pp. 1685–1692 (2014)
6. Sagonas, C., Antonakos, E., Tzimiropoulos, G., Zafeiriou, S., Pantic, M.: 300 faces in-the-wild challenge: database and results. Image Vis. Comput. **47**, 3–18 (2016)
7. Zhang, Z., Luo, P., Loy, C.C., Tang, X.: Facial landmark detection by deep multi-task learning. In: European conference on computer vision. pp. 94–108. Springer, Cham. (2014)
8. Redmon, J., Farhadi, A.: Yolov3: An incremental improvement. arXiv preprint arXiv:1804.02767 (2018)
9. Liu, W., Anguelov, D., Erhan, D., Szegedy, C., Reed, S., Fu, C.Y., Berg, A. C.: Ssd: Single shot multibox detector. In European conference on computer vision. pp. 21–37. Springer, Cham (2016)
10. Howard, A.G., Zhu, M., Chen, B., Kalenichenko, D., Wang, W., Weyand, T., Adam, H.: Mobilenets: efficient convolutional neural networks for mobile vision applications. arXiv preprint arXiv:1704.04861 (2017)
11. Zhu, X., Lei, Z., Liu, X., Shi, H., Li, S. Z.: Face alignment across large poses: A 3d solution. In: Proceedings of the IEEE Conference on Computer Vision and Pattern Recognition. pp. 146–155 (2016)
12. Zhou, E., Fan, H., Cao, Z., Jiang, Y., Yin, Q.: Extensive facial landmark localization with coarse-to-fine convolutional network cascade. In: Proceedings of the IEEE International Conference on Computer Vision Workshops pp. 386–391 (2013)
13. Belhumeur, P.N., Jacobs, D.W., Kriegman, D.J., Kumar, N.: Localizing parts of faces using a consensus of exemplars. IEEE Trans. Pattern Anal. Mach. Intell. **35**(12), 2930–2940 (2013)
14. Zhu, X., Ramanan, D.: Face detection, pose estimation, and landmark localization in the wild. In: 2012 IEEE Conference On Computer Vision and Pattern Recognition. pp. 2879–2886. IEEE (2012)
15. Wu, Y., Ji, Q.: Facial landmark detection: a literature survey. Int. J. Comput. Vis. **127**(2), 115–142 (2019)
16. Tan, M., Le, Q.V.: Efficientnet: Rethinking model scaling for convolutional neural networks. arXiv preprint arXiv:1905.11946 (2019)
17. Howard, A., Sandler, M., Chu, G., Chen, L.C., Chen, B., Tan, M., Le, Q.V.: Searching for mobilenetv3. In: Proceedings of the IEEE International Conference on Computer Vision pp. 1314–1324 (2019)

Image Identification of Multiple Parrot Species Belonging to CITES Using Deep Neural Networks

Woohyuk Jang[1] , Si Won Seong[2] , Chang Bae Kim[3] , and Eui Chul Lee[4]([✉])

[1] Department of Computer Science, Sangmyung University, Seoul, South Korea
[2] Department of AI and Informatics, Sangmyung University, Seoul, South Korea
[3] Department of Biotechnology, Sangmyung University, Seoul, South Korea
[4] Department of Human-Centered AI, Sangmyung University, Seoul, South Korea
eclee@smu.ac.kr

Abstract. Recently, not only studies on inanimate objects, but also deep learning-based image recognition studies on animals and plants are being actively conducted. Animals and plants designated by CITES are protected internationally. At airports and ports, it takes a lot of time and money due to manual inspection at customs clearance. Using this vulnerability, smugglers illegally import animals and plants belonging to CITES. To solve this problem, in this paper, we propose a method for classifying parrot species belonging to CITES by detecting objects based on deep neural networks. The SSD model was used for object detection, and the data augmentation technique was also applied to prevent overfitting. Using the trained model, parrot classification performance was measured as the mAP of about 95.7%.

Keywords: Object detection · Deep neural network · Parrot classification · CITES

1 Introduction

The animals and plants designated by Convention on International Trade in Endangered Species of wild flora and fauna(CITES) are endangered and are protected internationally. Animals and plants to be protected are illegally imported due to the fact that customs clearance inspections at airports and ports take a long time and are not accurate. Among illegal trade, wildlife trade is increasing annually [1]. To solve this problem, an automatic customs clearance system can be a good tool. For that, object detection is a suitable method using artificial intelligence technology. As biological image recognition research using artificial intelligence, various species classification studies such as toxic plant detection [2], fish detection [3], and bird detection [4] are being conducted. The method using deep neural networks in object detection performs much better than machine learning methods. Thus, object detection models such as R-CNN [5], Fast R-CNN [6], Faster R-CNN [7], YOLO [8] and SSD [9] based on deep neural networks appeared. In this paper, we propose a method for classifying parrot species belonging to CTIES using

© Springer Nature Switzerland AG 2021
M. Singh et al. (Eds.): IHCI 2020, LNCS 12616, pp. 248–253, 2021.
https://doi.org/10.1007/978-3-030-68452-5_26

artificial intelligence technology. The proposed method extracts feature from the image based on CNN and determines the parrot region and classification.

2 Proposed Method

2.1 Dataset and Preprocessing

The dataset collected images from the internet and, through the cooperation of the zoo, collected RGB images. RGB images were created by capturing video from various angles with an RGB camera and extracting video frames. Because parrot species have different beaks, eyes, and crests, we used the head region to classify parrot species. It consists of 11 parrot species (scientific name: Cacatua goffiniana, Cacatua ducorpsii, Cacatua alba, Cacatua moluccensis, Cacatua galerita, Ara macao, Ara ararauna, Camelot macaw, Ara chloropterus, Trichoglossus haematodus, psittacus erithacus). We applied data augmentation such as horizontal flip, rotation, scaling and translation a to prevent overfitting. In here, padding used edge pixels. (see Fig. 1). The total number of data is 35,806.

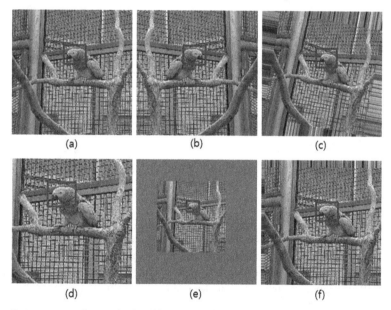

Fig. 1. Data augmentation method, red box (ground truth) (a) original image, (b) horizontal flip, (c) rotation, (d) scaling(zoom-in), (e) scaling(zoom-out), (f) translation

2.2 Deep Neural Network Model

Deep learning is known to robust not only image classification but also regression problems. Recently, research on object detection technology has been actively conducted.

Since the initial success of deep learning object detection technology was R-CNN [5], it developed into Fast R-CNN [6] and Faster R-CNN [7]. Also, YOLO [8] and SSD [9] are also object detection technologies using deep learning. Faster R-CNN uses feature maps obtained through the convolutional layer for the entire image, and is a 2-stage method with RPN (Region Proposal Network) and classifier that solves regression problems, so it is measured with high mean Average Precision(mAP) but low frames per second(FPS). YOLO and SSD models are a 1-stage method that solves regression and classifier at once, and is a model with high mAP and high FPS. Among the 1-stage models, YOLO is used as the last feature map from the convolutional layer and measured at 45 FPS and 63.4% mAP. On the other hand, the SSD model has a high FPS and mAP at 59 FPS and 74.3% mAP. We used SSD models to classify parrots (see Fig. 2). SSD is a model that classifiers (regression and classification) from feature maps that have passed through a convolutional layer for the entire image. Here, for feature extraction, VGGNet-16 [10] was used. Since the classifier is made at multiple feature maps, it has the advantage of being able to detect objects from small objects to large objects.

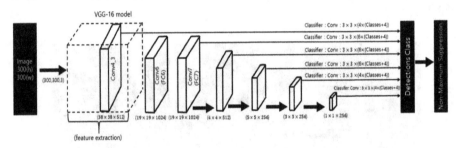

Fig. 2. parrot classification using deep neural network

3   Experimental Results

The SSD model was used for training to classify 11 types of parrots. As shown in Fig. 3, (a) and (b) are examples of correct and incorrect classification results, respectively. Incorrect classification results include cases where the parrot's head region is not detected due to occlusion and ambiguity between the background and the head region. Confusion matrix is provided to help understand the trained model performance. (see Fig. 4). It is normalized between 0 and 1, and the closer it is to 1, the better the prediction is, the x-axis is the predicted label, and the y-axis is the correct answer label. The part colored in blue tones on the diagonal is the result of the correct prediction, and the part colored in red tones is the result of the incorrect prediction. As you can see in the confusion matrix, the'background' is a case that was not detected. There were no misclassification results for the parrot species in row #10. The parrot species in rows #1 ~ #5 are included in the Cacatua genus, and are externally similar white species. In the cases, it was confirmed that many cases of misclassification occurred between each other. The trained model performance was achieved with mAP 95.7%.

Fig. 3. Examples of parrot species classification results (green: ground-truth, yellow: correct prediction, red: incorrect prediction). (a) Examples of correctly detected and classified. (b) Examples of incorrectly detected or classified. (Color figure online)

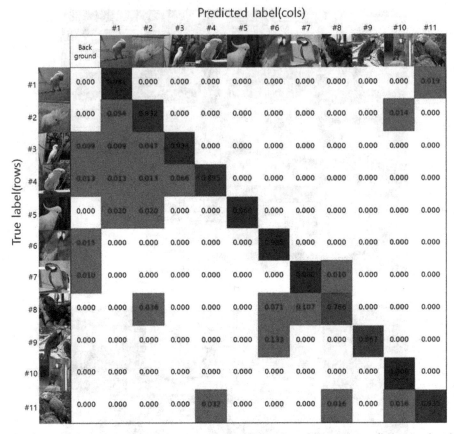

Fig. 4. Confusion matrix for 11 parrot species. On-diagonal results (correct predictions) colored in tones of blue, and colored in tones of red (incorrect predictions). (Color figure online)

4 Conclusion

In this paper, we proposed a parrot classification method belonging to CITES using a deep neural network method. We used the SSD model to differentiate between 11 parrot species. The network used for feature extraction was newly trained to make a model suitable for parrot classification without using the pretrained VGGNet-16. In the case of animal customs procedures at airports or ports for the protection of CITES species, if an automated system applying the proposed system is introduced, it will be possible to save a lot of time and simplify the procedure. Future research will add features to the torso and feet of each species, contributing to performance improvement.

Acknowledgement. This work was financially supported by a Grant (2018000210004) from the Ministry of Environment, Republic of Korea.

References

1. Bush, E.R., Baker, S.E., Macdonald, D.W.: Global trade in exotic pets 2006–2012. Conserv. Biol.Conserv. Biol. **28**(3), 663–676 (2014)
2. Kang, E., Han, Y., Oh, I.S.: Mushroom image recognition using convolutional neural network and transfer learning. KIISE Trans. Comput. Pract. **24**(1), 53–57 (2018)
3. Labao, A.B., Naval Jr., P.C.: Cascaded deep network systems with linked ensemble components for underwater fish detection in the wild. Ecol. Inf. **52**, 103–121 (2019)
4. Jang, W., kim, T., Nam, U., Lee, E.C.: Image segmentation and identification of parrot by using Faster R-CNN. In: Proceeding of ICNCT 2019, 12–14, pp. 91–92 (2019)
5. Girshick, R., Donahue, J., Darrell, T., Malik, J.: Rich feature hierarchies for accurate object detection and semantic segmentation. In: Proceedings of the IEEE Conference on Computer Vision and Pattern Recognition. pp. 580–587 (2014)
6. Girshick, R.: Fast r-cnn. In: Proceedings of the IEEE International Conference on Computer Vision. pp. 1440–1448 (2015)
7. Ren, S., He, K., Girshick, R., Sun, J.: Faster r-cnn: Towards real-time object detection with region proposal networks. In: Advances in Neural Information Processing Systems. pp. 91–99 (2015)
8. Redmon, J., Divvala, S., Girshick, R., Farhadi, A.: You only look once: Unified, real-time object detection. In Proceedings of the IEEE Conference on Computer Vision and Pattern Recognition. pp. 779–788 (2016)
9. Liu, W., et al.: Ssd: Single shot multibox detector.In: European Conference on Computer Vision, pp. 21–37. Springer, Cham (2016)
10. Simonyan, K., Zisserman, A.: Very deep convolutional networks for large-scale image recognition. ICLR, CoRR arXiv:1409.1556 (2015)

Automatic Detection of Trypanosomosis in Thick Blood Smears Using Image Pre-processing and Deep Learning

Taewoo Jung[1]([✉]), Esla Timothy Anzaku[2,4], Utku Özbulak[2,4], Stefan Magez[3,5], Arnout Van Messem[2,6], and Wesley De Neve[2,4]

[1] Ghent University Global Campus, Incheon, Republic of Korea
taewoojung7@gmail.com
[2] Center for Biotech Data Science, Ghent University Global Campus, Incheon, Republic of Korea
{eslatimothy.anzaku,utku.ozbulak,arnout.vanmessem, wesley.deneve}@ugent.be
[3] Center for Biomedical Research, Ghent University Global Campus, Incheon, Republic of Korea
stefan.magez@ugent.be
[4] Department of Electronics and Information Systems, Ghent University, Ghent, Belgium
[5] Department of Biochemistry and Microbiology, Ghent University, Ghent, Belgium
[6] Department of Applied Mathematics, Computer Science and Statistics, Ghent University, Ghent, Belgium

Abstract. Trypanosomosis, which is caused by the *Trypanosoma* parasite, is an infectious disease that affects both humans and animals. Today, the microscopic examination of a Giemsa or Wright stained blood smear from an infected individual is the standard procedure for diagnosis because of the straightforward nature of sample preparation. Unfortunately, this method is labor-intensive and prone to error, particularly resulting in false-negative scoring when parasite levels are low during chronic infections. Automating the detection of parasites in blood smear images can overcome sensitivity limitations related to a microscopic examination. We therefore propose a deep learning approach that aims at automatically classifying microscope images in terms of parasite presence or absence. To that end, we applied a ResNet18 model using a pre-processed dataset derived from microscope videos of unstained thick blood smears, with the blood smears originating from a mouse infected with *Trypanosoma brucei*. Our pre-processing strategy mainly involved image cropping and the application of a thresholding algorithm for facilitating effective model training. Moreover, our thresholding approach made it possible to observe a positive correlation between the percentage of parasite-related pixels in an image and the classification effectiveness.

Keywords: Deep learning · Image classification · Image pre-processing · Trypanosomosis · Unstained blood smears

© Springer Nature Switzerland AG 2021
M. Singh et al. (Eds.): IHCI 2020, LNCS 12616, pp. 254–266, 2021.
https://doi.org/10.1007/978-3-030-68452-5_27

1   Introduction

Trypanosomosis, an infectious disease caused by the *Trypanosoma brucei* (*T. brucei*) parasite, spreads mostly on the sub-Sahara African continent, where it is transmitted by the tsetse fly [9]. Subspecies of *T. brucei* comprise *T. b. brucei* and the human-infecting *T. b. gambiense* and *T. b. rhodesiense* [9]. Human trypanosomosis, which is also referred to as sleeping sickness, is characterized by two stages. The first stage of illness involves fever, headache, body ache, and an inflammation of the lymph nodes, with parasites being mainly present in the blood circulation system of the victim. The second stage involves the penetration of the parasites into the central nervous system, resulting in behavioral changes, sleep disturbances, and severe lethargy [17].

There is currently no viable vaccination against trypanosomosis, with treatment solely relying on chemotherapy. Indeed, a number of drugs are available that can be used to deal with the different stages of the infection [15]. Often, combinations of these drugs are administrated in order to obtain an effective treatment. In spite of the effectiveness of chemotherapies, administration of anti-trypanosomal drugs should be heedful because of possible severe side effects, including encephalitis and even mortality [15]. To avoid unnecessary treatment, reliable diagnostic methods should be able to find both true positives and true negatives with a high certainty.

When performing a diagnosis through microscopic examination, all blood smear fields are to be examined manually. Even with Giemsa or Wright staining [18] of the smears, errors can occur, which can be detrimental for a patient. In order to minimize microscopic examination labor and errors, several attempts have been made to automatically detect *Trypanosoma* parasites from blood smears via machine learning approaches. However, these attempts still require cumbersome sample preparation steps. Therefore, we prepared in-house unstained images of *T. brucei* blood samples to build computational models that can automatically detect parasites.

This paper is organized as follows. In Sect. 2, we review a number of state-of-the-art computational models for the automatic detection of protozoan parasites, including *Trypanosoma*, *Plasmodium*, and *Leishmania*. In Sect. 3, we propose a deep learning framework for detecting trypanosomosis directly from unstained thick blood smears. We subsequently discuss our experimental setup and the results obtained in Sect. 4. Finally, we present our conclusions and areas that need further study in Sect. 5.

2   Related Work

Traditional machine learning and deep learning approaches have been proposed to automate the detection of protozoan parasites in microscope images, with Giemsa and Wright staining [18] of the blood smear samples being a crucial step in the preparation of the underlying image datasets.

In traditional machine learning approaches, the majority of methods used segmentation-based feature extraction prior to the training of computational models. Both [4] and [13] exploited differing intensities between color channels of stained microscope images to discriminate the foreground (a *Trypanosoma* parasite or a specific feature of a *Trypanosoma* parasite) from the background of a blood smear. In [4], the difference between the a and b channels of the Lab color space was leveraged to obtain the foreground. Likewise, the difference between the blue and green channels of the RGB color space was used to locate pixels belonging to the kinetoplast and the nucleus of the *Trypanosoma cruzi* (*T. cruzi*) parasite in [13]. Combinations of thresholding techniques were then used to perform binary segmentation [4] or to select segmented sub-images of stained regions that belong to a parasite [13]. In case of *Plasmodium*, [12] used gray-scale images and the green channel of RGB images of Giemsa-stained blood smears to perform binary segmentation. As a final step, a wide range of features was extracted from binary images or sub-images. These extracted features were then used as parameters to build a classification model to detect *Trypanosoma* or *Plasmodium* parasites [4,12,13]. [4] used both support vector machines (SVMs) and K-nearest neighbors (KNN) for the detection of *Trypanosoma* parasites, where KNN outperformed the SVM-based approach by achieving an accuracy, sensitivity, and specificity of 0.983, 0.990, and 0.975, respectively. [13] also implemented KNN for *T. cruzi* classification, obtaining a sensitivity and specificity of 0.98 and 0.81, respectively. [12] employed an SVM to count red blood cells (RBC) infected with *Plasmodium*, obtaining a sensitivity of 0.931 and a specificity of 0.932. In general, segmentation-based feature extraction showed promising results.

Motion-based feature extraction introduced by [16] used neither staining nor segmentation. The objective was to automate the counting of *T. cruzi* parasites in video images. The pre-processing of video images of cultivated *T. cruzi* parasites without any host cells involved frame differencing and Otsu thresholding [6,16]. Frame differencing of successive image sequences made it possible to remove static background noise and to identify moving objects. The model was derived via multiple linear regression analysis of the cubic higher order local auto-correlation (CHLAC) feature to predict parasite counts [16]. The linear correlation, denoted as R^2, between hand-counted and predicted parasite counts was greater than 0.97. Even if host cells were present in the video images, the model was sufficiently sensitive to discriminate between the movement of host cells and *T. cruzi* parasites [16].

Traditional machine learning approaches require combinations of manual feature extraction, sophisticated pre-processing, and segmentation. On the other hand, deep learning facilitates the extraction of features from given raw or pre-processed data automatically (feature learning) [14], and the subsequent classification thereof (end-to-end learning). Since we are dealing with microscope images, it is important to mitigate imbalance problems, with these problems resulting from the region of interest (ROI) containing the parasites being significantly smaller in area compared to the background, and where the latter is of less interest to the use case at hand. To resolve these problems, image cropping can be used as a pre-processing method [2,7,19].

Thanks to their high effectiveness in the field of image understanding, convolutional neural networks (CNNs) have been used for both classification and segmentation purposes. A U-Net model [10] was employed for the segmentation of *T. cruzi* and *Leishmania* parasites in [2,7]. In both papers, the high imbalance, as mentioned above, was the limiting factor in feeding the raw data as input to the used network, given that it can drive the network to simply predict the most common class in the training set [2]. Therefore, image cropping was applied. Zero padding was used to maintain the input dimension of sub-images of stained blood smears containing *T. cruzi* parasites [7]. For *Leishmania* parasite segmentation, *Leishmania* species were co-cultured with macrophages and Giemsa-stained microscope images were used [2]. Apart from image cropping, a two-stage non-uniform sampling scheme was used in [2]. For the first 40 epochs, the model was trained with sub-images that have at least 40% of their pixels represent parasites, and the model was subsequently trained with all sub-images for the remaining epochs [2].

For the classification of *Plasmodium* parasites, a truncated VGG-19 network [6] was used in [19]. The truncated CNN architecture consists of the first six CNN layers of VGG-19, having three maxpool layers after every two CNN layers. For pre-processing, [19] proposed a cropping technique called Iterative Global Minimum Screening (IGMS) [19], making it possible to generate candidate sub-images with *Plasmodium* parasites from Giemsa-stained thick blood smear images.

Given that these segmentation- and motion-based traditional machine learning approaches involve staining and *in-vitro* cultivation, sample preparation is still cumbersome. Likewise, current deep learning models were trained from stained blood smear samples, and approaches were mostly done via segmentation. To the best of our knowledge, no attempt has thus far been made to build a deep learning model that can classify the presence or absence of *Trypanosoma* parasites, or any other protozoan parasites for that matter, from unstained blood smears, as we set out to do in this paper.

CNN-based deep learning approaches do not require complex pre-processing or segmentation-based feature extraction. If we can successfully build a computational model that can automatically detect *T. brucei* parasites in unstained blood smears, the model can contribute significantly to reducing the manual labor and the cost needed for sample preparation and diagnosis.

3   Proposed Approach

3.1   Motivation and Overall Framework

By reviewing already existing CNN-based computational models that were proposed to detect protozoan parasites, we are confident that deep learning models are capable of learning for this particular problem. In this paper, we aim at eliminating the task of sample preparation (such as Giemsa or Wright staining) by building a CNN-based classification model from unstained thick blood smears. Figure 1 illustrates our overall framework for the diagnosis of trypanosomosis via microscopic examination, utilizing a computational model for parasite detection. To ensure that our classification model is able to learn efficiently from the available dataset, we pay substantial attention to the development of an optimal image pre-processing strategy. The Residual Network (ResNet) [3] has proven its capability of learning through a number of experiments on the ImageNet and CIFAR-10 datasets. Thus, we made use of a ResNet as our CNN model for building a classification model. The proposed research effort can be considered the first attempt to apply a deep learning model for detecting *Trypanosoma* parasites in unstained microscope images.

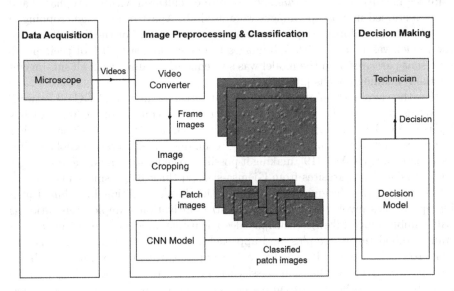

Fig. 1. Outline of our overall framework.

3.2 Data

The dataset for our experiments originated from microscope videos of unstained thick blood smears. First, we prepared two C57BL/6 mice, one of which was infected with *T. b. brucei* (the other one was healthy). We then collected a blood sample from each mouse, creating two slides from each blood sample for thick blood smear examination without any staining. The 20X magnification and 1360 × 1024 resolution settings on an IX83 inverted microscope (Olympus) were used to record microscope videos from each blood smear. When collecting the microscope videos, we randomly selected 12 and 11 different locations for the infected mouse and the healthy mouse, respectively, and we recorded each selected location for, on average, five seconds.

Only the odd-numbered frames of these videos were extracted as RGB images, so to be able to reduce the excessive overlap between successive images. In the end, we obtained 619 positive microscope images (containing at least 1 parasite) and 560 negative microscope images (containing no parasites).

The frames from our microscope videos were pre-processed and used as an input for the training and evaluation of the classification model. Several experiments were conducted to find the best pre-processing strategy.

In one of our attempts, we resized the 1360×1024 raw microscope images into 224×224 patches for training. This approach showed an outstanding training and validation accuracy with values close to 0.900. However, our model failed to generalize as confirmed by the test set: every image was predicted as positive.

After analyzing the obtained results, we found two possible explanations: (1) the number of pixels belonging to the foreground (i.e., the number of pixels belonging to visible parasites) is significantly lower than the number of background pixels in the raw microscope images and (2) the model might be learning features that are not exclusive to the presence or absence of parasites, but that are strongly related to the presence of an infection [2], such as the occurrence of irregularly shaped red and white blood cells. To address the observed lack of generalization, we decided to generate training images by applying image cropping as a pre-processing technique to the positive images only. Indeed, even though the resulting patches are all originating from positive images, many of these patches do not contain any visible parasites, making it possible to use patches without parasites as negative images and the other patches as positive images. As the use of cropping requires information about the location of each parasite, we needed binary masks for the positive images. We therefore selected six positive videos and manually generated binary masks using Labelbox, an online annotation platform, and Photos, an application available on iPadOS 13.5.1. With the help of binary masks and the size of the region of interest, we sorted the patches, as cropped from the positive images, into positive and negative patches. More details about this approach are provided in Sect. 4.

4 Experiments

4.1 Experimental Setup

Dataset Preparation. We cropped each 1360×1024 positive RGB image into 88 different 224×224 RGB patches in such a way that it allows for both a vertical and horizontal overlap of 112 pixels (see Fig. 2). When cropping the positive RGB images, we made use of the previously obtained binary masks. Each 1360×1024 binary mask is also cropped into 224×224 binary patches co-located with the aforementioned RGB patches. These binary patches were used as a reference to manually label the 224×224 cropped RGB image patches as positive or negative. For instance, if 2% or more of the pixels have a value of 255 (white) in a 224×224 binary patch, the co-located 224×224 RGB patch will be labeled as positive. Indeed, at 2%, we were able to clearly discriminate between positive and negative 224×224 RGB patches.

The pixels that belong to an image region where a parasite is clearly visible belong to the so-called region of interest. In what follows, the minimal ratio of the number of ROI pixels to the total number of pixels used to label a patch

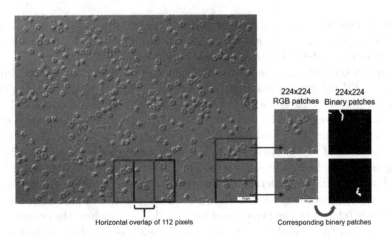

Fig. 2. Our approach for cropping an RGB image of size 1360×1024 that contains parasites into 224×224 patches.

as positive will be referred to as the ROI threshold, with Eq. 1 showing the calculation of this ratio:

$$ROI(\%) = \frac{Total\ number\ of\ white\ pixels\ in\ the\ binary\ patch}{Total\ number\ of\ pixels\ in\ the\ binary\ patch} \times 100\%. \quad (1)$$

In [2], it was suggested that the imbalance between the number of background pixels and the number of foreground pixels could trigger a model to predict the most abundant pixels as the most common feature of the class. We thus hypothesized that the use of a higher ROI threshold could help us in selecting patches that facilitate more effective learning by our classification models, given a higher availability of foreground pixels. To test this hypothesis, we prepared five balanced datasets (T1, T2, T3, T4, and T5), generated by empirically selecting different ROI thresholds for each dataset (see Table 1).

When setting the ROI thresholds for each dataset, we took into account that each video had a varying number of parasites. For instance, higher ROI thresholds were assigned to videos 4 and 5 because they contained a high number of parasites. In contrast, lower ROI thresholds were assigned to videos 2, 3, and 6, since they had less parasites. As shown in Table 1, the ROI thresholds used increase from dataset T1 to dataset T5, in the order of T1, T2, T3, T4, and T5. Moreover, since the number of negative patches is far greater than the number of positive patches, we matched the number of negative patches to the number of positive patches through random selection, thus obtaining balanced datasets. After generating the balanced datasets, we randomly split each dataset into three sets: a training set, a validation set, and a test set (see Table 2). The training, validation, and test sets represent 64%, 20%, and 16% of each dataset, respectively.

Table 1. ROI thresholds used for each dataset.

Dataset	ROI threshold (%)					
	Video 1	Video 2	Video 3	Video 4	Video 5	Video 6
T1	5.50	3.60	2.80	9.30	9.40	2.40
T2	6.00	4.00	3.20	10.30	10.00	3.00
T3	6.50	4.50	3.60	11.30	10.75	3.60
T4	7.13	5.10	4.00	12.50	11.50	4.00
T5	8.00	5.85	4.50	14.65	12.50	4.70

Table 2. Number of images (i.e., patches) in the training, validation, and test set of each dataset.

Dataset	Training	Validation	Test	Total
T1	4,794	1,506	1,204	7,504
T2	4,014	1,262	1,010	6,286
T3	3,172	998	800	4,970
T4	2,510	790	632	3,932
T5	1,724	546	436	2,706

ResNet18. In our experiments, we trained the classification models using the ResNet18 architecture. In this architecture, convolutional layers are used in combination with batch normalization layers, rectified linear unit (ReLU) activation layers, maxpool layers, average pooling layers, and one fully connected layer [3]. We applied the default parameters provided by PyTorch [8]. As shown in Fig. 3, the overall architecture of ResNet18 consists of a convolutional layer (with kernel size 7), eight basic residual blocks, and a fully connected layer. Since each basic residual block contains two 3×3 convolutional layers, the model consists of 17 convolutional layers and one fully connected layer in total. The main difference between ResNet and a plain network is the presence of an additional feedforward loop, called the shortcut connection, on each building block of ResNet (see Fig. 3) [3]. During training, the model takes as input a $224 \times 224 \times 3$ image and generates a logit for each class. Given that our datasets have two classes, namely positive and negative, the model will generate two logits in total. The obtained logits, after normalization by the softmax function, are subsequently used to make a prediction for every input image [8].

Model Training. The initial parameters of each convolutional layer were randomly sampled from a uniform distribution over the interval $[-\sqrt{k}, +\sqrt{k}]$, where k is computed as follows [8]:

$$k = \frac{1}{Number\ of\ input\ channels \times kernel\ size}. \tag{2}$$

Since the cross-entropy loss function is known to be efficient when a model needs to deal with a classification task, it was used as our loss function of choice [8].

Before training, the individual pixel values of the input images were re-scaled to the range $[0, 1]$, after which normalization took place using the mean and standard deviation of the re-scaled pixel values. We made use of two optimizers: Adaptive Momentum (Adam) [5] with a learning rate of 0.001 and Stochastic Gradient Descent (SGD) [11] with a learning rate of 0.01, a momentum of 0.9, and a gradient decay of 1×10^{-4} [1]. For SGD, we employed a learning rate scheduler called ReduceLROnPlateau, as available in PyTorch [8]. This learning rate scheduler divides the learning rate by a factor of 10 when the validation cross-entropy does not improve for 10 successive epochs. Furthermore, we used a batch size of 128 and trained each model for 200 epochs. We then selected the best model based on validation accuracy.

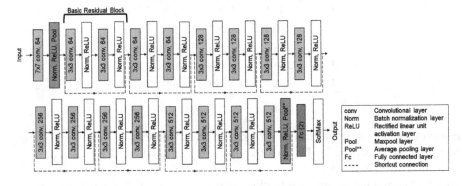

Fig. 3. ResNet18 architecture [3].

$$\text{Accuracy} = \frac{TP+TN}{TP+FP+TN+FN} \quad \text{Sensitivity} = \frac{TP}{TP+FN} \quad \text{Specificity} = \frac{TN}{TN+FP}$$

$$\text{Positive Predictive Value} = \frac{TP}{TP+FP} \quad \text{Negative Predictive Value} = \frac{TN}{TN+FN}$$

Fig. 4. Description of performance metrics used.

4.2 Experimental Results

For the evaluation of our classification models, we made use of five performance metrics: accuracy, sensitivity, specificity, negative predictive value (NPV), and positive predictive value (PPV). These metrics were calculated using the number of true positives (TP), true negatives (TN), false positives (FP), and false negatives (FN) (see Fig. 4). For our classification models, both PPV and NPV are of utmost importance to characterize the effectiveness of diagnostic tools, given the toxicity of chemotherapy against trypanosomosis.

Quantitative Results. Each dataset was used to train two models using different optimizers: Adam and SGD. That way, we obtained a total of 10 models, and for each optimizer, we selected the model with the highest validation accuracy. As shown in Table 3 (highlighted in gray), the T5 Adam and T5 SGD models are characterized by a high validation accuracy and these models were selected as our best classification models. The values of the different performance metrics can be found in Table 3. Compared to T5 Adam, T5 SGD has a higher sensitivity and a higher NPV, but a lower specificity and a lower PPV. It is thus difficult to conclude that T5 SGD outperforms T5 Adam.

Table 3. Results obtained by our classification models.

Model	Validation	Test				
	Accuracy	Accuracy	Sensitivity	Specificity	PPV	NPV
T1 Adam	0.679	0.651	0.721	0.581	0.633	0.676
T2 Adam	0.661	0.649	0.703	0.594	0.634	0.667
T3 Adam	0.666	0.626	0.665	0.588	0.617	0.637
T4 Adam	0.662	0.633	0.636	0.630	0.632	0.634
T5 Adam	**0.686**	0.677	0.688	0.665	0.673	0.681
T1 SGD	0.665	0.626	0.674	0.578	0.615	0.640
T2 SGD	0.668	0.632	0.697	0.566	0.617	0.652
T3 SGD	0.649	0.675	0.690	0.660	0.670	0.680
T4 SGD	0.677	0.685	0.690	0.680	0.683	0.687
T5 SGD	**0.708**	0.679	0.706	0.651	0.670	0.689

During dataset preparation, the five datasets were designed using different ROI thresholds when selecting 224×224 RGB positive patches. Dataset T1 used the smallest ROI thresholds and dataset T5 used the largest ROI thresholds. When comparing the test results obtained by T1 and T5 SGD, we could observe increases in validation accuracy of 0.043 (+6.47%), in test accuracy of 0.053 (+8.47%), in sensitivity of 0.032 (+4.75%), in specificity of 0.073 (+12.63%), in PPV of 0.055 (+8.94%), and in NPV of 0.049 (+7.66%). For the T1 and T5 Adam models, there were increases in validation accuracy of 0.007 (+1.03%), test accuracy of 0.026 (+3.99%), specificity of 0.084 (+14.46%), PPV of 0.040 (+6.32%), and NPV of 0.005 (+0.74%), but also a slight decrease in sensitivity of 0.033 (−4.58%). Nevertheless, T5 Adam seems a more reliable model than T1 Adam. Furthermore, we noticed that all models seemed to favor positive predictions, but this imbalance seems to decrease with increasing threshold values (from 57.0% positive predictions to 51.1% for Adam and from 54.8% to 52.8% for SGD). Figure 5 shows that the cross-entropy loss curves for the training and

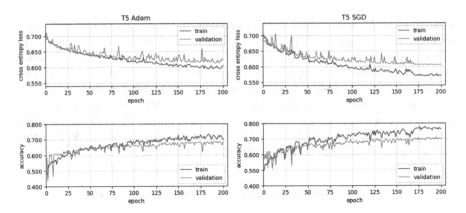

Fig. 5. Cross-entropy loss and accuracy obtained for every epoch during the training and validation of the T5 Adam and T5 SGD models.

validation sets of the T5 Adam and T5 SGD models differ by 8.6% and 8.1%, respectively, which is a good indicator that there was no overfitting during the training of our models.

Qualitative Results. A qualitative analysis was conducted to observe patches that were classified correctly and incorrectly by the T5 models. We observed that our models were able to classify several challenging patches correctly. However, our models still found it difficult to predict certain patches (see Fig. 6).

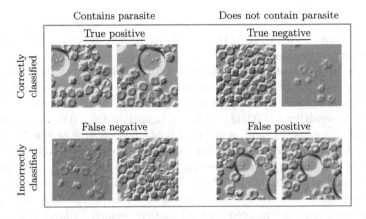

Fig. 6. Qualitative results obtained for the test set. Regions encircled in red represent parasite locations. (Color figure online)

5   Conclusions and Future Work

In this paper, we presented a first deep learning approach for detecting *Trypanosoma* parasites in unstained microscope images, leveraging a ResNet18 architecture and a pre-processing strategy that consists of image cropping and the application of a thresholding technique for facilitating effective model training. Given the experimental results obtained, we can conclude that CNN-based classification models are capable of classifying patches of unstained blood smears according to parasite presence and absence. In addition, the application of an ROI thresholding strategy resulted in the observation that there is a positive correlation between the percentage of parasite-related pixels in an image and the effectiveness of deep learning-based classification models for parasite detection. Considering the performance of the segmentation-based classification models proposed in [4,13] (see Sect. 2), further optimization is required of the data collection and pre-processing techniques to enhance our model performance.

During data acquisition, we collected positive videos from a single blood smear sample. As a result, the raw microscope images extracted from these videos come with limited representative power. As an important future work item, we should collect more microscopy videos from several blood smears from different infected mice. Additionally, we can improve our image cropping strategy by generating patches that are smaller than 224×224, and where these patches are non-uniformly depending on the shape of a parasite. That way, we should be able to maximize the number of foreground pixels in each patch and minimize the influence of background pixels during training.

References

1. Goodfellow, I., Bengio, Y., Courville, A.: Deep Learning. MIT Press, Cambridge (2016)
2. Górriz, M., et al.: Leishmaniasis parasite segmentation and classification using deep learning. In: Perales, F.J., Kittler, J. (eds.) AMDO 2018. LNCS, vol. 10945, pp. 53–62. Springer, Cham (2018). https://doi.org/10.1007/978-3-319-94544-6_6
3. He, K., Zhang, X., Ren, S., Sun, J.: Deep residual learning for image recognition. In: Proceedings of the IEEE Conference on Computer Vision and Pattern Recognition, pp. 770–778 (2016)
4. Faria, J.R.N.: Automatic detection of trypanosomes on the blood stream. Ph.D. thesis, University of Porto (2016)
5. Kingma, D.P., Ba, J.: Adam: a method for stochastic optimization. CoRR abs/1412.6980 (2014)
6. Krig, S.: Computer Vision Metrics. Survey, Taxonomy, and Analysis. Apress, Berkeley (2014). https://doi.org/10.1007/978-1-4302-5930-5
7. Ojeda-Pat, A., Martin-Gonzalez, A., Soberanis-Mukul, R.: Convolutional neural network U-Net for Trypanosoma cruzi segmentation. In: Brito-Loeza, C., Espinosa-Romero, A., Martin-Gonzalez, A., Safi, A. (eds.) ISICS 2020. CCIS, vol. 1187, pp. 118–131. Springer, Cham (2020). https://doi.org/10.1007/978-3-030-43364-2_11
8. Paszke, A., et al.: Automatic differentiation in PyTorch (2017)

9. Radwanska, M., Vereecke, N., Deleeuw, V., Pinto, J., Magez, S.: Salivarian Try-panosomosis: a review of parasites involved, their global distribution and their interaction with the innate and adaptive mammalian host immune system. Front. Immunol. **9**, 2253 (2018)

10. Ronneberger, O., Fischer, P., Brox, T.: U-Net: convolutional networks for biomed-ical image segmentation. In: Navab, N., Hornegger, J., Wells, W.M., Frangi, A.F. (eds.) MICCAI 2015. LNCS, vol. 9351, pp. 234–241. Springer, Cham (2015). https://doi.org/10.1007/978-3-319-24574-4_28

11. Rumelhart, D.E., Hinton, G.E., Williams, R.J.: Learning representations by back-propagating errors. Nature **323**(6088), 533–536 (1986)

12. Savkare, S.S., Narote, S.P., et al.: Automatic detection of malaria parasites for estimating Parasitemia. Int. J. Comput. Sci. Secur. (IJCSS) **5**(3), 310–315 (2011)

13. Soberanis-Mukul, R., Uc-Cetina, V., Brito-Loeza, C., Ruiz-Piña, H.: An automatic algorithm for the detection of Trypanosoma Cruzi parasites in blood sample images. Comput. Methods Programs Biomed. **112**(3), 633–639 (2013)

14. Stevens, E., Antiga, L., Viehmann, T.: Deep learning with PyTorch essential excerpts. Manning (2019)

15. Steverding, D.: The development of drugs for treatment of sleeping sickness: a historical review. Parasites Vectors **3**, 15 (2010). https://doi.org/10.1186/1756-3305-3-15

16. Takagi, Y., Nosato, H., Doi, M., Furukawa, K., Sakanashi, H.: Development of a motion-based cell-counting system for Trypanosoma parasite using a pattern recognition approach. Biotechniques **66**(4), 179–185 (2019)

17. The Editors of Encyclopaedia Britannica: sleeping sickness (2020). https://www.britannica.com/science/sleeping-sickness

18. Vilchez, C.: Examination of the peripheral blood film and correlation with the complete blood count. Rodak's Hematology-E-Book: Clinical Principles and Appli-cations, pp. 201–205 (2019)

19. Yang, F., Poostchi, M., Yu, H., Zhou, Z., Silamut, K., Yu, J., Maude, R.J., Jaeger, S., Antani, S.: Deep learning for smartphone-based malaria parasite detection in thick blood smears. IEEE J. Biomed. Health Inform. **24**(5), 1427–1438 (2019)

Adaptive Margin Based Liveness Detection for Face Recognition

Gabit Tolendiyev⬤, Mohammed Abdulhakim Al-Absi⬤, Hyotaek Lim⬤,
and Byung-Gook Lee^(✉)⬤

Dongseo University, Busan 47011, South Korea
{d0165114,d0185123}@kowon.dongseo.ac.kr
{htlim,lbg}@dongseo.ac.kr

Abstract. Face recognition is currently is becoming the hotspot in the area of deep learning, pattern recognition, and computer vision where it has been broadly utilized in many fields. Facial feature extraction is a key link in the face recognition system. The texture features of human faces are highly discriminative, so extracting the texture features of face images can often get a good classification and recognition effect. Image texture feature extraction methods can generally be classified into four categories: statistical methods, model methods, structural methods, and signal processing methods. Recently, face recognition based person authentication systems have been popular among other biometrics. However, hacking methods are also developed with this methodology. In this paper, we present a margin based liveness detection method (MLDM) for the face recognition system based on texture feature analysis. The fake images captured from a video that has edges generated by differences among different face images of real and fake people images. Moreover, we exploit a convolutional neural network to extract these features and differentiate real and fake face images. Experimental results show that our model has higher accuracy and can efficiently classify real faces and spoofed faces compare with the existing model. The outcome shows that our approach is better than the existing work which is experimentally proven.

Keywords: Face recognition · Face liveness detection · Texture feature · Margin based method · 2D spoofing attack

1 Introduction

Bio-metrics are being broadly used in person authentication systems in the last few decades. However, the most popular bio-metrics are fingerprint identification [7], palm recognition [6], retina recognition [3], IRIS recognition [4], face recognition and etc. are being used at airports, access control systems, payment verification systems and other various fields. Among these bio-metrics face recognition techniques are being used in various fields because of its contactless, fast speed, high accuracy and user-friendliness. Especially in the COVID-19

© Springer Nature Switzerland AG 2021
M. Singh et al. (Eds.): IHCI 2020, LNCS 12616, pp. 267–277, 2021.
https://doi.org/10.1007/978-3-030-68452-5_28

pandemic contactless bio-metric face recognition system is replacing other traditional bio-metrics, such as fingerprint identification and being widely deployed as a person authentication system. However, with the development of this technology, hacking methods are also developing as well. For example, an unauthorized person might use an authorized person's photo to attempt to be authenticated. In computer security it is called a spoofing attack or representation attack. Some examples of the 2D face spoofing attack are illustrated in Fig. 1. For that reason, it is compulsory to develop a new technology that distinguishes a real face image of authorized person and spoofed or fake image is needed.

face spoofing attack examples.PNG

Fig. 1. 2D face spoofing attack examples.

The research work contribution is as follows:

- A database for face liveness detection model is constrained containing real face image and replay spoofing attack image.
- A novel margin based liveness detection method for face recognition is proposed.

The paper organization is as follows. An extensive survey is carried out in Sect. 2. and Sect. 3, proposed a Margin based Liveness Detection Method (MLDM). Section 4 is devoted to the results and experiments. Section 5 is the discussion. Future works is described in Conclusions section.

2   Related Work

2.1   Face Recognition

Face recognition (FR) is more popular than other biometric systems such as fingerprint, palm vein and eye iris recognition. A big reason for using FR is its contactless, non-invasiveness and secureness.

As a face recognition model pretrained OpenFace 0.2.0 is utilized. OpenFace is a free and an open source library for facial recognition with deep convolutional neural networks. It is a Python and PyTorch implementation of facial recognition with deep convolutional neural networks and is based on Computer Vision and Pattern Recognition Conference 2015 (CVPR) [12]. PyTorch allows the convolutional neural network to be carried in the central processing unit (CPU) or with a parallel computing platform CUDA [1]. Even though OpenFace model is train on the public datasets have orders of magnitude less than private industry datasets, the accuracy of the model is remarkably high on the standard Labeled Faces in the Wild (LFW) public benchmark (Table 1).

Table 1. Accuracy evaluation of OpenFace models on LFW benchmark

Model	Accuracy	AUC
nn4.small2.v1 (Default)	0.9292 ± 0.0134	0.973
nn4.small1.v1	0.9210 ± 0.0160	0.973
nn4.v2	0.9157 ± 0.0152	0.966
nn4.v1	0.7612 ± 0.0189	0.853
FaceNet Paper (Reference)	0.9963 ± 0.009	Not provided

2.2 Liveness Detection

There are several approaches of liveness detection for face recognition, including:

- **Texture analysis techniques**, including computing the Local Binary Patterns (LBPs) over face regions and using an support vector machines (SVMs) to classify the real and fake face images [8].
- **Frequency analysis techniques**, a method of liveness detection by examining the Fourier domain of the face [8].
- **Variable focusing**, a method of liveness detection for 2D fake face images by examining the pixel values variation among two consecutive frames captured in different focuses [9].
- **Heuristic-based algorithms**, a liveness detection method based on blink detection, lip movement and eye movement. These algorithms attempt to track blinks and eye movement to make sure the authenticating person is not holding a printed photo of an authorized person [14].
- **3D face shape**, a method that distinguishes between real faces and printouts, photos, and images of another person by comparing its 3D meshes [10].
- **Combinations of the above methods**, face recognition engineers choose face liveness detection models appropriate to their particular applications.

3 Proposed Approach

In this work, we focused on face liveness detection method against replay spoofing. Because of widely use of smartphone and its availability adversaries possibly attack with their smartphone rather than using printed face image or 2D mask of the authorized person.

In this paper, the liveness detection model (MLDM) is treated as a binary classification model. It classifies the given face images as real face image or fake spoofed face image. We trained the model on a dataset containing real and spoofed face images (Fig. 2).

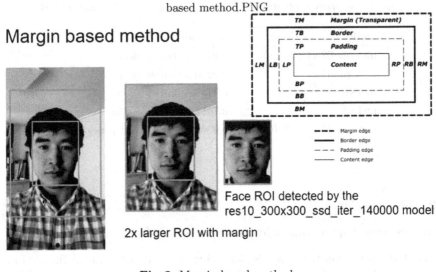

Fig. 2. Margin based method.

Flowchart of the face liveness detection method is depicted in Fig. 3.

Table 2. Model-architecture of MLDM.

LivenessNet64-structural details												
#	Input image			Output			Layer	Stride	Kernel		In	Out	Parameter
1	64	64	3	32	32	16	conv3-16	1	3	3	3	16	448
2	32	32	16	32	32	16	conv16-16	1	3	3	16	16	2320
	32	32	16	16	16	16	maxpool	2	2	2	16	16	0
3	16	16	16	16	16	32	conv16-32	1	3	3	16	32	4640
4	16	16	32	16	16	32	conv32-32	1	3	3	32	32	9248
	16	16	32	8	8	32	maxpool	2	2	2	32	32	0
5	1	1	2048	1	1	64	FC		1	1	2048	64	131136
6	1	1	64	1	1	2	FC		1	1	64	2	130
Total													147,922

Our liveness detection model is extended from [11]. Network architecture of the liveness detection model is presented in Table 2. Input image resized to 64 × 64 and all the pixel intensities are scaled from original range to the range 0 to 1. Data augmentation techniques are also used. A data augmentation object which will generate new face images with random rotation, zoom, width shifting, height shifting, channel shift, shear intensity, horizontal flop and vertical flips. The model is implemented with OpenCV built-in face detection library, Tensorflow and Keras framework. As an optimizer Adam optimization algorithm is used. Training parameters: Learning rate - 1e-4, batch size - 8, number of epochs - 50. Training and testing datasets separated as 75% and 25% respectively. As you can see in Fig. 4, texture features of real and fake images are different. This texture feature difference helps to distinguish real and fake face images (Fig. 5).

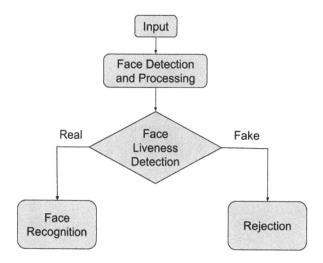

Fig. 3. Flowchart of face liveness detection method (MLDM).

4   Experiments and Results

4.1   Dataset Preparation

The amount of data and its distribution are crucial for training deep learning models. Insufficient data and not well distributed data might affect on generalization ability of the model. As a result, classification accuracy might reduce when model receives as a input a new data. For liveness detection model our dataset has to contains fake and real face images. In order to generate training data for real face images, we recorded a selfie video with the long 20 s of members of our laboratory. Using ResNetSSD face detection model [2] face area detected, cropped with the size of 64 × 64 and stored on the local disk. By skipping every 4 frame consisting face image, 300 face images which is extracted from each video. 20 volunteers are participated in data preparation. Around 300 images

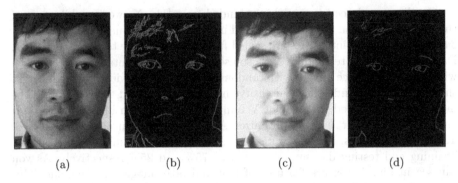

Fig. 4. Samples of input images (a) real image and (b) its edges, (c) fake image and (d) its edges.

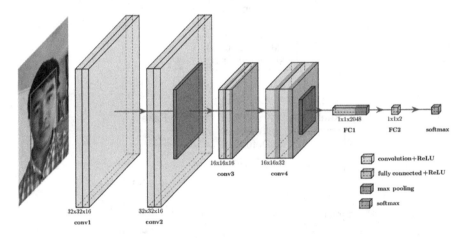

Fig. 5. Model architecture in 3D

Fig. 6. Some of the collected real and fake face images examples for training. The first row shows the real face images and the second row is the fake face images.

were extracted from each persons video. Our dataset contains (total of 15,849 images) 8,328 of real face images and 7,521 of fake face images.

Using the recorded videos, we extracted 3,000 real face images. In order to extract face images, we applied ResNetSSD face detection model [2] described in the preceding section to the whole dataset. Real and fake face images are stored in the separate folder. As a result, 286 real and 3856 fake face images of 20 different people were obtained. The input size of the network was 64 × 64 × 3, therefore all the images are resized to match the input layer of the network. Some samples of the training images are illustrated in Fig. 6.

4.2   Training Model

We trained the our liveness detection model from scratch using the dataset described in the previous section. The dataset was split into a testing set and training set within a proportion of 75% and 25% respectively. To train the liveness detection network, we applied the Adam optimization algorithm with Binary Cross-Entropy Loss function, with the starting learning rate of 0.0004. Training was done on NVIDIA GeForce RTX 2070 16 GB GPU with a batch size of 8. We trained the network for 50 epochs. In prepossessing step, image pixels intensity values are scaled to the range from 0 to 1. Real and Fake label names are strings, they are transformed to integers and the on-hot encode function applied. Also, before training the network, data augmentation operation applied

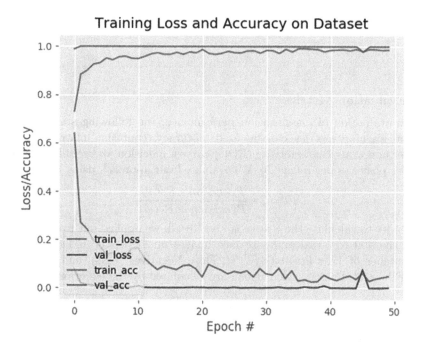

Fig. 7. Training loss and accuracy on dataset

with the following setting (rotation range = 20, the rang of the width shift = 0.2, height shift range = 0.2, zoom range = 0.15, shear range = 0.15, fill mode = "nearest", horizontal flip = True) in order to generalise well the model.Training Loss and Accuracy on training and testing datasets depicted in Fig. 7.

As can be seen from Fig. 8a and 8b, the model distinguishes real image from fake images accurately. For the convenience to differentiate, True face image is shown in blue color, while fake face image in red color. In the detection result window shown the label name, its confidence, ROI covering the face image.

4.3   Experimental Setup

In Table 3, listed experimental requirements.

Table 3. Experimental requirements

Hardware & Software requirements	Programming language	Packages
PC	Python 3.6	*Tensorflow 1.12*
Web camera		*OpenCV 3.4*
Windows OS or Linux		*Keras 2.2.4*
		scipy 1.1.0
		playsound
		pygame
		numpy
		imults
		nose
		gtts

4.4   Evaluation Metrics

For the evaluation of classification performance, the following statistical and machine learning metrics can be used: accuracy, confusion matrix, log-loss, Receiver operating characteristic (ROC) curves, precision and recall, F1-scores, and false positives per image [5]. We evaluated our approach using F1-score:

$$F_1 = 2 * \frac{Precision * Recall}{Precision + Recall} \tag{1}$$

In order to calculate the Precision and Recall, we applied our liveness detector on the 400 test real and fake face images of 20 person and counted the total number of True Positives (TPs), False Negatives (FNs), and False Positives (FPs) [13]. Precision and Recall are calculated by the equations:

$$Precision = \frac{TP}{TP + FP} \tag{2}$$

$$Recall = \frac{TP}{TP + FN} \tag{3}$$

4.5   Results

As a result of training, the accuracy of the classifier on the entire dataset was 99.8%. The experimental results obtained by applying face liveness detection model to the image stream received from a web-camera as an input image as shown in Fig. 8b. Living face images is shown with red color rectangle and recognized it with accuracy of 93.99%, whiles fake image (i.e spoofed image) displaying on smartphone is classified as a fake image and shown in a red color. From these these data, we calculated precision and recall values. After that, F1-score were calculated by Eq. 1 and added to the last column of Table 4.

5   Discussion

Our findings suggests that larger region of interest and training on various different people generalize the model. The model can distinguish new people that unseen before. The existing work in [11] cannot classify the new person's face image, while our model can distinguish correctly which is experimentally proven.

(a) Prints attack test.   (b) Replay attack test.

Fig. 8. Experimental results. Liveness detection model running on PC.

Accuracy comparison between our model with the existing model is shown in the Table 4. However, the margin based liveness detection method (MLDM) has higher accuracy than the existing model (Tables 5 and 6).

Table 4. Accuracy comparison

Model	Precision	Recall	F1-score
[11]	0.709	0.675	0.6915
Our model (MLDM)	0.996	0.996	0.996

Table 5. Confusion matrix of the our trained convolutional neural network

True label	Real	117	0
	Fake	4	142
		Real	Fake
	Predicted label		

Table 6. Confusion matrix of [11]

True label	Real	64	57
	Fake	67	75
		Real	Fake
	Predicted label		

6   Conclusions

In this study, the authors presented a novel methodology for liveness detection against replay spoofing in face recognition (MLDM). We look into the dissimilar nature of imaging variability from a real face or a fake photograph face image based on the analysis of Margin based face liveness detection model, which leads to a new method to exploit the additional information contained in the given image (i.e edges and fingers). We show that phone edges and fingers also contribute to learning fake face image features, which helps to distinguish real face images and fake face images captured from smartphone display. Experiments on a real and fake face images database show that the proposed method promising replay spoofing detection performance, with advantage of real-time testing.

This is the first paper that used margin based liveness detection learning techniques to distinguish whether the given static face images are from a real live human or printed/displayed photos. In order to take measure the robustness of the proposed model, the authors are collecting more person's face images to enlarge their database to further improve and import the performance for future work.

Acknowledgment. This work was supported by Institute for Information and Communications Technology Promotion (IITP) grant funded by the Korea government (MSIT) (No.2018-0-00245, Development of prevention technology against AI dysfunction induced by deception attack). And it was also supported by the National Research Foundation of Korea grant funded by the Korea government(MSIT) (No. 2020R1A2C1008589).

References

1. Amos, B., Ludwiczuk, B., Satyanarayanan, M.: Openface: a general-purpose face recognition library with mobile applications. Technical report CMU-CS-16-118, CMU School of Computer Science (2016)

2. Balu, G.: Resnetssd face detector, March 2018. https://github.com/gopinath-balu/computer_vision/blob/master/CAFFE_DNN/res10_300x300_ssd_iter_140000.caffemodel. Accessed 9 Sept 2020

3. Choraś, R.S.: Retina recognition for biometrics. In: Seventh International Conference on Digital Information Management (ICDIM 2012), pp. 177–180. IEEE (2012)

4. Daugman, J.: How iris recognition works. In: Bovik, A.C. (ed.) The Essential Guide to Image Processing, pp. 715–739. Elsevier, Amsterdam (2009)

5. Flach, P.A.: The geometry of roc space: understanding machine learning metrics through roc isometrics. In: Proceedings of the 20th International Conference on Machine Learning (ICML2003), pp. 194–201 (2003)

6. Hadi, A.H., Abd, Q.: Vein palm recognition model using fusion of features. Telkomnika **18**(6), 2921–2927 (2020)

7. Hoshino, S.: Personal identification authenticating with fingerprint identification, US Patent 6,636,620, 21 October 2003

8. Kim, G., Eum, S., Suhr, J.K., Kim, D.I., Park, K.R., Kim, J.: Face liveness detection based on texture and frequency analyses. In: 2012 5th IAPR International Conference on Biometrics (ICB), pp. 67–72. IEEE (2012)

9. Kim, S., Yu, S., Kim, K., Ban, Y., Lee, S.: Face liveness detection using variable focusing. In: 2013 International Conference on Biometrics (ICB), pp. 1–6. IEEE (2013)

10. Lagorio, A., Tistarelli, M., Cadoni, M., Fookes, C., Sridharan, S.: Liveness detection based on 3D face shape analysis. In: 2013 International Workshop on Biometrics and Forensics (IWBF), pp. 1–4. IEEE (2013)

11. Rosebrock, A.: Liveness Detection with OpenCV, March 2019. https://www.pyimagesearch.com/2019/03/11/liveness-detection-with-opencv/. Accessed 9 Sept 2020

12. Schroff, F., Kalenichenko, D., Philbin, J.: Facenet: a unified embedding for face recognition and clustering. In: Proceedings of the IEEE Conference on Computer Vision and Pattern Recognition, pp. 815–823 (2015)

13. Seidaliyeva, U., Akhmetov, D., Ilipbayeva, L., Matson, E.T.: Real-time and accurate drone detection in a video with a static background. Sensors **20**(14), 3856 (2020)

14. Singh, A.K., Joshi, P., Nandi, G.C.: Face recognition with liveness detection using eye and mouth movement. In: 2014 International Conference on Signal Propagation and Computer Technology (ICSPCT 2014), pp. 592–597. IEEE (2014)

A Study on a Mask R-CNN-Based Diagnostic System Measuring DDH Angles on Ultrasound Scans

Seok-min Hwang, Hee-Jun Park, and Jong-ha Lee[✉]

Department of Biomedical Engineering, Keimyung University, Daegu, South Korea
segeberg@gmail.com

Abstract. Recently, the number of hip dysplasia (DDH) that occurs during infant and child growth has been increasing. DDH should be detected and treated as early as possible because it hinders infant growth and causes many other side effects In this study, two modelling techniques were used for multiple training techniques. Based on the results after the first transformation, the training was designed to be possible even with a small amount of data. The vertical flip, rotation, width and height shift functions were used to improve the efficiency of the model. Adam optimization was applied for parameter learning with the learning parameter initially set at $2.0 \times 10e^{-4}$. Training was stopped when the validation loss was at the minimum. No significant difference in angle measurements was found between the model and doctor. The differences in the maxi-mum, minimum, and mean base angles were 6.93, 0, and 1.81°, respectively. The differences in the maximum, minimum, and mean α angle were 19.78, 0.01, and 3.56°, respectively. The differences in the maximum, minimum, and mean β angle were 25.92, 0.01, and 4.51°, respectively A novel image overlay system using 3D laser scanner and a non-rigid registration method is implemented and its accuracy is evaluated. By using the proposed system, we successfully related the preoperative images with an open organ in the operating room.

Keywords: DDH · Developmental dysplasia of the hip · Medical image · EOS · AI · CNN

1 Introduction

The incidence of developmental dysplasia of the hip (DDH), which occurs during the developmental period of young children, has been increasing. DDH does not cause external deformities in young children. About 1–2% of young children on average have DDH, and DDH can be passively detected by medical imaging in 15% of DDH cases. DDH can be easily treatment if detected within the first year after birth but can incur serious problems if left untreated. Although infants are regularly screened for DDH in the United States, there is not yet a standardized and clear method of DDH screening. South Korea has included DDH in the National Health Screening Program for Infants since 2006. Research has also been started on DDH in South Korea but without clear

© Springer Nature Switzerland AG 2021
M. Singh et al. (Eds.): IHCI 2020, LNCS 12616, pp. 278–287, 2021.
https://doi.org/10.1007/978-3-030-68452-5_29

research standards as is the case in the United States. DDH is manually detected by a physician using ultrasonography (US) as a standard practice. However, factors such as the examiner and the system used to perform US can affect the interpretation of the US results. In US-based DDT screening, a medical team measures the acetabulum-femoral head angle on ultrasound scans. A physician then diagnoses DDT based on the angle measurement. However, such measurements are subject to errors, which can affect the diagnosis results. US resolution can also have a significant impact on DDT screening results. Since DDT is manually detected with the eyes, the resolution and accuracy of ultrasound images can significantly affect screening results. Since DDT is manually screened by a human without a standard guideline, poor resolution can cause large errors. Additionally, a diagnosis is not just made based on an ultrasound image but also on length and angle measurements in different sites. Currently, a medical team manually obtains and provides a physician with these measurements. This manual process takes over 10 min per image, challenging health professionals in their endeavors to efficiently provide medical services.

Computational speed has been rapidly increasing due to recent technological advances. The improved performance of graphics processing units (GPU), used in image and video processing, has led to the increasing use of machine learning algorithms used for image processing in industries as well as in the fields of science and medicine. Visual interpretation of medical images is important in the field of medicine. Machine learning algorithms such as convolutional neural networks (CNNs) and improved Resnet50 are increasingly used in ultrasound image analysis. Related literatures have demonstrated that artificial intelligence (AI)-based image analysis has significantly improved objectivity and productivity in medical settings and that these machine learning algorithms may be used to develop a computer-aided DDH diagnostic system using ultrasound images.

CNNs are a machine learning algorithm used in image analysis and are divided into machine learning and deep learning depending on the level of human intervention. The lower the level of human intervention, the closer an algorithm is to deep learning and the greater the amount of data required. However, it is often the case that there is not

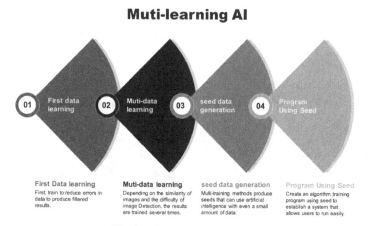

Fig. 1. The goal of the system.

enough data to even just train a mode l. If more data are required for deep learning, then human intervention may be increased to resolve the data shortage problem. Thus, we propose a multi-training technique using a user filter.

In this study Fig. 1, we propose a MASK R-CNN-based DDH diagnostic system that can be trained on a small amount of ultrasound image data. The system uses a deep leaning algorithm to automatically calculate the acetabulum-femoral head angle and classify the hip bones of children as normal or abnormal.

2 Method

This study is aimed at developing a system that uses a multi-training method to process unprocessed data to increase the user accessibility. Previously, DDH was diagnosed by detecting the areas containing the ilium and acetabular roof, measuring the angles of the ilium and acetabular roof, and using a machine learning algorithm to make a prediction based on these angle measurements. Instead of measuring the angles from the ilium and acetabular roof following area detection, the areas containing the α and β angles of the ilium and acetabular roof are detected to measure the angles.

The Mask R-CNN method is used to first segment the shapes on ultrasound images formed by the ilium and acetabular head. Two points on the ilium and three points for calculating the α and β angles (four points in total with one overlapping point) are used to determine the acceptability of an image. The points are connected to obtain meaningful values for DDH diagnosis.

Based on a previous study in which segmentation and key points were used to extract a large amount of information from a given amount of data and achieved higher machine learning performance, the four points on the ilium and acetabular roof and segmentation were used to produce a deep learning model (Fig. 2).

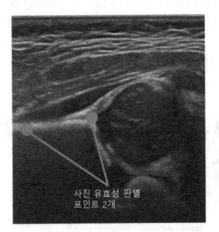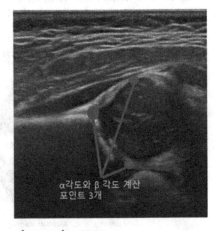

Fig. 2. Four points on ultrasound scans

2.1 Segmentation Environment

Mask R-CNN was used for segmentation. Mask R-CNN is an extension of Fast R-CNN with the addition of a Mask-head. The Mask head is first used to detect the areas corresponding to the ilium and acetabular roof. A ResNet algorithm consisting of 50 layers modified using the background algorithm of Mask R-CNN. The ROIAlign and Feature Pyramid Network (FPN) were used for feature extraction.

2.2 File Conversion

I The US data obtained from the participants were in the DICOM format. The DICOM files included information about ultrasound scans (date, type of equipment, imaged area, and patient) and image data. For personal information protection and memory efficiency reasons, this information was removed, and the ultrasound images were converted to 256-bit greyscale images.

2.3 Layer Extraction

A model was trained on ultrasound scans for segmentation. An area to be masked for the first training phase must be marked. The area is marked using the CVAT program developed by the MIT (Fig. 3).

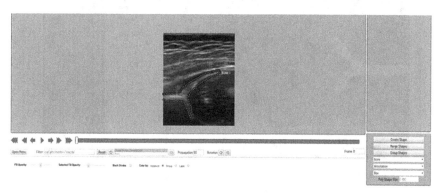

Fig. 3. Model training using CVAT

2.4 First Training

Polypoints were assigned to the areas on an image that contain the ilium and acetabular roof using the point selection tool. Points were assigned across a sequence of images. An extraction tool can be used to obtain XML or JSON files containing point records. The obtained XML and image files were used to train the first model (Fig. 4).

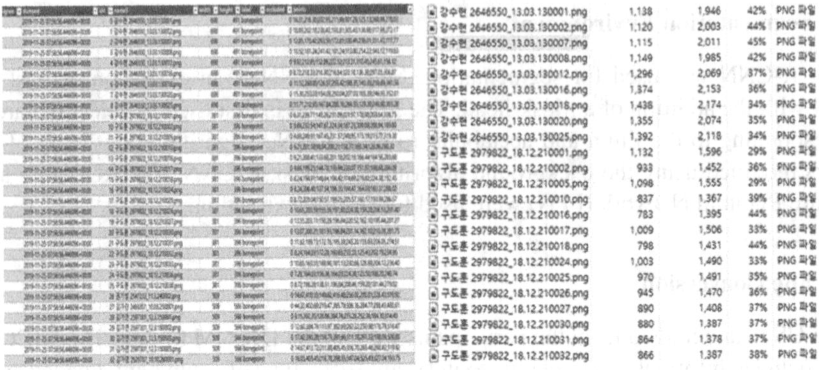

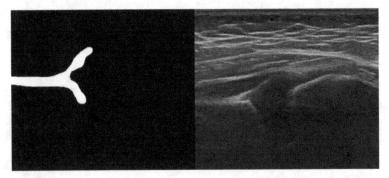

Fig. 4. Image files processed using CVAT (Left) XML file containing location data

2.5 First Training Results and Image Conversion

XML files in which the areas corresponding to the ilium and acetabular roof were masked and image files are used to train a model. The test dataset was then run on the trained model. The pixel values of the areas c orresponding to the ilium and acetabular roof were converted to 255 (white) and the rest to 0 (black) (Fig. 5).

Fig. 5. Original image (Left). The image after the first conversion (Right)

2.6 Final Results

Once the second training was complete, the first and second models were used to predict the angles and lengths of the ilium and acetabular roof. Hence, an AI system was developed that could assist physicians in diagnosing DDH based on these predicted values (Fig. 6).

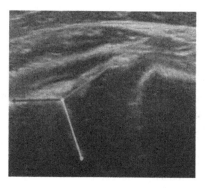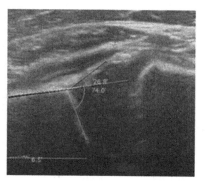

Fig. 6. Final results of the multi-AI system

3 Results

The following data were used:

Total data: 2,692 images
Valid data: 1,241/2,692 images
Dataset for model training and testing: 321/1,241 images
Training dataset: 288/321 images
Test dataset: 33/321 images
Test dataset: 920/1,241 images

A total of 2,692 ultrasound scans collected from 2016 and 2019 at Dongsan Medical Center were obtained. Of these, 1,242 were acceptable for use in the analysis. 321 of these scans were randomly selected and then arbitrarily divided into a training dataset containing 288 scans and a test dataset containing 33 scans.

To produce a machine learning model, the multi-training algorithm proposed in this study. The training parameter was set to $2.0e^{-3}$, and the model was trained until the validation loss was at the minimum. The trained model was then tested using 920 scans.

The 322 ultrasound scans obtained from Dongsan Medical Center were arbitrarily divided into a training dataset containing 288 scans and a validation set containing 33 scans. Of the total 920 scans, 389 were deemed acceptable (could be clearly interpreted by an orthopedic doctor at Dongsan Medical Center). The system correctly detected DDH in 369 out of the 389 scans, achieving a detection accuracy of 94.86%.

3.1 Performance Classification

We performed face recognition under various directions of the face. The experiment was conducted on 50 people (23 to 33 years old, 33 men and 17 women). The recognition rate for this system was 96.0% on the front and 86.0% on the face side. Also, the recognition rate for irregular lighting was 94.0%.

The heart rate measurement experiment was done in bright and dark light, the first to fifth experiment was done under bright fluorescent light, and the sixth to tenth experiment was done in dark incandescent light. Heart rate measurements (measured after accurate ambient light value measurement) were processed from the peak value of 30 to 35 s with images taken under bright fluorescent lamps, resulting in the same difference in heart rate values. (or not more than one) measured on each RGB channel. The G- channel was filtered the same as other channels but showed a uniform shape. The experiment was conducted 10 times per person and the average heart rate accuracy was 99.1%.

Additionally, we conducted an experiment on the situation that occurs when the subject of measurement uses a mask. This experiment was conducted with 10 people (46 to 88 years old, 5 men and 5 women) who visited a medical center. Using this system, we tested the accuracy of heart rate when a part of the face was covered by wearing a mask and when the mask was removed. As a result of the experiment, when the system was used, when the mask was worn, the accuracy was 97% compared to when the mask was not worn.

3.2 Non-contact Oxygen Saturation Measurement

To measure the performance of the system, the model's output was classified as appropriate detection, fail detection, different result, false image, inappropriate detection, or insufficient image. 'OK' indicated that the system produced normal outputs for normal data. 'Error' indicated that although an output was produced, an image was deemed inappropriate since the base angle value was not within the normal range. 'N/A' indicated that an output could not be produced due to non-available values. The detection results of the total 920 scans were as follows: 'Appropriate detection' for 413 scans, 'Fail detection' for 98 scans, 'Different result' for 30 scans, 'False image' for 336 scans, 'Inappropriate detection' for 9 scans, and 'Insufficient image' for 34 scans. The Fig. 7 below shows a bar graph the detection results starting with 'Appropriate Detection' from the far left followed by 'False Image', 'Different Result', 'Insufficient Image', 'Fail Detection', and 'Inappropriate Detection'.

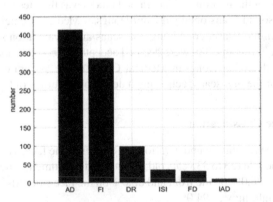

Fig. 7. Detection results

Performance parameters were used to assess the systems' performance. Since the answer must be divided into true or false, the classifier was also divided into OK (True) and error. A 2 × 2 matrix could then be created. Of the total 510 scans excluding the ones classified as 'N/A', 320 were true positives (TPs), 48 were false positives (FPs), 49 were false negatives (FNs), and 93 were true negatives (TNs).

3.3   System Performance Parameters

Based on the above classification, performance parameters could be calculated. The equations below Table 1 were used to calculate precision, recall, accuracy, F1 score, and fall-out. Precision was calculated at 0.87, recall at 0.87, accuracy at 0.81, F1 score at 0.87, and fall-out at 0.34.

Table 1. The values of the performance parameters

	Value
Precision	$8.796e^{-3}$
Recall	$8.672e^{-3}$
Accuracy	$8.098e^{-3}$
F1 score	$8.684e^{-3}$
Fall-out	$8.404e^{-3}$

The vertical flip, rotation, width and height shift functions were used to improve the efficiency of the model. Adam optimization was applied for parameter learning with the learning parameter initially set at $2.0 \times 10e^{-4}$. Training was stopped when the validation loss was at the minimum. n addition, in order to check the recognition rate in various situations, an experiment was also conducted on the error rate when rotation or noise is generated or when the target object is covered. In the deformation due to rotation, the measurement angle was changed from 5° to 20° at intervals of 5°, and the error was tested when it was deformed. The noise generation generated random noise in all the feature points, and the ratio was adjusted at 10% intervals for experiments. In the process of extracting the image sequence, the cover generated an obstacle at an arbitrary location to hinder the extraction of the feature point at the corresponding location, and then the feature point was extracted and compared with the database, and the error was calculated (Fig. 8).

No significant difference in angle measurements was found between the model and doctor. The differences in the maximum, minimum, and mean base angles were 6.93, 0, and 1.81°, respectively. The differences in the maximum, minimum, and mean α angle were 19.78, 0.01, and 3.56°, respectively. The differences in the maximum, minimum, and mean β angle were 25.92, 0.01, and 4.51°, respectively.

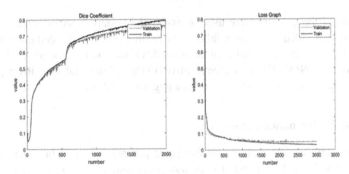

Fig. 8. Dice and Loss values for the validation and training dataset

4   Conclusion

IRecently, the number of hip dysplasia (DDH) that occurs during infant and child growth has been increasing. DDH should be detected and treated as early as possible because it hinders infant growth and causes many other side effects. DDH's standard screening method is to utilize Ultrasonography (US). However, the Ultrasonography (US) may differ in the reading results depending on the proficiency of the physician. It also takes a lot of skill to define DDH levels depending on the quality of the Ultrasonography (US). In order to improve the doctor's proficiency, many patient cases should be encountered and large-capacity Ultrasonography (US) should be obtained. However, due to the nature of infants and toddlers, only 2 to 10% of the photos taken can be used as meaningful photos. Currently, significant photographs are used by doctors to determine DDH in ultrasonic images, such as the angle of the Acetabulum-Femoral head for ultrasonic image reading. However, this method causes a lot of errors because the doctor works manually. Recently, machine learning techniques using Convolutional Neural Networks (CNN) and improved Resnet50 have been widely used for ultrasound image analysis. Research shows that computer-aided image analysis greatly improves objectivity and productivity at medical sites. This shows that the DDH computer-aided diagnostic algorithm through ultrasound images can solve difficult challenges in orthopedics.

In this paper, we proposed a computer-aided diagnostic algorithm that can automatically measure and diagnose DDH using CNN. In this study, the algorithm was designed to automatically calculate the angle of the acetyaboulum-femoral head for normal/nonnormal reading of the infant hip joint and to perform diagnosis automatically by deep learning of existing images. Experiments have shown that the speed and accuracy of diagnosis are improved compared to doctor.

Acknowledgement. This work is supported by the Foundation Assist Project of Future Advanced User Convenience Service" through the Ministry of Trade, Industry and Energy (MOTIE) (R0004840, 2020) and Basic Science Research Program through the National Research Foundation of Korea (NRF) funded by the Ministry of Education (NRF-2017R1D1A1B04031182).

References

1. Webster, J.G.: Encyclopedia of Medical Devices and Instrumentation. John Wiley & Sons, New York (1988)
2. Geddes, L.A., Baker, L.E.: Principles of Applied Biomedical Instrumentation. John Wiley & sons, New York (1989)
3. Cameron, J.R., Skofronick, J.G.: Physics of the Body Medical Physics Publishing Corporation (1999)
4. Perloff, D., et. al.: Human blood pressure determination by sphygmomanometry. J. Am. Heart Assoc. **88**, 2460–2470 (1993)
5. O'Grady, N.P., et al.: Guidelines for the prevention of intravascular catheter-Related infection, CDC (2011)
6. Drzewiecki, G.M., Melbin, J., Noordergraaf, A.: The Korotkoff sound. Ann. Biomed. Eng. **17**, 325–359 (1989)
7. Moraes, J.C.T.B., Cerulli, M., Ng, P.S.: Development of a new oscillometric blood pressure system. IEEE Comput. Cardiol. Conf. **26**, 467–470 (1999)
8. Wax, D.B., Lin, H.M., Leibowitz, A.B.: Incasive and concomitant noninvasive intraoperative blood pressure monitoring. Anesthesiology **115**, 973–978 (2011)

Comparison of SVM and Random Forest Methods for Online Signature Verification

Leetesh Meena[1], Vijay Kumar Chaurasiya[1], Neetesh Purohit[1],
and Dhananjay Singh[2(✉)]

[1] Indian Institute of Information Technology Allahabad, Allahabad, India
{ism2014008,vijayk,np}@iiita.ac.in
[2] Hankuk (Korea) University of Foreign Studies (HUFS), Gyeonggi-do, South Korea
dan.usn@ieee.org

Abstract. Signatures are widely used for authentication purposes. Signature verification has many applications in banking, in crossing international borders, in boarding of planes etc. For verifying identity of a person, signature is legally and widely accepted biometric trait. This work presents two simple and efficient methods for online signature verification. The paper proposes Support Vector Machine (SVM) and Random Forest Method for verifying online signatures. To measure the performance of algorithm f1 score is used and experiments were performed on SUSIG dataset. In Method-1, data after preprocessing is taken as feature set and in Method-2 feature vector are made by concatenation of bins of different attributes of the signature. The attributes taken are k^{th} derivative of x and y coordinates, and k^{th} derivatives of pressure. The classification is performed on extracted features using SVM and Random Forest. The performances of proposed methods were evaluated by using confusion matrix on SUSIG dataset. Results show that the proposed methods are capable of verifying online signatures with acceptable level of accuracy.

Keywords: SVM · Random forest · Classification · Confusion matrix · Feature extraction · Online signature.

1 Introduction

Signature verification has been a challenging and exciting problem for biometric researchers from many years [2]. In recent times signature forgery crimes have been increased many folds. The signature verification system becomes handy in government offices for document validation, banking applications, student mark sheet verification etc. Signatures can be broadly divided into two categories:

1. Offline Signatures
2. Online Signatures

Supported by Ministry of Education, Government of India.

© Springer Nature Switzerland AG 2021
M. Singh et al. (Eds.): IHCI 2020, LNCS 12616, pp. 288–299, 2021.
https://doi.org/10.1007/978-3-030-68452-5_30

Offline signatures take place on paper, and analysis can be carried out when information of a signature on the document is extracted through image processing techniques. The verification process is based on the signatures that are stored digitally. This information can be termed as features and primarily used to distinguish between two signatures.

Online Signatures, on the other hand, are signatures that take place on an electronic device that are capable of recording the movement of the signature at a fixed interval of time, digitally. The x and y coordinate of the user's signatures along with some other attributes like pressure, time and pen up/down are also acquired to represent an online signature. Because of this vast range of dynamic features, the online signature verification system usually achieves better performance accuracy than the offline method and contains more information than an offline signature. Online signatures are more secure and reliable as compare to offline signatures due to dynamic features [3]. The objectives of this work are to propose an approach for verification of online signatures:

1. Deriving set of features using two different methods
2. Classifying signatures using SVM and Random Forest
3. Determining the performance of model, to analyze the performance using different set of confusion matrices

The rest of this paper is organized as follows: Section 2 presents literature review, Sect. 3 presents methodology, Sect. 4 presents experiments and discussions, Sect. 5 presents the results and analysis, followed by conclusion and future scope in Sect.6 of this paper.

2   Literature Review

Online signature verification is among the most acceptable, spontaneous, rapid and cost-effective system for user privacy (authentication). It requires some dynamic features like quickness, pressure, direction, length of the stroke and pen-ups/pen-downs to verify user's identity. Many researches are working in this field to improve feature-based approaches, dynamic time wrapping, Hidden Markov Model (HMM), Gaussian Mixture Models (GMM), Support Vector Machine (SVM), and Neural Networks for its applicability in this domain. All the above approaches have some advantages and disadvantages.

Dolfing et al. [4] implemented online signature verification using the HMM. They have concentrated more on the pan-tilt using a digitizer tablet and investigated the authentication accuracy based on distinct types of forgeries. They analyze the discriminatory cost of the different features based on a linear type discriminate analysis (LDA) to found that pen-tilt as a critical feature. Based on the HMM, their methodology reached rates of the error to some equal extent between 1.0%–1.9%. The forged access to both the spatial and dynamic characteristics of genuine signatures made the verification more difficult [5].

Liang Wan et al. [6] have used an Artificial Neural Network using a back propagation algorithm for signature verification. To calculate the model's accuracy and effectiveness, they took parameters as a false reject rate, equal error rate, and false accept rate. They used signatures of 20 individuals consisting of 400 signature samples of forged and genuine signatures. They aim to limit computer singularity in the signature verification process. The type of Neural Network they have used is the Feed Forward Neural Network. They have attained comparable speed, throughput, and accuracy to the benchmark algorithms in signature verification.

Pallavi V. et al. [7] have used MCYT and SVC2004 signature verification datasets. They have used the Longest Common Subsequence (LCS) to find the similarity of online signatures. They have also used the Support Vector Machine (SVM) to classify signatures, whether they are forged or genuine. They have also explored SVM with other kernel algorithms like dynamic time wrapping and concluded that the SVM with the longest common subsequence is more accurate. They also considered the inclination angles of the pen while doing the signature on a graphic tablet.

Christian Gruber et al. [8] acquired online signatures by using a digital tablet. Information provided by the signature includes the whole trajectory of the signature and pressure with respect to time. This paper proposed a new online signature verification system for both forgeries and genuine signatures. Here in the place of standard features, a combination of features is used and verification is performed using several classifiers. This method applied on several signature databases for a performance check and it gives high verification performance.

Vahab Iranmanesh et al. [9] mainly focused on how to find the most distinctive features for the online verification system. For dimensionality reduction, Principal Component Analysis (PCA) is used and verification is performed using a Multi Layer Perceptron (MLP). A comparison of previous approaches is also done to better view the effect of the selected features from PCA analysis on recognition results. The comparison done in the paper shows that the method proposed here resulted in a subset of 50 features that is more efficient than previous methods proposed.

For signature verification, two types of approaches can work. The first one is based on the features and the second one is based on the classification functions. Both of the approaches are discussed below. Since feature extraction plays a vital role in the classifier's performance, therefore, the main focus of the proposed work is on feature extraction. Before classification, one needs to create a proper set of features so that classifiers can perform well on these feature vectors. Therefore this work focuses on both the feature extraction technique as well on the classification method optimization.

3   Methodology

This section describes the steps involved in the proposed methodology. In this work two methods of feature extraction have been proposed and both are explained in the subsequent section. The proposed methodology has following steps:

1. Data Acquisition
2. Preprocessing the dataset
3. Feature Extraction
4. Feature Modelling
5. Classification

The above steps of the proposed approach are elaborated in the next section.

4   Experiments and Discussions

This section elaborates the steps proposed in the methodology.

4.1   Data Acquisition

SUSIG is an online dataset which consists of signatures of 100 people containing 3000 genuine and 2000 skilled forgery signatures. The dataset contains following information i.e., X, Y, Time Stamp, Pressure, Pen Up/Down. Where X and Y are x and y coordinates, Time stamp is the time at that instant from start time, Pen Up/down indicates whether finger is touching screen or not at that instant, Pressure indicates its usual meaning [1].

4.2   Preprocessing the Dataset

Since any type of data which are collected may not be perfect or it may not be as per the requirements of the classification techniques, therefore, data preprocessing becomes necessary before extracting features from dataset. It essentially implies changing crude information to a clean dataset. This gives improved outcomes when passed to the classification model. It is proposed to apply following three steps to make the dataset suitable for feature extraction.

1. Filtering
2. Interpolation
3. Normalization

Filtering. It is not desired to change the shape of the signature after the filtering process therefore RDP filtering algorithm has been used because it takes the subset of signature coordinates to represent a signature after filtering, which does not change the shape of the signature after filtering. With filtering, the number of coordinates in the signature will decrease and the signature looks crisp. After filtering, the noise is removed and data becomes free of noise.

It can be observed that there is no difference between the two signatures, i.e., the RDP algorithm removes noise and decreases points, but it does not change the actual shape or features of the original signature, which can be clearly noticed from the above figures.

Interpolation. After filtering, interpolation was performed because some of the points have been removed. Since it is required that the dataset needs to be in constant length; therefore, the missing values are filled using interpolation. Here 256 points are considered in the signature of every individual. Homogeneity is to be considered for the dataset to take 256 points which are equally spaced in a signature to maintain equal number of points for every signature of a particular user. To maintain constant length of feature vector the coordinates representing a signature are interpolated. Linear interpolation is used to interpolate points. Linear Interpolation actually finds unknown value at a point given the values in its surrounding points. After applying linear interpolation all the obtained values are then floored to integers to avoid complexity and for convenience.

Normalization. Since signatures can be of any size and they can have different time of signing so they are normalized with respect to location, size and time respectively. To make the dataset time invariant, time data in the signature is removed. Size normalization is used to make the signature size independent. Size normalization is essential because any individual cannot make the signatures which are exactly equal in size. Size normalization is performed on the signatures such that coordinates after normalization lies between 0 and 1 irrespective of how lengthy the signature could be. Normalization is important in verification process.

$$x_i = (x_i - minimum)/(maximum - minimum) \qquad (1)$$

where '$minimum$' denote minimum value of feature and '$maximum$' maximum value of feature 'x_i' respectively.

4.3   Feature Extraction

After data preprocessing 256 x coordinates, 256 y coordinates, 256 pressure coordinates and pen up and downs are collected. The time coordinate is removed. Finally feature vector will contain the X and Y coordinates and P pressure coordinates along with pen up and downs. The present data is considered as the feature vector with the above mentioned columns as features.

4.4   Feature Modeling

Two methods for feature modeling is proposed here. In **method 1** after feature extraction, 10 genuine and 10 forgery signature files for each individual are created. Thereafter, all the genuine and forgery files are combined into one file and add an additional feature at the end for each user which corresponds to genuine or forgery signature (setting 1 or 0) for that feature. The final feature vector will contain the features- X, Y, P, Pen-up/down, 1/0. This will be passed to classifier to classify the signatures. In Method-1, only size normalization was performed but location normalization was not considered.

In **method -2**, location normalization was also performed. We are making the system location invariant means independent of location change by using differences of coordinated which is making it position independent. For this, feature vector is concatenation of bins of different attributes of the signature. The attributes taken are k^{th} derivative of x and y coordinates, as well as k^{th} derivatives of pressure. The kth derivative of each attribute is taken to make the feature vectors position invariant (i.e. a user cannot start the signing the signature from the same point and in the same orientation all the times he/she signs). The kth derivatives of the points are used to solve this problem. Now each attribute is made in to bins of regular intervals using max and min values of every attribute. For above attributes min and max values are found from the data directly. Now this entire range is divided into regular intervals called as bins. Initially the counts of all the bins are zero. Then by iterating through each tuple from the data, counts in the bins are incremented by checking to which bin that particular value belongs. Then the bins of all attributes were concatenated to form a feature vector of a single signature.

$$F = |B_1||B_2|||B_3||....|B_k| \tag{2}$$

After extracting feature vectors, these will be given to the classifier to classify the signatures.

4.5 Classification

After forming feature vector, the final step is the verification process for which classification algorithm will be used. This can be done by many machine learning algorithms like SVM, Neural Networks etc. In this work, SVM and Random Forest are used for classification and verification purpose. The details of both of the classifiers with their implementation are discussed in the following sections.

1. Support Vector Machine

SVM is a supervised learning model that is used as a classification algorithm. SVM takes data sample and give output the hyper plane that separates the data samples optimally. SVM try to finds a hyper-plane with the maximum margin in the higher dimensionality which is linearly separable. The equation of hyper-plane can be written as

$\mathbf{w}^T\mathbf{x}=0$. hyper-plane can be written mathematically as for x satisfying $\mathbf{w}.\mathbf{x}+b=0$.

(Decision rule: if ≥ 0, then positive samples).

Considering 2-dimensional vectors, let us assume that the two hyper-planes h_1 and h_2 both of them separates the data points so that h_0 (h_0 is the optimal hyper-plane) is equidistant from h1 and h2. The hyper-planes that meet the following constraints are selected where for each vector xi either: 1 or −1.

$$(\mathbf{w}.\mathbf{x}_i + b) \geq 1 \tag{3}$$

for class 1 and

$$(\mathbf{w}.\mathbf{x}_i + b) \leq -1 \tag{4}$$

for class -1.

In the case of online signature verification which is a binary classification problem (i.e., forgery or genuine), radial basis function kernel (RBF) is used. This can be written as

$$K(\ x_i,\ \ x_j) = exp(-\gamma(\|x_i - x_j\|)^2), \gamma > 0. \tag{5}$$

2) Random Forest

Decision tree is the building block of Random Forest Method. Random forest grows several classification trees. It consists of a number of decision trees which are not dependent on each other and work as a whole. It works as follows.

A. Let D be the training Dataset and it generates k bootstrap samples of D. Each bootstrap sample D(i) consists same number of tuples as that of D by sampling and replacing from D(i.e, some original tuples and some duplicated tuples of D).

B. For each D(i) a decision tree is constructed and k decision trees are formed.

C. Decision tree algorithm uses Gini index for tree generation.

$$Gini\,(D) = 1 - \sum_{i=1}^{m} p_i^2 \tag{6}$$

Where, pi is the probability of a tuple that belong to a class in training dataset D. Impurity of a dataset D is given by Gini index.

$$Gini_A\,(D)(= \frac{|D1|}{|D|}\,Gini\,(D1) + \frac{|D2|}{|D|}\,Gini\,(D2)) \tag{7}$$

This gives the Gini index of d by binary split on A (attribute) and splits into D1 and D2.

$$(A) = Gini\,(D) - Gini)_A \tag{8}$$

This gives the reduction in impurity by splitting D on attribute A. To classify a given test tuple X, each tree classification result is counted and X is classified with a class having the maximum count. Finally, compare the different classification results based on precision, recall and accuracy.

4.6   Implementation

The generated feature vectors, as explained above, are used for the classification process. Firstly, an SVM classifier is used to classify whether the given signature is genuine or forged. The data in feature vectors are divided into training and testing sets. The classifier gets trained using the training dataset. It draws a hyper-plane that divides the genuine and forges classed after training the classifier. Then the testing dataset was used to obtain the accuracy of the predictions. For Random Forest after, preprocessing data is divided into train and test data for train data. Decision trees were created for a randomly selected subset of data for bootstrapping. After iterating through all the trees, the final decision is made based on voting. The final class was the one that has more votes and considered as the predicted class. In this model, hyper-parameter tuning was not done; default parameters are used. Random Forest gives the probability and in the case of SVM, it gives the support vectors which need to be converted to probability.

5   Results and Analysis

This section discusses the results obtained after the experiments. The performance of classifiers is evaluated using confusion matrix.

For method-1 only overall accuracy is calculated. For method-2 separate classifiers were implemented for all 100 users. Table 1 and 2 are summarizing the results obtained for method 1 and method 2. Both the tables represent the average values obtained for all the 100 users in two separate cases. Where the first row represents the results obtained before the hyper parameter tuning and the second row represents the results obtained after the hyper parameter tuning.

Table 1. Accuracy obtained using method − 1 for feature extraction

For Method 1	
SVM	Random forest
Before hyper tuning	**Before hyper tuning**
Accuracy = 80.33 F1 Score = 62.0	Accuracy = 83.25 F1 Score = 79.0
After hyper tuning	**After hyper tuning**
Accuracy = 80.75 F1 Score = 79.85	Accuracy = 83.3 F1 Score = 83.34

Table 2. Accuracy obtained using method - 2 for feature extraction

For Method 2	
SVM	**Random forest**
Before hyper tuning	**Before hyper tuning**
Accuracy = 83.0 F1 Score = 73.0	Accuracy = 88.0 F1 Score = 83.0
After hyper tuning	**After hyper tuning**
Accuracy = 88.28 F1 Score = 87.44	Accuracy = 91.57 F1 Score = 91.29

This section will discuss the results obtained during the hyper parameter tuning of both the models.

SVM hyper-parameter tuning Plots

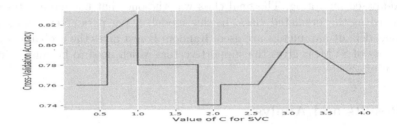

Fig. 1. Output of hyper parameter tuning C parameter of SVM

It can be observed from the Fig. 1 that the accuracy is maximum at $C = 1$. Therefore, 1 is considered as the best value of C parameter for the proposed model.

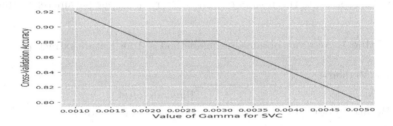

Fig. 2. Output of hyper parameter tuning Gamma parameter of SVM

It can be observed from the Fig. 2 that the accuracy is maximum at 0.001 and after that it is decreasing so the best value for gamma is 0.001 for the proposed model.

Random Forest hyper-parameter tuning Plots

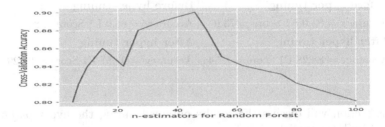

Fig. 3. Output of hyper parameter tuning of random forest

For this parameter high value is generally preferred because more trees will be better to learn from data. Unfortunately high value can decrease the training process so hyper-parameter tuning was required. It can be concluded from the Fig. 3 that parameter value 46 is good because the accuracy is maximum at this value.

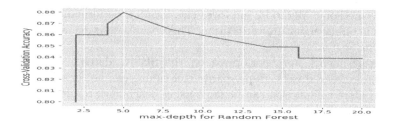

Fig. 4. Output of max_depth parameter tuning of random forest

max_depth parameter indicates the depth of each tree in the forest. The deeper the tree then it will have more number of splits so it will contain more information. Depending on the above discussion, the range from 1 to 32 was used to fit the decision trees and it can be observed from the Fig. 4 that depth = 5 is a good choice.

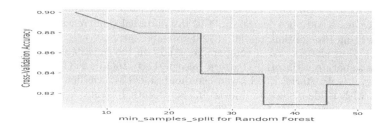

Fig. 5. Output of hyper parameter tuning of random forest

min_samples_split parameter tells about the requirement of minimum number of samples for splitting on every node. From the Fig. 5 it can be concluded that value of 5 is giving good amount of accuracy so it will be considered. In this way hyper-parameter tuning is done for both SVM and Random Forest and best parameters are obtained using grid search.

Hence, from the above figures and tables it can be concluded that the F1 Score and accuracy for Random Forest is more as compare to SVM in both of the cases. This is after cross-validation so it is generalized. It should perform well on other datasets too.

6   Conclusion and Future Scope

The work proposed a simple and effective online signature verification system. For classification purposes, SVM and Random Forest have been used. From the results it can be concluded that the Random Forest outperforms SVM in both the proposed methods of feature extraction. Therefore it is better to use the Random Forest algorithm for online signature verification. For feature extraction out of two methods; Method-2 improves classification performance for both the classifiers in online signature verification. This conclusion is derived on the basis of F1 Score.

The experiments were performed on the SUSIG dataset, which can be extended to other datasets. Hyper-parameter tuning can be considered for classifiers and more comparison can be done using benchmark results of previously used verification models.

The collection of the Signature dataset can be achieved by using mobile devices. The verification of signature for authenticating the user on mobile devices has unique aspects that are not shown in those datasets whose acquisition is done under proper supervision. Experimentation can be done in this field; for verification purposes, the same methodology can be used as proposed in this work.

References

1. Kholmatov, A., Yanikoglu, B.: SUSIG: An on-line signature database, associated protocols and benchmark results. J. Pattern Anal. Appl. **12**(3), 227–236 (2009)
2. Impedovo, D., Pirlo, G., Plamondon, R.: Handwritten signature verification: new advancements and open issues. In: 2012 International Conference on Frontiers in Handwriting Recognition, pp. 367–372, September 2012
3. Lei, H., Govindaraju, V.: A comparative study on the consistency of features in on-line signature verification. Pattern Recogn. Lett. **26**(15), 2483–2489 (2005)
4. Dolfing, J.G.A., Aarts, E.H.L., van osterhout, J.J.G.M.: On-line signature verification with hidden Markov Models. In: Fourteenth International Conference on Pattern Recognition, 20 August 1998
5. Wan, L., Wan, B., Lin, Z-C.: On-line signature verification with two-stage statistical models. In: Eighth International Conference on Document Analysis and Recognition (ICDAR 2005), 31 August 2005
6. Wan, L., Wan, B., Lin, Z-C.: On-line signature verification with two stage statistical models. In: Eighth International Conference on Document Analysis and Recognition (ICDAR 2005), pp. 282–286 (2005)
7. Hatkar, P.V., Prof Salokhe, P.T., Malgave, A.A.: Handwritten signature verification using neural network. In: International Journal of Innovations in Engineering Research and Technology [IJIERT], vol. 2, no. 1, January 2015
8. Gruber, C., Gruber, T., Krinninger, S., Sick, B.: Online signature verification with support vector machines based on LCSS kernel functions. IEEE Trans. Sys. Man Cybern. - Par: B Cybern. **40**(4), 1088–1100 (2010)
9. Iranmanesh, V., Ahmad, S.M.S., Adnan, W.A.W., Yussof, S., Arigbabu, O.A., Malallah, F.L.: Online handwritten signature verification using neural network classifier based on principal component analysis. The Scientific World Journal 2014 Article

10. Hassaïne, A., Al-Maadeed, S.: An online signature verification system for forgery and disguise detection. In: Huang, T., Zeng, Z., Li, C., Leung, C.S. (eds.) ICONIP 2012. LNCS, vol. 7666, pp. 552–559. Springer, Heidelberg (2012). https://doi.org/10.1007/978-3-642-34478-7_67
11. Feng, H., Wah, C.C.: Online signature verification using a new extreme points warping technique. Elsevier **24**(16), 2943–2951 (2003)
12. Afsar, A., Arif, M., Farrukh, U.: Wavelet transform based global features for online signature recognition. In: Pakistan Section Multitopic Conference. Karachi, pp. 1–6 (2005)

Human-Centered AI Applications

Audio Augmented Reality Using Unity for Marine Tourism

Uipil Chong$^{(\boxtimes)}$ ⓘ and Shokhzod Alimardanov ⓘ

University of Ulsan, 93 Daehak-ro, Mugeo-dong, Nam-gu, Ulsan, Republic of Korea
upchong@ulsan.ac.kr, alimardanovshohzod@gmail.com

Abstract. The development of marine tourism content is a much-needed challenge at this point. Audio augmented reality (AAR) is a new and fast-growing technology. We assume that AAR will improve the quality of services provided by marine tourism. In this research, the actual potential of AAR in marine tourism was studied by developing an AAR prototype using the Unity engine. We have used whale images with 3D sounds. The prototype can localize several 3D sounds in real-world locations. We conducted a listening test comparing the normal whale audios with the 3D whale audios generated by AAR. 81% of the participants were positive for 3D sounds and whale images. Interviewees felt better engagement in the service. They considered that the technology can be useful for receiving more information easily and making them more physically active. We conclude that the main contributions of AAR may be giving insightful knowledge about marine life in an interesting way and stimulating people to engage in nature closer.

Keywords: Audio augmented reality (AAR) · Unity · 3D sounds · ARCore · Marine tourism

1 Introduction

The tourist attraction is a major point of interest in the marine tourism industry. There are a wide variety of factors that have a direct impact on users' choice. Visitors always have a desire to find the uniqueness of the destination while having fun. Proper navigation and guidance are some of the things that could help to meet those expectations. Current systems in marine tourism rely on professional guides and hardware-dependent technology to satisfy their needs. This causes some other problems such as language barriers and high costs. Although English is the most widespread language in the world, there is always an insufficiency of English speakers in some destinations. Hiring professional guides as many as needed is costly for both consumers and sellers in the industry. The study analyzed the role of AAR as a solution to the problems.

Audio augmented reality (AAR) is a new-rising technology that augments objects with virtually spatialized sounds. The concept of AAR has been proposed a few decades ago but remained less explored in the field of augmented reality (AR) research, of which a majority has focused on visual augmentation [1–3]. With AAR, users can perceive the virtually synthesized sounds by hearing them via normal headphones.

© Springer Nature Switzerland AG 2021
M. Singh et al. (Eds.): IHCI 2020, LNCS 12616, pp. 303–311, 2021.
https://doi.org/10.1007/978-3-030-68452-5_31

AAR is a voice-oriented part of Augmented Reality (AR). AR resembles a bridge that connects the virtual and real worlds. It makes real life more interesting and interactive by adding extra information. People can see real-size whales wherever they want, mechanics may be given extra information about the current problem as well as safety warnings, education can bring real jungle in the classroom. In fact, we have seen AR's assistance in many areas of our lives [4].

As a result of AR's significant development over the last years, we can find its implementations in various industries. Experiments in construction [5], education [6], gaming, retail, marketing, medicine [7], tourism [8], and automotive industries [9] have noted positive improvements.

The study in reference [10] proposed a synergy between the human body development, computing machines, and neural environment. Their focus was to motivate children toward nature. Pokémon Go, which was developed by Niantic, is an AR product that would encourage exercise, but not in a heavy-handed way [11]. AR games make players engage in nature physically.

Another study [12] investigated AR's impact on sociability, entertainment, and learning. There were two groups of people using either an AR app or non-AR apps. AR app users noted the entertainment value and learning opportunities. They also acknowledged that they were better engaged in the exhibition.

After researching the impact of AR in other industries, we assume that it might also be profitable to create software that makes people more physically active by being better engaged in the service of the marine tourism industry.

We have developed an application that augments several 3D sounds on top of real-world objects. The service provided by our AAR application can be experienced using headphones and smartphones.

This paper is organized as follows. Section 2 presents background and related works. Section 3 explains important 3D sound implementation tools of Unity with C# coding algorithm. It also consists of the prototype development process. Experimental results are given in Sect. 4. Conclusions are presented in Sect. 5.

2 Background and Related Works

2.1 Augmented Reality

Augmented Reality (AR) combines virtual objects and real-world observations. Virtual objects not only come into view as part of the user's environment, but users can also interact with the objects. Typical AR systems comprise technologies, such as computing, motion tracking hardware, and an event registration interface [5].

AR is classified based on various criteria. For example, depending on the kinds of hardware that users use there are stationary AR systems, spatial AR systems, desktop AR systems, mobile AR, head-mounted displays, contact lenses.

Stationary AR systems use powerful cameras and recognize real objects and scenes. They are used in virtual fitting rooms.

Spatial AR systems are also called projection mapping and video mapping. They bring real life-sized virtual objects to real life. Spatial systems are much popular in the automotive industry. Automakers are practicing spatial AR systems for showing their new models to their clients.

Desktop AR systems take advantage of webcams and project digital data onto real-world objects and display them on the desktop screen. Online stores are implementing desktop AR systems for virtual fitting rooms.

Mobile AR systems are the most popular AR system among others due to ubiquitous mobile devices. Usually, mobile devices receive information from GPS to extract the user's position and show specific data that has a direct connection with user locations. Mobile AR systems play an important role in navigating people in tourism cities.

Head-mounted displays' most popular representative is smart glasses. They can be either video see-through or optical devices depending on the technology behind.

Contact lenses are considered the future of AR since it is still under development.

The next classification depends on the identification of scenes or objects in the real world. There are vision-based AR and location-based AR.

Vision-based AR mainly focuses on the recognition and interpretation of images, objects, and scenes in the reality.

Location-based AR anchors virtual data to certain locations persistently with the help of GPS. In tourism, this persistently localized data helps tourists to find their destinations and collect more information about different facilities.

2.2   3D Sounds and AAR

Adding 3D sound effects with the relative listener's position will make the sound systems more realistic and suitable. A detailed description of 3D sounds and technologies behind are given in reference [13, 14].

Audio Augmented Reality (AAR) enhances the real audio environment by superimposing virtual audios [15]. It's a part of augmented reality that concentrates mainly on 3D sounds. In the research, we use AAR to generate 3D sounds.

2.3   Unity and ARCore

Developed by Unity Technologies, Unity is one of the most popular game engines for both 2D and 3D cross-platform game development. It was launched in 2005. C# and JavaScript are used as the main scripting in Unity. While installing Unity, it is required to have a Unity account. Once Unity is installed, you will be able to access the game engine from the icon [16].

ARCore is Google's platform for creating AR applications for Android and iOS. Its first version was launched in February 2018. It makes use of different capabilities, such as motion tracking, environmental location, and light estimation. Targeted at Android 7.0 and above, not all devices in the market support this technology, although this number has increased greatly since the first SDK version [17].

3   Development of Audio Augmented Reality Prototype

3.1   3D Sound Coding in Unity

The development process begins by installing the `ARFoundation` package in Unity (see Fig. 1). `ARFoundation` is a cross-platform API that helps to create augmented reality for both Android and iOS. Developed by Google and Apple, `ARCore` and `ARKit` are special platforms to build AAR projects on Android and iOS respectively. We can create one AR app and run it on both `ARCore` and `ARKit` with the help of `ARFoundation`. Since the scope of the current study centralizes mainly on Android, the android AR development process will be covered only. However, the main part of the prototype is the same, except for the deployment process on a targeted operating system. The prototype can be implemented on the iPhone without much effort.

The next stage is setting up Unity for android development. We import several components in the Game Engine.

Initially, we import `ARSessionOrigin`. It is the parent for an AR setup. `ARSessionOrigin` contains an `ARCamera` that detects features, such as planes or point clouds, and listens to 3D sounds relative to the position of a phone's Camera.

We have used Resonance Audio SDK to spatialize audios in AR. Created by Google, Resonance Audio gives AAR developers powerful technology for delivering high fidelity spatial audio at scale, to users across the top mobile and desktop platforms. One of its main script files is `ResonanceAudioListener` which is attached to `ARCamera` and listens to the 3D sounds in real-time.

To make the experience interesting and visible, we have imported two 3D whale models and attached the audios.

There is an important component inside the Unity Editor, `AudioSource`. The `AudioSource` manages sound behaviors through the AAR experience. It plays back the sound in either 2D or 3D or in mixed. We import the `AudioSource` component. It comes with several special properties. These properties include enabling sound spatialization tools and minimum/maximum distances which determine audio listening zones.

Inside the `AudioSource` component, there is a property called Spatial Blend. It takes values from 0 to 1 which calculate how much the 3D engine is utilized. Increasing Spatial Blend to 1 means the sound will be played back fully in 3D.

The other properties are min/max distances. Minimum distance specifies how close users and audio source should be to listen to the sound. Reversely, maximum distance is for specifying the borders of the virtual sound wives can reach at most. Users hear the sound if there are in this determined sphere. The unity engine simulates the sound to give the feeling of real audios.

Degrees of fading audios or increasing the volume are regulated by Rolloff Mode in the game engine. We applied linear rolloff mode. The further away we get from the sound, the less we hear it.

C# script file was created to manage audios and put them to positions in the sphere. We have defined several variables to assign sound objects as well as to keep track of the number of 3D sounds that a user can localize in the real world. The program inputs values for each variable and then waits for a user to touch the phone screen.

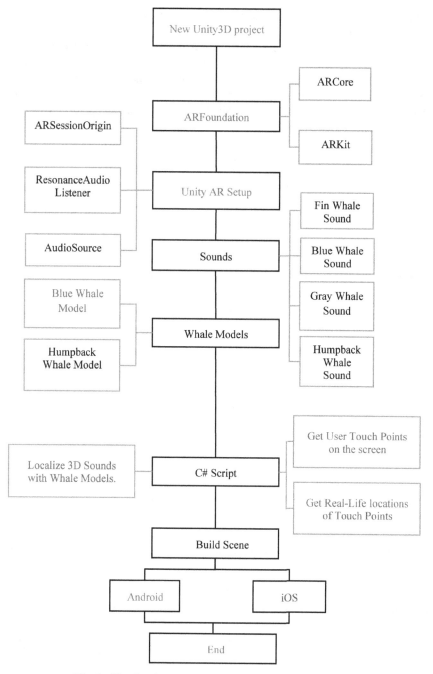

Fig. 1. The development process of AAR apps using Unity.

The Function in Algorithm 1 gives information about the user and phone interactions and checks if a user touches the screen. It stores touch positions, returns `true` when touch inputs exist. In the reverse case, the function returns `false` and gives the default value to the `touchPosition` variable. The `true`/`false` values are essential to make decisions for 3D sound localization. It should be clarified that the function in Algorithm 1 obtains only the touchpoint on the phone screen, not the location in the real world. Real-world positions (3-dimensional coordinates of the screen touchpoint) are taken by the `Update` function (see Algorithm 2).

```
// Get Vector2 touch positions.
BOOL TryToGetPosition(OUT Vector2 touchPosition)
        // Obtain a screen touch struct.
        IF (Screen touch phase == Beginning of screen touch)
                touchPosition = Screen touch position
                RETURN TRUE
        ENDIF

        touchPosition = DEFAULT
        RETURN FALSE
```

Algorithm 1. Function declaration that obtains information about the user and phone interactions.

Looking more closely inside the `Update` function in Algorithm 2, there are two if-statements. The first ensures nobody is touching the device. Second if-statement casts ray and saves hit information.

```
VOID UPDATE ()

        IF (!TryToGetPosition(OUT Vector2 touchPosition))
                RETURN
        ENDIF

        IF (raycastManager.Raycast(touchPosition, hits, TrackableType.PlaneWithinBounds)
                VAR hitPose = hits[0].pose

                IF (Placed 3D sounds < Maximum limit)
                        // Instantiate 3D sounds at the position and orientation.
                        Instantiate (3DSounds, hitPose.position, hitPose.rotation)
                ENDIF
        ENDIF
```

Algorithm 2. The `Update` function declaration that places 3D sounds with whale images.

To withdraw the real-world coordinates of the points on a mobile phone screen and implementing the above-stated ray casting methods, we have used the `Raycast()` function. `Raycast()` is a function that gets real-world positions of screen points by casting virtual rays toward real-world objects. Furthermore, it gathers hit information. The hit information contains the obstacles that the virtual ray comes across as it travels. These can be walls, the ground, chairs, etc.

In the nested if-statement in Algorithm 2, the number of placed objects is calculated and checked if it is more than the allowed limit. In the end, the program localizes 3D sounds in world space.

We used the `Instantiate()` function that takes an object, the position and rotation (see Algorithm. 2). The function clones game objects like the duplication in the editor. Duplicated game objects are placed at the location specified by position and rotation arguments.

3.2 Building the AAR Scene on Android

We used an Android phone to build the prototype. Figure 2 demonstrates the full process starting from the creation of the Android developer environment, goes through build settings inside Unity, and ends at deploying the app on a phone.

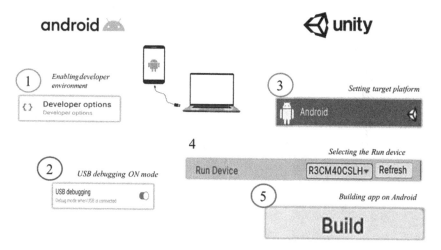

Fig. 2. Building the AAR app on an android phone.

Firstly, we enable developer options on Android. Developer options can be set to visible by going to `Settings > About > Device` and tap seven times on `Build number`. Secondly, we allow USB debugging that is located inside the `developer options`. We are giving access to a computer for building apps directly on the phone. The process continues with connecting a smartphone to a computer via a USB cable.

The final part is done inside Unity. We build the scene on the target android phone by implementing Unity's special building tools. Figure 2 presents configurations in Unity. The word `Android` at the third step is the target platform. It should be set to the desired type of operating system before building the scene. The fourth step is showing the device that executes the AAR application. Finally, the `build` command starts the actual building process, after which we can practice the AAR experience through the smartphone.

4 Experimental Results

The experiment was conducted successfully on a whale tour ship. We implemented a listening test on August 15, 2020, comparing the normal whale audios with the 3D whale audios. Participants in the experiment listened to 2D sounds at first. Then they listened to 3D sounds.

Thirty-six people from teens to 50 s participated in the experiment, which had a 50/50 female-male ratio. The test results of the experiment are shown in Fig. 3. 81% of people answered positively when they were asked whether 3D sounds felt better than normal sounds. 67% of people gave positive answers to the question about the potential of AAR for improving the quality of the services in marine tourism.

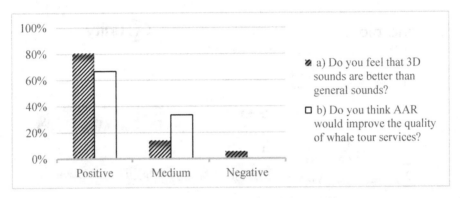

Fig. 3. The survey results on the whale tour ship.

5 Conclusions

We have tested the role and potential of AAR in marine tourism. 81% of participants confirmed positive effects. Using AAR, we may motivate people to interact with nature in a physically active way. We have confronted some limitations of the technology as well. When it comes to the surface that moves continuously on the ship, AAR has shown weak points to persistently locate 3D models.

Acknowledgments. This research is supported by the 2020 National Research Foundation (No. 2017R1D1A3B05030815) of the Korean Government.

References

1. Patil, P.P., Ronald, A.: Cross-platform application development using unity game. Int. J. Adv. Res. Comput. Sci. Manage. Stud. 3(4), 19–27 (2015).
2. Bederson, B., Druin, A.: Computer augmented environments: new places to learn, work, and play. Adv. Hum. Comput. Interaction, 5, 37–66 (1995).

3. Kim, K., Billinghurst, M., Bruder, G., Duh, H.B., Welch, G.F.: Revisiting Trends in Augmented Reality Research: A Review of the 2nd Decade of ISMAR, pp. 2947–2962 . IEEE TVCG (2018).
4. Rovithis, E., Moustakas, N., Floros, A., Vogklis, K.: Audio Legends: Investigating Sonic Interaction in an Augmented Reality Audio Game. Information MDPI (2019).
5. Mutis, I., Ambekar, A.: Challenges and Enablers of Augmented Reality Technology for in Situ Walkthrough Applications. Journal of Information Technology in Construction - ISSN 1874-4753 (2020).
6. Petrov, P.D., Atanasova, T.V.: The Effect of Augmented Reality on Students' Learning Performance in Stem Education. Information, MDPI. ISSN 2078-2489 (2020).
7. Parekh, P., Pate, S., Patel, N., Shah, M.: Systematic review and meta-analysis of augmented reality in medicine, retail, and games. Visual Computing for Industry, Biomedicine, and Art (2020)
8. Williams, M., Kelvin K.K., Yao, J.R.C.: Nurse: Developing an Augmented Reality Tourism App through User-Centered Design. arXiv:2001.11131 (2020).
9. Čujan, Z., Fedorko, G., Mikušová, N.: Application of virtual and augmented reality in automotive. Open Eng. **10**(1) (2020).
10. Wirza, R., Nazir, S., Khan, H.U., Garcı́a-Magariño, I., Amin, R.: Augmented reality interface for complex anatomy learning in the central nervous system: a systematic review. J. Healthcare Eng. Hindawi (2020)
11. Bueno, S., Gallego, M.D., Noyes, J.: Uses and Gratifications on Augmented Reality Games: An Examination of Pokémon Go. Applied Science, MDPI (2020)
12. Savela, N., Oksanen, A., Kaakinen, M., Noreikis, M., Yu, X.: Does Augmented Reality Affect Sociability, Entertainment, and Learning? A Field Experiment. Applied Science, MDPI (2020)
13. Yanghan Kim, Youngjin Park: 3D Sound Manipulation: Theory and Application. ICSV15, Daejeon, Koreapp, pp. 1–18 (2008).
14. Usman, M., Kamal, K., Qayyum, R., Akram, S., Mathevan, S.: 3D sound generation using kinect and HRTF. In: IEEE 2nd ICSIP, pp. 307–310 (2017)
15. Harma, A., Jakka, J., Tikander, M., Karjalainen, M.: Augmented reality audio for mobile and wearable appliances. J. Audio Eng. Soc. (2004).
16. Jing, Y., Gabor, S.: Audio Augmented Museum Experiences using Wearable-Inertial Odometry. MUM 2019, Pisa, Italy (2019)
17. Geronazzo, M., et al.: Creating an Audio Story with Interactive Binaural Rendering in Virtual Reality. Hindawi Wireless Communications and Mobile Computing (2019).

A Prototype Wristwatch Device for Monitoring Vital Signs Using Multi-wavelength Photoplethysmography Sensors

Nguyen Mai Hoang Long[1], Jong-Jin Kim[2], and Wan-Young Chung[1(✉)]

[1] Department of AI Convergence, Pukyong National University, Busan, South Korea
wychung@pknu.ac.kr
[2] Department of Electronic Engineering, Pukyong National University,
Busan, South Korea
kimjj@pknu.ac.kr

Abstract. Photo-plethysmography (PPG) is among hot research topics today for developing wearable devices for human healthcare. In this paper, we introduce a wristband device using multi-wavelength PPG (mW-PPG) that permits us to use it for multi-purpose monitoring. By taking advantage of ultra-low-power analog front end technology, our design is minimized and saves energy. Through the gadget, we investigated three points along the radial artery at the wrist hand namely Cun, Guan, and Chi, and defined Guan point is the best place for collecting PPG with Near-infrared (NIR) wavelength. Also, using state-of-the-art blood pressures (BP) estimation from the reflective pulse time transition (R-PTT) method, we make a demonstration of estimating BP which employed a non-linear regression technique. Moreover, a phenomenon namely the invert-phase effect is discovered from mW-PPG techniques which can open a new research direction to detect illness on this.

Keywords: PPG · Multi-wavelenth · Blood pressure · Invert-phase effect · Wearable · Pulse transition time

1 Introduction

Vital signs are essential information that indicates health status and supports the doctor in detecting many problems in the patient. PPG could be employed for non-invasive monitoring vital signs such as body temperature, pulse rate, respiration, SpO2, BP, glucose concentration (GC), Heart rate variable (HRV), and other important health information [1]. Recently, the mW-PPG method, which combines visible and NIR lights to estimate BP and GC, has attracted more consideration for its versatility. Based on measuring shifting time between

This work was supported by an National Research Foundation grant of Korea Government (2019R1A2C1089139).

M. Singh et al. (Eds.): IHCI 2020, LNCS 12616, pp. 312–318, 2021.
https://doi.org/10.1007/978-3-030-68452-5_32

two pulses from different blood vessels [2] or reflective pulse transition time (R-PTT) between an incident and a reflective wave of blood flow [3], BP can be estimated. Likewise, NIR-spectroscopy at several bands in 750–2500 nm is used for calculating GC [4]. Our previous work, BP and GC has been investigated but mainly focused on discrete circuits [5,6]. In this paper, a wearable device with four-wavelength, three-site locating along the left-hand wrist radial artery is proposed. The reason for choosing a multi-sensor array is to improve BP and GC estimation accuracy through PTT. These selected locations are inspired by Chinese Traditional Medicine (TCM), where the doctor uses to diagnose the patient's health status and organ [7].

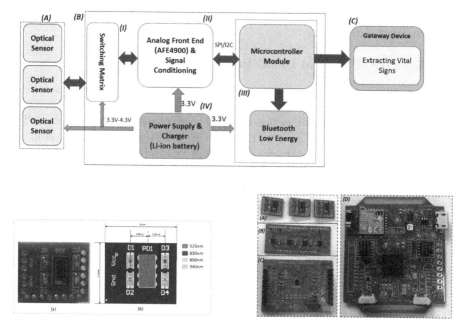

Fig. 1. Overview of the system. (a) Architecture of the system, (b) Sensor Shape, (c) Main parts of wearable device.

2 System Design

Design work includes two parts: creating a custom PPG-sensor array with four wavelengths and building the system with an analog front end (AFE), and data acquisition. The diagram for the system as can be seen in Fig. 1a. Because almost commercial sensors today often use three wavelengths, specifically green (530 nm), red (640 nm) and NIR (940 nm), to address BP and GC measurement, our sensor as in Fig. 1b takes one more wavelength (850 nm). All LED in 0402 SMT packages helps us design optimally around a photodetector (PD) VEMD8080. For interfacing the sensor, AFE AFE4900 is selected because it supports four LED phases and three multiplex-time PDs. However, to control

12 LEDs from 3 optical sensors, a switching matrix based on TS5A3399 with low on-state resistance is added. Next, MSP432P401R, a 32-bit ARM Cortex-M4F for interfacing to AFE is selected because it is an ideal wireless host MCU with ultra-low-power consumption 80 uA/MHz and has a well-supporting resource. In order to transfer data wirelessly, Bluetooth low energy (BLE) 4.2 protocol based on CC2650 core is preferred and configured a maximum throughput at 300 kbps. Finally, a charger and DC/DC converter circuit are adopted with TPS63031DSKT and TPS61251DSGT for DC output 3.3 V and 4.2 V respectively and BQ21040DBVT for recharging the battery with an output current up to 500 mA. All printed circuit boards with full components after soldering could be seen in Fig. 1c.

3   Methodology

3.1   Reflective PPG and Blood Pressure Estimation

First, transmitted light from the source will penetrate the human body, and then some portions are absorbed, scattering and reflected following Beer-Lambert Law. The optical path's depth depends on the wavelength of light and generally in the form of a banana [3]. Then, electric voltage signal sensing from PD through the trans-impedance circuit will include AC and DC components, as in Fig. 2a. The appearance of the pulsatile part is caused by changing of the blood volume in vessels periodically from contracting and relaxing of heart operation. The more blood flow comes in, the less intensity of reflected light is.

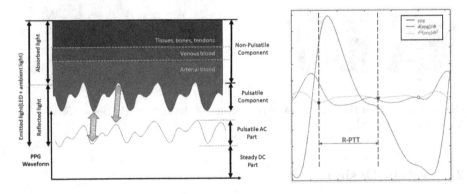

Fig. 2. (a) PPG waveform responses by reflected light intensity, (b) R-PTT based on max-peak of 2nd derivative of PPG signal

In this study, we demonstrate in single site mode at wrist to collect mW-PPG signals and estimate BP using model regression from work of Kao et al. [3] based on reflective PTT (R-PTT) as shown in Fig. 2b. For estimating SBP and DBP, Eq. (1) and Eq. (2), respectively.

$$SBP = DBP + K_a \cdot \frac{1}{R\text{-}PTT^2} \tag{1}$$

$$DBP = K_b + \frac{2}{0.031} ln(\frac{K_c}{R\text{-}PTT}) - \frac{1}{3}\frac{K_a}{R\text{-}PTT^2}. \tag{2}$$

With K_a, K_b, K_c are calibrated coefficients

3.2 Experiment

The device firstly is set up to measure at three points to define which kinds of wavelength is the clearest one and where is the best place to get the signal between three locations called Cun, Guan and Chi, according to TCM [8]. The device is configured to sample 100 Hz and send back to the host computer through Bluetooth. Then, three healthy young males without any cardiac history problems were invited to participate in the experiment. The process for mW-PPG data collection passed seven stages, as shown in Fig. 3 and reference BP points were obtained by Oscar 2 (SunTech Medical, USA). Within each step, $2^{nd}-7^{th}$, the measurement was repeated three times with a duration time of 2 min for the resting stage and 30 s for Valsalva Maneuver and took average after that. As a result, 7 BP points were archived, including 4 (2-min) and 3 (30-s) sections per subject for model training.

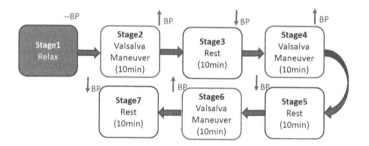

Fig. 3. Process of collecting data for training

4 Result

The result of signals at three sites is illustrated in Fig. 4. From our study, the Cun position gives the clearest signals for all wavelengths, as shown in Fig. 4a, while NIR wavelengths at Guan are superior in shape. All signals at Chi point have many mixed noise variations, and amplitude is the smallest compared to two other sites.

Table 1. Correlation between Cun, Guan and Chi points

Site	R-PTT ± STD	AC/DC ratio (%)			
		850 (nm)	630 (nm)	940 (nm)	530 (nm)
Cun	20.81 ± 0.04	0.31	0.36	0.21	5.53
Guan	20.54 ± 0.39	0.29	2.47	0.84	7.11
Chi	20.46 ± 0.37	0.14	0.71	0.16	17.04

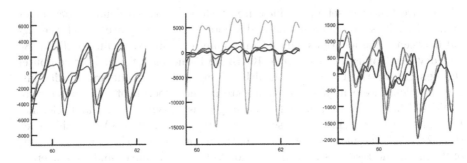

Fig. 4. Multi-wavelength PPG signals at Cun (a), Guan (b) and Chi (c) points; *Note: 850 nm (blue) - 630 nm (red) - 940 nm (yellow) - 530 nm (violet) (Color figure online)

Table 1 shows that greenlight gives the best of AC/DC ratio while the variation of R-PTT at Cun point is the smallest versus other sites and the average difference between R-PTTs is insignificant. Guan point gives the better shape in NIR wavelength permit us to select Guan position, and this wavelength for the next experiment to estimate blood pressure.

After collecting mW-PPG signals and reference points, model regression is applied from Eq. (2) for DBP and gets the parameters Ka, Kb, Kc are {99.8306, 80.7283, 16.6247} with a losses is 12.574. The estimation less accurate because the BP reference points are asynchronously obtained with mW-PPG. Figure 5a displays the result of the estimation after the trained model. Noticeably, during the process of collecting PPG data, an invert phase effect is discovered, which also announced in some studies on remote PPG imaging recently by Kamshilin et al. [9]. This result may be a hint for research to extract the ballistic signal for detecting health status using PPG. Figure 5b shows a typical invert effect from our experiment.

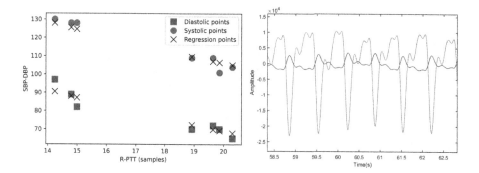

Fig. 5. (a) Reference points and estimation points (X) after training, (b) Invert phase effect between 940 nm vs. 850 nm PPGs.

5   Conclusion

In this paper, a wearable device using mW-PPG is introduced with a demo to measure BP is illustrated. We confirm that NIR wavelength lights at 850 nm and 940 nm are superior to visible light in the case of stationary condition experiment. Moreover, we discovered that mW-PPG signals at the wrist have an invert phase effect, which also is informed by some researches on remote PPG imaging recently. Lastly, we define that the Guan site is the best place to get an mW-PPG signal, which matching with the pilot research from Kim et al. [8]. Because we aim to make a device to monitor vital signs, the next direction in our research is to estimate GC and apply modern machine learning algorithms to cancel calibration and hand-features engineering, and our results are promised definitely to be improved.

References

1. Castaneda, D., Esparza, A., Ghamari, M., Soltanpur, C., Nazeran, H.: A review on wearable photoplethysmography sensors and their potential future applications in health care. Int. J. Biosens. Bioelectron. **4**(4), 195 (2018)
2. Scardulla, F., et al.: A novel multi-wavelength procedure for blood pressure estimation using opto-physiological sensor at peripheral arteries and capillaries. In: International Society for Optics and Photonics, Design and Quality for Biomedical Technologies XI, vol. 10486, p. 1048614 (2018)
3. Kao, Y., Paul, C.-P.C., Chin-Long, W.: Design and validation of a new PPG module to acquire high-quality physiological signals for high-accuracy biomedical sensing. IEEE J. Sel. Topics Quant. Electron. **25**(1), 1–10 (2018)
4. Yadav, J., Asha, R., Vijander, S., Bhaskar, M.M.: Prospects and limitations of non-invasive blood glucose monitoring using near-infrared spectroscopy. Biomed. Signal Process. Control **18**, 214–227 (2015)
5. Rachim, V.P., Toan Huu, H., Wan-Young, C.: Wrist photo-plethysmography and bio-impedance sensor for cuff-less blood pressure monitoring. In: 2018 IEEE Sensors, pp. 1–4. IEEE (2018)

6. Rachim, V.P., Chung, W.Y.: Wearable-band type visible-near infrared optical biosensor for non-invasive blood glucose monitoring. Sensors and Act. B: Chem. **286**, 173–180 (2018)

7. Tang, A.C.Y.: Review of traditional Chinese medicine pulse diagnosis quantification. Complementary therapies for the contemporary healthcare, p. 61 (2012)

8. Kim, J.U., Lee, Y.J., Lee, J., Kim, J.Y.: Differences in the properties of the radial artery between cun, guan, chi, and nearby segments using ultrasonographic imaging: a pilot study on arterial depth, diameter, and blood flow. Evid.-Based Complement. Altern. Med. (2015)

9. Kamshilin, A.A., et al.: A new look at the essence of the imaging photoplethysmography. Sci. Rep. **5**, 10494 (2015)

User Perception on an Artificial Intelligence Counseling App

Hyunjong Joo[1]([✉]), Chiwon Lee[2], Mingeon Kim[3], and Yeonsoo Choi[3]

[1] Hyundai Motor Securities, Seoul, South Korea
jamesjoo5764@gmail.com
[2] Yonsei University, Seoul, South Korea
[3] Cogito, Seoul, South Korea

Abstract. Mental health is a serious issue amongst college students in the Republic of Korea. The stigma related to mental health issues renders college students to be reluctant when it comes to seeking help regarding it. Moreover, college students often seek alternative means to rely on rather than in-person counseling sessions due to shortage of time, money, and lack of access to resources. Existing research has revealed that the utilization of technology can be conducive when it comes to alleviating mental health concerns. The development of a mobile application that provides its users with a counseling session via Artificial Intelligence (AI) could be a helpful resource for college students as it could be accessed regardless of the time and location of the students that wish to access the service. This research aims to conduct user research on college students to learn the stigma related to mental health, and to derive implications from the user research to develop a counseling app prototype to serve as a guideline for the development of a counseling app for college students in the Republic of Korea that could potentially be widely used.

Keywords: Artificial Intelligence · Counseling · Mental health · Multimedia human communication · Intelligent mobile interfaces

1 Introduction

Despite substantial attention to the problem of suicide amongst college students over the past several decades, suicide remains as a leading cause of death amongst college and university students in a numerous OECD countries, such as the Republic of Korea and the United States [1, 2]. In particular, according to data from the Korea Suicide Prevention Center, from 2011 to 2018, the major cause of death at juveniles in the Republic of Korea is death due to deliberate self-harm caused by mental health issues such as depression [3]. 2,017 youth committed suicide in 2018 alone.

College administrations have acknowledged the struggle of college students regarding this matter and have deployed counseling centers and mental health professionals within college campuses; nonetheless, a majority of students have not been able to reap the benefits of this system. Unfortunately, students distressed by these illnesses often do not seek out services or choose to fully engage in them. One factor that impedes care seeking and undermines the service system is stigma related to mental health issues [4].

© Springer Nature Switzerland AG 2021
M. Singh et al. (Eds.): IHCI 2020, LNCS 12616, pp. 319–325, 2021.
https://doi.org/10.1007/978-3-030-68452-5_33

This study was conducted to develop a prototype that could serve as a guideline for the creation of a mobile environment in which students can receive more help if there is a case where it is difficult for students to benefit from the services provided by the institution for any reason.

2 Previous Work and Existing Studies

2.1 Existing Studies of Suicide Prevention

The World Health Organization (WHO) reported that national suicide prevention strategies has brought numerous benefits, such as public recognition of mental health as a necessity, provision of guidance from authorities with evidence-based methodology, and identifying the significant issues with existing legislation, data collection, and service provision. It also recommends that governments should implement such strategies in a cultural and historical context; effective solutions for those who go under mental health problems should not be universal, as such problems are subjective [5]. Barriers such as scarcity of specialized health workers, guardians' reluctance to discuss mental health issues, and inadequate data collection hinders individuals from receiving psychological assistance. The WHO further suggests that to overcome these issues, governments should train and assess health workers with rigorous criterion, work closely with teachers and educational institutes, and strengthen surveillance systems for suicide attempts [6]. On the other hand, mental health workers of the Ministry of Health and Welfare of the Republic of Korea reported their occupation as "the sewer of emotions" and "the front most line of cannon folders," reflecting their stressful work conditions [7]. Additionally, installing surveillance systems are imperfect as despite the current systems of Korea, suicide rates per 100,000 population is 20.2%, which is twice the global average of 10.2% as reported by the WHO [8].

2.2 Existing Products for Suicide Prevention

Following up with the technology being feasible to get integrated with the mobile phone application for suicide prevention, there already exist about 27 applications related to it. However, there are only a limited number of services available to adolescents or young adults who are attending educational institutions [9]. For instance, in the Republic of Korea, the application called "Trost" was invented in 2015 with a goal of lowering the bar that exists as a stigma towards psychological treatment. With the record of being downloaded 200,000 times and having approximately 70% of them as the official user, it proves to be the one of the most influential mobile applications in Korea in the department of healthcare [10]. However, its target group is the general public who is seeking psychological assistance which can seem broad. According to the 2019 Statistics on the Death Cause published by Statistics Korea, it is shown that in the age group of 20–29, self-harm is listed as the most prominent factor comprising 51% of mortality causes [11]. Therefore, the mental issues which often lead individuals to render radical decisions can be better tackled with the application that specializes at empathizing with the college or university students.

3   Methods

This study investigates users' demand and perception of psychological comfort via two methods: the implementation of a survey and a focus group interview (FGI). In order to comprehend the pain points and set our design goals which users have, we gathered both quantitative and qualitative data through a survey, and a FGI.

3.1 Survey

An online survey was conducted on 124 college students in the Republic of Korea to learn more about their experiences and perceptions of utilizing counseling sessions. With the goal of alleviating mental health concerns along with perceptions, the survey mainly focused on whether the students would be willing to utilize a mobile application for counseling.

3.2 Focus Group Interview

A FGI session was carried out on August 9th, 2020 to observe prospective user's viewpoints on the design goals and the corresponding functional and nonfunctional requirements at a more accurate level. There were a total of 7 participants recruited to participate in the FGI; they were 3 males and 4 females in their 20s. All participants attend a college in Seoul, the Republic of Korea, and students of diverse grade levels and majors were recruited.

4   Results

The 7 participants shared the sentiment that the main reason a student would not go to therapy nor seek psychological counseling is the inaccessibility of services for the average college student in the Republic of Korea. The problem of inaccessibility can be further specified into: the social stigma attached, the constraint of time, the cost of therapy sessions, and the lack of awareness of where these services are located or how these services are to be reached.

4.1 Difficult to Utilize School Counseling Services Because of Stigma: Survey

According to 124 responses gathered from the survey, more than 77% student participants were aware of the college's counseling center. This indicates that the students acknowledged that they could attain help from the institution if they needed psychological assistance. However, when asked about the intention to use the college provided counseling service, more than 90% of students answered they would not use the counseling service provided by the college. Most of the students (76%) stated that they are afraid of people around them finding out that they used therapy.

"I went to the counseling office for an appointment, and there I met a student that I kind of know. I was immediately scared that the person might spread rumors. It was a very uncomfortable memory and I never want to feel it again." P4.

"I am worried that my counseling records and usage details will remain at the school and negatively affect me in the future." P1.

Students [P1, P2, P4, P5, P6] were reluctant to use the school's counseling program because of stigma. They also voiced concerns over leaving a counseling record as they stated that the societal perception is still generally negative towards those who receive counseling.

4.2   Restraint of Time as a Factor of Discouraged Counseling Usage

Among the students who participated in the survey, 62% answered that they would not use college counseling services because they find it difficult sparing time to register for counseling services. Participants [P2, P3, P6] stated that college counseling service centers are usually booked, and that it is difficult to carve out time to allocate to counseling services.

"If it is not a service that is available at night, I cannot make time for counseling. There is no time for counseling at the end of the school day due to school hours, time for group work and assignments, and part-time jobs." P2.

"External counseling agencies take a long time and are expensive. For example, 20 sessions of 1 h each Wednesday night is not that feasible as I have exams, immediate assignments, and a part-time job." P3.

"I wanted to get counseling at night, but most of the counseling services for students in the college were in office hours, where all the students and faculty were around the campus." P6.

Participants thought that the counseling services provided by the college did not meet their own time needs. Particularly, during interviews, responses there were repeating comments on the need for evening counseling sessions which are not provided by the colleges.

4.3   Desire for Calm and Solitary Experience

While providing distinct insights to the research topic, one of the counseling elements that participants [P1, P2, P7] concurred was that they would prefer a serene counseling session conducted by a counseling entity that is not judgmental.

P1 stated that receiving service at the counseling center requires in-person meeting and time arrangement, which is often challenging for students. Therefore, friends and acquaintances at college often resort to negative coping methods such as smoking or drinking. Since they could resort to such methods at a time of their convenience, which is in-between classes and commitments and at night, students would further avoid visiting the counseling center.

Participants [P2, P7] further expressed the preference for a counseling service mobile application as it would not make them appear as if they are receiving counseling which allows them to utilize the service regardless of the situation that they are in.

"It would be nice if I could quietly concentrate on myself at the end of the day." P2.

"In case someone looks over at my screen, I want it to appear that I'm just using a cool and interesting app rather than receiving counseling because I am having a difficult time." P7.

5 Implications

Via the research, a prototype was developed that would be apt to fulfill student needs regarding counseling. The design was developed based on the results from the user research conducted. The "Moon" theme was derived based on the FGI results as keywords such as "quietness," "calm," "solitary," "light," "darkness," "night," "the time they preferred," "the time they are able to use," and more. The moon phase has also been the key concept of Microsoft's social media facilitated tool for emotional reflection and wellness: Moon Phrases [12] (Figs. 1 and 2).

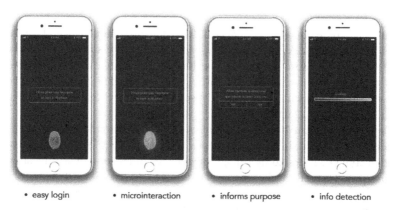

• easy login   • microinteraction   • informs purpose   • info detection

Fig. 1. Prototype screen for a secure login and user consent collection

6 Conclusion

The conducted survey and FGI revealed that college students in the Republic of Korea have mental health concerns that they find it difficult to alleviate due to the lack of access

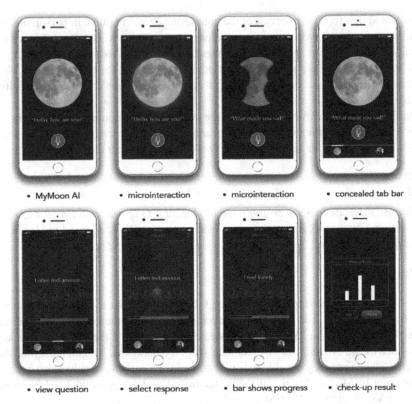

- MyMoon AI   • microinteraction   • microinteraction   • concealed tab bar

- view question   • select response   • bar shows progress   • check-up result

Fig. 2. Prototype for counseling and check-up post-counseling

to resources and time or monetary concerns. Research also figured that college students are often reluctant to seek professional in-person counseling services because of the social stigma against those with mental health issues. Regardless of a general understanding that such social stigma exists, participants were highly interested in receiving counseling for wellness purposes if it could resolve the concerns that they have regarding conventional counseling methods.

We were able to learn that a mobile application that is powered by Artificial Intelligence could be conducive in mitigating mental health concerns of students since such an application could be accessed by college students regardless of time and location. The application prototype reflects the interest of the college students in the Republic of Korea regarding psychological wellness. The application would foment the sense of security, create an environment where the user would be comfortable to discuss concerns, and become an interface that does not stigmatize the user.

We invite further research to be done in the area as mental health amongst college students is an issue that should be addressed. There are high hopes that further research in this field could enable more college students to obtain access to resources that are needed for them to achieve mental well-being.

Acknowledgement. We would like to thank our research participants for their generous contribution of time and insight that was utilized for this research.

References

1. Corrigan, P.W., Druss, B.G., Perlick, D.A.: The impact of mental illness stigma on seeking and participating in mental health care. Psychol. Sci. Public Interest **15**(2), 37–70 (2014). https://doi.org/10.1177/1529100614531398
2. De Choudhury, M., Gamon, M., Hoff, A., Roseway, A.: Moon Phrases: A Social Media Facilitated Tool for Emotional Reflection and Wellness (2019). https://www.microsoft.com/en-us/research/publication/moon-phrases-a-social-media-facilitated-tool-for-emotional-reflection-and-wellness. Accessed 27 Sept 2020
3. Human Company Co., Ltd. Trost. https://trost.co.kr/. Accessed 29 Oct 2020
4. Kennard, B.D.: Developing a brief suicide prevention intervention and mobile phone application: a qualitative report. J. Technol. Hum. Serv. **33**(4), 345–357 (2016)
5. Korea Suicide Prevention Center: Suicide state in the past decade (2019). https://spckorea-stat.or.kr/korea01.do. Accessed 15 Sept 2020
6. Kim, O.: Why do public officials at the ministry of health and welfare self-educate themselves (2020). https://news.khan.co.kr/kh_news/khan_art_view.html?artid=202006201530001. Accessed 30 Oct 2020
7. Statistics Korea: 2019 Statistics on the Death Cause (2020). https://kostat.go.kr/assist/synap/preview/skin/doc.html?fn=synapview385219_2&rs=/assist/synap/preview. Accessed 30 Oct 2020
8. Suicide Prevention Resource Center: Suicide among College and University Students in the United States. Education Development Center, Inc. (2014)
9. The Korea Herald: Suicide No. 1 Cause of Death for S. Korean Teens, Youths, May 1, 2019 (2019). https://www.koreaherald.com/view.php?ud=20190501000216
10. World Health Organization: National Suicide Prevention Stratgies (2019)
11. World Health Organization: Suicide in the world (2019)

Authentication of Facial Images with Masks Using Periocular Biometrics

Na Yeon Han[1] , Si Won Seong[1] , Jihye Ryu[1] , Hyeonsang Hwang[2] ,
Jinoo Joung[3] , Jeeghang Lee[3(✉)] , and Eui Chul Lee[3(✉)]

[1] Department of AI and Informatics, Graduate School, Sangmyung University,
Hongjimun 2-Gil 20, Jongno-Gu, Seoul 03016, Republic of Korea
[2] Department of Computer Science, Graduate School, Sangmyung University,
Hongjimun 2-Gil 20, Jongno-gu, Seoul 03016, Republic of Korea
[3] Department of Human-Centered AI, Sangmyung University, Hongjimun 2-Gil 20,
Jongno-gu, Seoul 03016, Republic of Korea
{jeehang,eclee}@smu.ac.kr

Abstract. Due to COVID-19 pandemic, wearing a mask rapidly becomes a new social norm that people in the society should comply. Although it is altruistic behavior preventing all people from serious infections, it brings about hassles for individuals. For example, when a face is partly covered with a mask, identification using a face recognition is going to be malfunctioning. To this end, we propose a novel computational framework that enables the personal authentication with a partial face image, a face covered with a mask. For the experiments, we constructed the datasets of facial images containing the periocular regions only, extracted from full facial images covered with the mask. Given the datasets, we trained our framework, a variant of a Siamese network, with various configuration of hyperparameters. As a result, RMSprop optimizer with the learning rate 1×10^{-5} trained from periocular datasets showed the highest accuracy for the personal authentication. Next, we conducted a comparative experiment with our proposal and the model trained with datasets containing the full facial regions. When testing with the periocular region images, our proposal is superior in the authentication accuracy to that of the model trained with the full facial regions. This result raises the optimistic expectation that in the era of COVID-19, facial images covered with mask can still be used for the authentication using face recognition at a nearly same level of accuracy. This means that people can use the face recognition applications without taking off the mask, which provides the safe circumstances against the infections.

Keywords: Periocular biometric · Siamese network · Face biometric · Masked face recognition

1 Introduction

COVID-19 pandemic across the entire world brings about a steep increase in the use of facial masks. To prevent people from infections, wearing a mask rapidly becomes

---

N. Y. Han and S. W. Seong—These authors contributed equally to this work.

M. Singh et al. (Eds.): IHCI 2020, LNCS 12616, pp. 326–334, 2021.
https://doi.org/10.1007/978-3-030-68452-5_34

a normative behavior everywhere in public places. Despite this altruistic behavior can make sure all the people's safety, it also causes some drawbacks. For example, most smart phones have a biometric authentication method using a face recognition technique. It is likely to fail when using this application with a face partially covered with the mask. To be successful on the function, people would take off the mask every time when using smart phone. This significantly diminished the user experience – not only is it inconvenient, but also is there the high risk of infections.

Periocular region is a small region around the eye and may include eyebrows. However, the periocular region is not defined specifically, it is defined slightly different by researchers and papers. In some papers, an enlarged image of one eye is called a periocular region and in some other papers, an area including both eyes is called a periocular region. There is a research presented a standalone modular biometric system based on periocular information to authenticate towards device. Researchers in the study applied three well known feature extraction techniques, SIFT, SURF and BSIF independently in the proposed periocular based authentication system. It indicated the applicability of the proposed periocular based mobile authentication system in a real-life scenario [1]. However, in recent years, most of these methods are replaced by convolutional neural networks [2]. When extracting features, CNN is well known to be more advantageous than other conventional methods.

There are some benefits of using a periocular region than using a full face when the face including mouth and nose is occluded with objects. Researchers in the study said that the use of periocular reduces the discomfort to the scanner and is more stable to the variation of the face than the iris. A classification algorithm was used for the image set with various modulations for one person's periocular to compare the recognition accuracy for various factors. Experiment results showed the accuracy of periocular areas with eyebrows was higher than the case of without eyebrows. In addition, the local matching method extracts and matches local characteristics of photos. It is more robust to occlusion or deformation than the global matching method, which matches all pixels in the photo area with feature values [3]. Periocular region is less sensitive to facial expressions, age, beards, pose changes than face recognition biometrics using full faces. Thus, it would be a good candidate for personal authentication having the genuine information. If the eye area is not covered, face recognition is possible even if the lower part of the chin or face is covered [4].

In this study, we propose a novel computational framework that enables personal authentication with a partial face image, a face covered with a mask. To this end, we will crop the regions of both eyes to use as ROI (Region of Interest) images instead of full face images used in general face recognition systems. In addition, we suggest a method to utilize periocular recognition as an alternative to existing facial recognition systems by comparing the accuracy of entire face and periocular authentication.

2 Proposed Method

2.1 Datasets

We first collected public databases on face recognition composed of unmasked face images for the purpose of model training. With these, we constructed two datasets: one is the identical to the set of original unmasked face images, and the other is the set of masked face images. In total, five different databases were used. RFW (11430 images) [5], Face Spoofing DB (272 images) [6], and IAS-Lab RGB-D Face (312 images) [7] are color images, AT&T (400 images) [8] and Cas-Peal pose (15600 images) [9] are gray scale images. Also, color images were converted to gray scale images.

We also collected masked face images in order to test model performance. Masked face dataset consists of 300 images (50 people, 6 per each). We considered two scenarios subject to the way people are wearing masks: (i) entirely wearing masks that covered the entire nose, and (ii) partly wearing masks that opened the nose tip and covered elsewhere. Also, we took pictures for each person from above, front, and low in each scenario.

2.2   Face Image Pre-processing

Through the pre-processing of the collected data, both periocular region and entire face image were obtained. Pre-processing consists of three steps: (i) to detect facial landmarks, (ii) to get ROI images, and (iii) to resize.

Fig. 1. Regions of interest [11].

First, facial landmarks were detected using Adrian's method [10]. Six landmarks out of the total 68 were used to capture the periocular region, an area around the eyes. These six landmarks were used to crop the ROI, a periocular region, in the second stage.

In Park's study, experiments were done to investigate the use of periocular region for biometric recognition with GO, LBP, and SIFT local feature extractors. Also, various combinations of datasets were used to study how much eyebrow, facial expressions, occlusion, iris, and other factors affect accuracy. Experiments indicated that datasets with eyebrows showed 10.66% better performance than datasets without eyebrows in rank-one accuracy [3].

After extracting the facial landmarks, Full, Region 1, Region 2, and Region 3 were successively cropped as shown in Fig. 1. The entire face, the area between the eyes and the mouth, the area between the eyes and the nose, and the eye area were cropped and set as Full, Region 1, Region 2, and Region 3, respectively [11]. In this study, Region 2, which includes both eyes, eyebrows, and nose, was chosen for a periocular region.

Afterwards, the area around the eyes, a periocular region, was cropped and resized to 91 × 64 size. The same pre-processing was carried out in order to obtain the entire face images as obtaining periocular images. Four of the 68 facial landmarks were used to crop the entire face images and resized to 120 × 120 size (see Fig. 2).

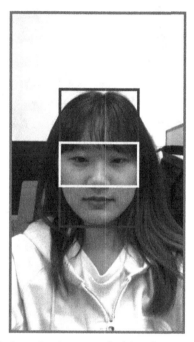

Fig. 2. Terminology description related to unmasked face image pre-processing. (red: original image, blue: unmasked face image, yellow: unmasked periocular image) (Color figure online)

Then the pre-processed images were divided by a ratio of 14:3:3 for train, validation, and test data. Train data contains 923 classes (persons), validation and test data each contains 198 classes (persons). Each class contains 2–21 images.

The masked dataset constructed in this study followed the same pre-processing procedures to obtain masked face images and masked periocular images then used as test data (see Fig. 3). After pre-processing procedures, the biometric authentication model goes through training and other stages shown in Fig. 4 to determine if two input images are the same person.

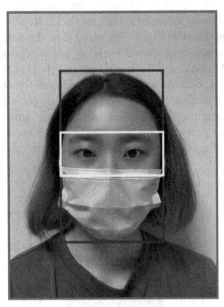

Fig. 3. Terminology description related to masked facial image pre-processing. (red: original image, blue: masked face image, yellow: masked periocular image) (Color figure online)

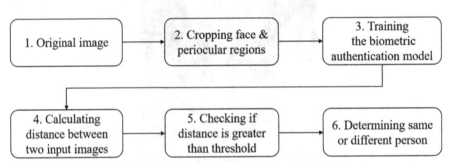

Fig. 4. The whole process of the proposed method

2.3 Algorithms

Siamese network showed an excellent performance in face recognition among deep learning networks [12]. The network takes two images as input, vectorizes them through convolution, and then calculates the distance between the two vectors. Additionally, we use the concept of one-shot learning for efficient learning about classes with small data. This is because in a class containing only two data, only one pair is available as an input to the Siamese network.

$$L\left(W, Y, \overrightarrow{X_1}, \overrightarrow{X_2}\right) = (1 - Y)\frac{1}{2}(D_W)^2 + (Y)\frac{1}{2}\{max(0, m - D_W)\}^2 \qquad (1)$$

The model in this study used a contrastive loss function as Eq. (1) because the goal is embedding, not classification. W is weight, Y is label, $\overrightarrow{X_1}$ and $\overrightarrow{X_2}$ are images respectively,

D_W is the Euclidean distance between each feature vector for two images. If two images $\overrightarrow{X_1}$ and $\overrightarrow{X_2}$ are in the same class (genuine pair), $Y = 0$ and if they are different (imposter pair), $Y = 1$. And the margin denoted by m is set to 2.0.

2.4 Model Training

Datasets containing both color and gray scale images were preprocessed as described in Sect. 2.2, and were resized in (105, 105, 1) to train the model afterwards. Siamese network extracts feature vectors from the two input images through convolutions. The images are reduced to 100×1 size through a fully connected layer after training. The following Table 1 gives a summary of network architecture. After extracting and shrinking the dimensions of input images, the model updates weights to have the least contrastive loss. In the training process, we used contrastive loss function and Adam, RMSprop optimizers.

Table 1. Network architecture

Layer	Type	Maps	Size	Kernel size	Stride	Padding	Activation
In	Input	1	105×105	–	–	–	–
C1	Convolution	64	96×96	10×10	1	0	ReLU
S2	Max Pooling	64	48×48	–	2	–	–
C3	Convolution	128	42×42	7×7	1	0	ReLU
S4	Max Pooling	128	21×21	–	2	–	–
C5	Convolution	128	18×18	4×4	1	0	ReLU
S6	Max Pooling	128	9×9	–	2	–	–
C7	Convolution	256	6×6	4×4	1	0	ReLU
F8	Fully Connected	-	4096	–	–	–	ReLU
F9	Fully Connected	-	1000	–	–	–	ReLU
Out	Fully Connected	-	100	–	–	–	–

3 Experiment Results

After training the model, the result of prediction was determined by threshold based on the distance. If the distance between the two images is smaller than the threshold, the pair is determined as the same person and vice versa. There are two ways to determine the threshold. One is to make the accuracy-score higher, and the other is to make the f1-score higher. The accuracy-score method is a good metric when the distribution of class is similar. However, using only accuracy-score method can make measurement results biased depending on the configuration of the test data. On the other hand, the f1-score method is a better metric when the distribution of class is imbalanced.

In experiments, we adjusted the number of genuine matching pairs and the number of imposter matching pairs similarly then set the threshold to have the highest f1-score. After setting the threshold, we measured the performance of the model using the accuracy-score method, instead of f1-score method.

After training the Siamese network with unmasked images, we measured the performance of the model using both datasets with unmasked images and masked images. The following Table 2 gives a summary of test accuracy of Siamese network for both cases.

Table 2. Test accuracy of Siamese network with masked face and masked periocular images

Dataset	Siamese-face	Siamese-periocular
Unmasked images	0.930	0.923
Masked images	0.634	0.663

Siamese-face is a model trained using the unmasked face images, and Siamese-periocular is a model trained using only unmasked periocular images. In the case of not wearing a mask, the model trained with unmasked face images (accuracy = 0.930) showed 0.007 higher performance than the model trained with unmasked periocular images. Conversely, in the case of wearing a mask, the model trained using only unmasked periocular images (accuracy = 0.663) showed 0.029 higher accuracy than the model trained with unmasked face images. The result shows that the model trained with the unmasked face images is slightly better when people are not wearing masks, but the model trained with unmasked periocular images is superior to otherwise when people are wearing masks.

There are ROC (receiver operating characteristic) curves of testing Siamese-face and Siamese-periocular models with masked face images and masked periocular images respectively (see Fig. 5). The horizontal axis means False Positive Rate and the vertical axis means True Positive Rate. The AUC of Siamese-face (green) is 0.6936, and the AUC of Siamese-periocular (orange) is 0.7378. When testing with the masked images, the model trained with periocular images showed higher accuracy than the model trained with face images. (True = genuine matching, False = imposter matching).

We also calculated the ratio of a mask covering faces in each image. After analyzing masked face dataset constructed in this study, the ratio of areas covered by masks was higher in masked face images than in masked periocular images. Mean and standard deviation of masked face images is 0.200, 0.086 and those of masked periocular images is 0.086, 0.088. Based on the results, wide ROI was not as advantageous as general cases for extracting features. It can be assumed that the ratio of the area not covered by masks is an important factor for face recognition performance.

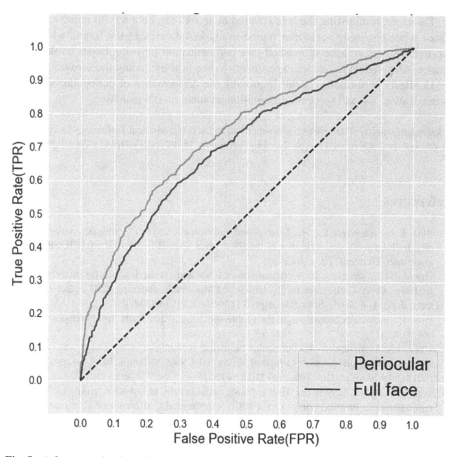

Fig. 5. A figure caption is ROC curve of Siamese network with masked face images (green) and masked periocular images (orange) (Color figure online).

4 Conclusion

In this study, we compared the biometric authentication performance of the entire face and periocular regions. Siamese network model trained using RMSprop optimizer with the learning rate 1×10^{-5} showed the highest accuracy. When testing with unmasked images, the model trained with unmasked face images showed higher performance (accuracy = 0.930) than the model trained with unmasked periocular images (accuracy = 0.923). However, when using masked images, the model trained with unmasked periocular images (accuracy = 0.663) showed higher performance than the model trained with unmasked face images (accuracy = 0.634). The accuracy of both models trained using unmasked face images and unmasked periocular images dropped when testing with masked images. Comparing the test performance of the unmasked images and masked images, the performance degradation of Siamese-face (0.296) was greater than that of Siamese-periocular (0.260).

Therefore, minimizing the area covered by masks can be a way to improve performance when obtaining periocular region in masked dataset. In the future, we plan to improve the performance of masked face authentication by using larger masked face dataset. Also, we will conduct experiments to see how excluding the area covered by masks affects performance. After improving the performance of periocular biometric authentication, we will solve the recognition problem of 1:N matching.

Acknowledgement. This research was supported by the Bio & Medical Technology Development Program of the NRF funded by the Korean government, MSIT (NRF-2016M3A9E1915855).

References

1. Raja, K.B., Raghavendra, R.: Smartphone authentication system using periocular biometrics. In: International Conference of the Biometrics Special Interest Group (BIOSIG) 2014, Darmstadt, Germany, pp. 1–8 (2014)
2. Özaydın, U., Georgiou, T.: A comparison of CNN and classic features for image retrieval. In: International Conference on Content-Based Multimedia Indexing (CBMI) 2019, Dublin, Ireland, pp. 1–4 (2019).https://doi.org/10.1109/cbmi.2019.8877470
3. Park, U.: Periocular biometrics in the visible spectrum. IEEE Trans. Inf. Forensics Secur. **6**, 96–106 (2011)
4. Uzair, M., Muhammad, A.: Periocular biometric recognition using image sets. In: IEEE Workshop on Applications of Computer Vision 2013, WACV, Tampa, FL, USA, pp. 246–251 (2013).https://doi.org/10.1109/wacv.2013.6475025
5. Wang, M.: Racial faces in the wild: reducing racial bias by information maximization adaptation network. In: Proceedings of the IEEE International Conference on Computer Vision, Seoul, Korea, pp. 692–702 (2019)
6. Bok, J.Y.: Verifying the effectiveness of new face spoofing DB with capture angle and distance. Electronics **9**(4), 661 (2020)
7. IAS-Lab RGB-D Face dataset. https://robotics.dei.unipd.it/reid/index.php/8-dataset/9-overview-face. Accessed 8 Sept 2020
8. AT&T dataset. https://git-disl.github.io/GTDLBench/datasets/att_face_dataset/. Accessed 8 Sept 2020
9. Cas-Peal face data base. https://www.jdl.ac.cn/peal/index.html. Accessed 8 Sept 2020
10. Bulat, A.-G.: Super-fan: integrated facial land-mark localization and super-resolution of real-world low resolution faces in arbitrary poses with GANs. In: Proceedings of the IEEE Conference on Computer Vision and Pattern Recognition, pp. 109–117. The University of Nottingham, United Kingdom, Salt Lake City, Utah, USA (2018)
11. Merkow, J., Jou, B.: An exploration of gender identification using only the periocular region. In: IEEE International Conference on Biometrics: Theory, Applications and Systems (BTAS) 2010, Washington, DC, USA, pp. 1–5 (2010). https://doi.org/10.1109/btas.2010.5634509
12. Koch, G., Zemel, R.: Siamese neural networks for one-shot image recognition. In: ICML Deep Learning Workshop 2015, vol. 2 (2015)

Analysis of User Preference of AR Head-Up Display Using Attrakdiff

Young Jin Kim and Hoon Sik Yoo[✉]

Seoul Media Institute of Technology, Seoul, South Korea
dogzone276@gmail.com, hsyoo@smit.ac.kr

Abstract. The head-up display, a necessity for military and civil aircraft, is growing rapidly with the emergence of a variety of new markets such as automobile, sports, and infotainment. The global automotive head-up display market is expected to grow from $3.71 billion in 2020 to an annual average growth rate of 28.54%, reaching $13.02 billion and 15.6 million units in 2025. And AR HUD (augmented reality-based head-up display) has been mainly developed in the head-up display industry. The purpose of this study is to define what elements should be strategically utilized for the future car AR head-up display development by using Attrakdiff in automotive head-up display design to analyze user feedback according to the graphic style of the head-up display. To conduct the study, we selected two graphic styles of 2.5D, AR head-up display, and surveyed 500 people (250 males, 250 females) using Attrakdiff. As a result, we were able to define what elements should be strategically utilized for the future car AR head-up display development as an early stage study.

Keywords: Head-up display · AR head-up display · User experience design · Graphical user interface · Attrakdiff

1 Research Background and Purpose

The head-up display, a necessity for military and civil aircraft, is growing rapidly with the emergence of a variety of new markets such as automobile, sports, and infotainment. The global automotive head-up display market is expected to grow from $3.71 billion in 2020 to an annual average growth rate of 28.54%, reaching $13.02 billion and 15.6 million units in 2025. Among them, AR head-up display is expected to increase from $2.123 billion in 2020 to an annual average growth rate of 30.30%, and is expected to reach $7.97 billion in 2025. In addition, the adoption rate of electric vehicles is expected to account for 16% of the total HUD market by 2025, and the introduction of HUD in addition to ADAS (Advanced Driver Assistance System) and connected technologies are expected to be implemented in electric vehicles which are integrated with various advanced technologies [1].

© Springer Nature Switzerland AG 2021
M. Singh et al. (Eds.): IHCI 2020, LNCS 12616, pp. 335–345, 2021.
https://doi.org/10.1007/978-3-030-68452-5_35

Fig. 1. Volkswagen ID.3 AR Head up display (Left), Mercedes-Benz 7th generation S-class AR Head up display (Right).

Recently, AR HUD (augmented reality-based head-up display) has been mainly developed in the head-up display industry. Volkswagen adopted augmented reality (AR)-based head-up display (HUD) developed by LG Electronics for 2019's new electric vehicle ID.3. As the first case in the automotive industry to adopt AR-based HUD for mass-production vehicles, LG Electronics achieved commercializing the world's first AR HUD [2]. And the AR-based head-up display mounted on the 7th generation S-Class (W223) that Mercedes-Benz will release in the near future, guides the navigation route along the actual road, and it is capable of warning the user of dangerous situations on the road. According to Mercedes-Benz, the experience is like viewing a 77-inch monitor from 10 m away [3] (Fig. 1).

Research has been conducted on AR head-up display cognitive response [4], element design of displaying information [5], whether visual demand is reduced [6], preference area of displaying information [7], and driver's viewpoint tracking technology [8] etc. Prior study showed the necessity to conduct the study from user experience perspective regarding the difference between the AR head-up display and the currently used 2.5D head-up display graphic style type. Therefore, in this study, in designing a car head-up display, the user's feedback according to the graphic style of the head-up display is analyzed using Attrakdiff to define what elements should be strategically utilized for the future car AR head-up display development.

2   User Research

To analyze user preferences of two types of car head-up display (2.5D, AR) graphic style, as shown in Table 1, 500 Koreans (250 males and 250 females) from 20s to 60s were surveyed. A survey was conducted to determine user preference regarding the two type of vehicular head-up display graphic styles (Fig. 2). The survey was conducted to measure user preference based on 28 factors in the following four categories: PQ (pragmatic quality) and HQ-I (hedonic quality-identity), HQ-S (hedonic quality-Stimulation), and

ATT (attractiveness). The result of the survey was analyzed, using Attrakdiff's 7-point scales (-3: negative words to 3: positive words) for 28 factors. The question items used in the survey are shown in Table 2 below.

Table 1. Experimental participant demographic information.

Item		Number	Ratio (%)
Gender	Male	250	50
	Female	250	50
	Total	500	100
Age	20s	100	20
	30s	100	20
	40s	100	20
	50s	100	20
	60s	100	20
	Total	500	100

Table 2. Attrakdiff usability evaluation survey questions.

Category	Pragmatic Quality (PQ)	Hedonic Quality-Identity (HQ-I)	Hedonic Quality-Stimulation (HQ-S)	Attractiveness (ATT)
Question	Technical - human	Isolating - connective	Conventional - inventive	Unpleasant - pleasant
	Complicated - simple	Unprofessional - professional	Unimaginative - creative	Ugly - attractive
	Impractical - practical	Tacky - stylish	Cautious - bold	Disagreeable - likeable
	Cumbersome - straightforward	Cheap - premium	Conservative - innovative	Rejecting - inviting
	Unpredictable - predictable	Alienating - integrating	Dull - captivating	Bad - good
	Confusing - clearly structured	Separates me - Brings me closer	Undemanding - challenging	Repelling – appealing
	Unruly - manageable	Unpresentable - presentable	Ordinary - novel	Discouraging - motivating

Fig. 2. 2.5D Head up display (Left), AR Head up display (Right).

3   Results

Table 3 shows the results of analyzing the user preference of each of the two types of head-up displays (2.5D, AR) using a 7-point scale (−3: negative words to 3: positive words) using the Attrakdiff usability evaluation method.

Table 3. Attrakdiff usability evaluation result by head up display type

Contents	Pragmatic Quality (PQ)	Hedonic Quality-Identity (HQ-I)	Hedonic Quality-Stimulation (HQ-S)	Attractiveness (ATT)
2.5D Head-up Display	0.81	1.00	0.99	1.19
AR Head-up Display	0.42	1.20	1.39	1.29

As shown in Table 3, according to the AttrakDiff evaluation results, the satisfaction of Pragmatic Quality (PQ) between 2.5D head-up display was greater than that of the AR head-up display, including evaluation criteria such as 'impractical - practical' and 'unruly - manageable'. In the case of AR head-up display, the Hedonic Quality-Identity (HQ-I) which includes evaluation criteria such as 'unprofessional - professional', 'tacky - stylish', and Hedonic Quality-Stimulation (HQ-S) including evaluation criteria such as 'unimaginative - creative', 'conservative - innovative', Satisfaction with Attractiveness (ATT), including evaluation criteria such as 'ugly - attractive' and 'repelling - appealing', was found to be higher than the 2.5D head-up display.

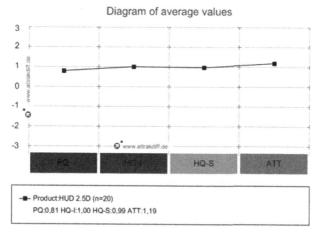

Fig. 3. 2.5D head-up display diagram of average values.

As shown in Fig. 3, 2.5D head-up display had the highest evaluation of Attractiveness (ATT) (1.19), followed by Hedonic Quality-Identity (HQ-I) (1.00), Hedonic Quality-Stimulation (HQ-S) (0.99), and Pragmatic Quality (PQ) (0.81).

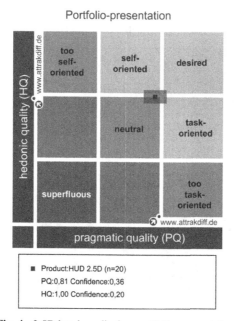

Fig. 4. 2.5D head-up display portfolio-presentation.

As shown in Fig. 4, the results of the Attrakdiff usability evaluation show that the 2.5D head up display is belongs to the region of 'neutral' close to the 'desired' direction in the correlation among Pragmatic Quality (PQ), Hedonic Quality-Identity (HQ-I), Hedonic Quality-Stimulation (HQ-S).

Diagram of average values
3
2
1
0
-1
-2
-3
www.attrakdiff.de
HQ-S
ATT
Product:HUD AR (n=20)
PQ:0,42 HQ-I:1,20 HQ-S:1,39 ATT:1,29

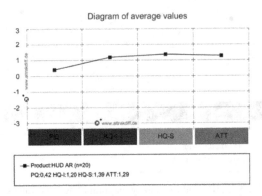

Fig. 5. AR head-up display diagram of average values.

As shown in Fig. 5, AR head-up display had the highest evaluation of Hedonic Quality-Stimulation (HQ-S) (1.39), followed by Attractiveness (ATT) (1.29), Hedonic Quality-Identity (HQ-I) (1.20), and Pragmatic Quality (PQ) (0.42).

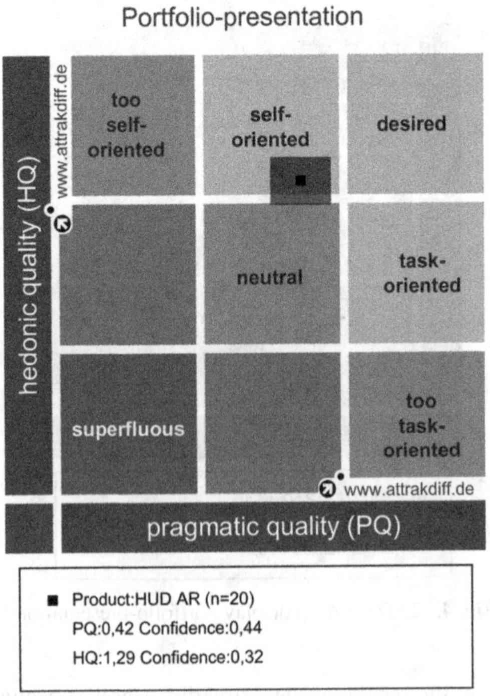

Portfolio-presentation
hedonic quality (HQ)
www.attrakdiff.de
too self-oriented
self-oriented
desired
neutral
task-oriented
superfluous
too task-oriented
www.attrakdiff.de
pragmatic quality (PQ)
Product:HUD AR (n=20)
PQ:0,42 Confidence:0,44
HQ:1,29 Confidence:0,32

Fig. 6. AR head-up display portfolio-presentation.

As shown in Fig. 6, the results of the Attrakdiff usability evaluation show that the AR head up display belongs to the region of 'self-oriented' close to the 'neutral' direction in the correlation among Pragmatic Quality (PQ), Hedonic Quality-Identity (HQ-I), Hedonic Quality-Stimulation (HQ-S).

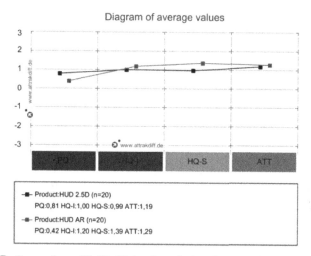

Fig. 7. Comparison of 2.5D-AR head-up display diagram of average values.

When comparing the 2.5D-AR head-up display types as shown in Fig. 7, the AR head-up display type had 0.20 points of Hedonic Quality-Identity (HQ-I), 0.40 points of Hedonic Quality-Stimulation (HQ-S), and 0.10 points of Attractiveness (ATT) higher than the 2.5D head-up display type, and the practicality was 0.39 points lower.

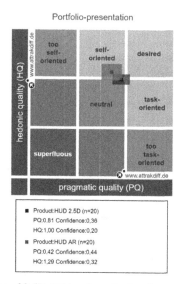

Fig. 8. Comparison of 2.5D-AR head-up display diagram of average values.

As shown in Fig. 8, when comparing the 2.5D-AR head-up display by using the Attrakdiff usability evaluation, the 2.5D head-up display type was more 'neutral' oriented than AR head-up display type in its connection to Pragmatic Quality (PQ), Hedonic Quality-Identity (HQ-I), and Hedonic Quality-Stimulation (HQ-S). AR head-up display type, on the other hand, belongs to the 'self-oriented' area more than 2.5D head-up display type.

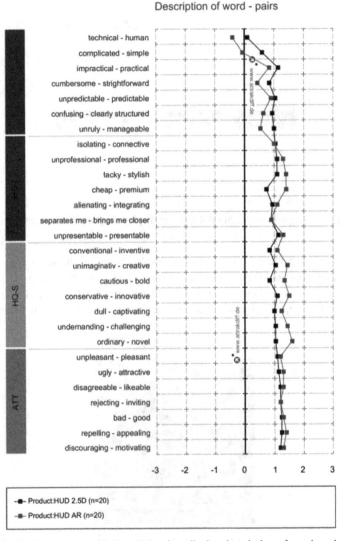

Fig. 9. Comparison of 2.5D-AR head up display description of word - pairs.

As shown in Fig. 9, the result of analyzing the details which indicated that 2.5D head-up display type was 0.39 points higher than the AR type in Pragmatic Quality (PQ), showed that based on the seven detailed evaluation criterion ('technical – human', 'complicated – simple', 'impractical - practical', 'cumbersome – straightforward', 'unpredictable – predictable', 'confusing - clearly structured', 'unruly – manageable') of Pragmatic Quality (PQ), 2.5D head-up displayed scored higher than AR head-up display type. Overall, the 2.5D head-up display type tends to be evaluated to be more practical than the AR head-up display type.

Table 4. Analysis results using the Attrakdiff usability evaluation method

Item	Attrakdiff
2.5D Head-up display	Attractiveness (1.19) > Hedonic Quality-Identity (1.00) > Hedonic Quality-Stimulation (0.99) > Pragmatic Quality (0.81)
AR Head-up display	Hedonic Quality-Stimulation (1.39) > Attractiveness (1.29) > Hedonic Quality-Identity (1.20) > Pragmatic Quality (0.42)
Comparison of 2.5D-AR Head-up display	The AR head-up display type was 0.20 points higher in Hedonic Quality-Identity (HQ-I), 0.40 points in Hedonic Quality-Stimulation (HQ-S), and 0.10 points in Attractiveness (ATT) than 2.5D type, and Pragmatic Quality (PQ) was 0.39 points lower than 2.5D type

A result of analyzing the details which indicated that AR head-up display type was 1.39 points higher than the 2.5D head-up display in Hedonic Quality-Stimulation (HQ-S), showed that based on the seven detailed evaluation criterion ('conventional – inventive', 'unimaginative – creative', 'cautious – bold', 'conservative – innovative', 'dull – captivating', 'undemanding – challenging', 'ordinary – novel') of Hedonic Quality-Stimulation (HQ-S), AR head-up display scored higher than 2.5D head-up display type. This is the biggest difference in the assessment amongst the four core characteristic elements (PQ, HQ-I, HQ-S, ATT) between AR head-up display and 2.5D head-up display. Overall, AR head-up display type has a special advantage over the 2.5D head-up display in terms of Hedonic Quality-Stimulation (HQ-S) criteria (Table 4).

In this study, the user preference in terms of graphic style of the head-up display was analyzed using Attrakdiff usability evaluation method. The analysis indicated that 2.5D head-up display type was 0.25 higher than AR head-up display type in terms of user preference with Pragmatic Quality (PQ), which includes evaluation criteria such as 'impractical - practical' and 'cumbersome - straightforward' for 2.5D type was 0.25 points higher than AR type. Due to the fact that 2.5D head-up display type is the current dominant form of head-up display type, the familiar head-up display type is considered to be more human than the AR type, which is heavily technologically oriented. In addition, 2.5D head-up display type, provides information in a simple cluster form and is considered to be simpler, more practical, straightforward, predictable, clearly structured and more manageable than its AR counterpart.

The AR head-up display type has a 0.20 points higher rating than 2.5D in terms of Hedonic Quality-Identity (HQ-I), including evaluation criteria such as 'unprofessional - professional' and 'tacky - stylish'. Hedonic Quality-Stimulation (HQ-S), which includes evaluation criterion such as 'conventional - inventive', 'unimaginative - creative' was 0.40 points higher. Attractiveness (ATT), which includes evaluation criterion such as 'ugly - Attractive', and 'disagreeable - likeable' was 0.10 points higher. It seems that the AR head-up display types were considered to be more professional, attractive, and significantly more expensive than the 2.5D head-up display type. This is so because users are able to know intuitively that the AR type requires advanced technology by displaying information display elements according to the situation of the road seen through the windscreen, and the advanced graphic elements were added than the 2.5D type.

Analyzing the detailed results of the Hedonic Quality-Stimulation (HQ-S), AR head-up display interface was more inventive, creative, bold, innovative, more captivating, and tended to be more challenging and novel than the 2.5D head-up display type. When analyzing the detailed results of Attractiveness (ATT), it was found that though the difference is not large, the AR head-up display type tends to be more attractive, likeable and appealing than the 2.5D head-up display type. It seems that this is due to the fact that many of the participants in the survey were experiencing a new type of head-up display graphic style rather than the head-up display format that they were familiar with, and in reality, the graphic style of AR head-up display tends to be more visually attractive than the pre-existing 2.5D head-up display type.

AR navigation system shows it's Hedonic Quality-Identity (HQ-I) to be 0.2 points, the Hedonic Quality-Stimulation (HQ-S) 0.4 points, and the Attractiveness (ATT) 0.1 points higher than the 2.5D navigation system. So, it is necessary to maintain the advantage of visually appealing (especially stylish, premium, bold, novel, attractive etc.) because it provides a new and innovative experience by displaying information display elements in a sophisticated graphic style according to the road situation. The 2.5D head-up display type's satisfaction with Pragmatic Quality (PQ) was 0.25 points higher than the AR type. It seems that this is so because of the AR head-up display type's unpredictable information display elements, which may prove to be difficult to discern. Indicating that the graphic design should be supplemented in a direction to minimize the disadvantages attributed to AR head-up display such as complicated, unruly, and cumbersome.

4   Conclusions

This study has significance and value in defining what elements should be strategically utilized for the future car AR head-up display development, where AR head-up display is actively being developed. However, this study is not a prototype in which the technology is actually implemented in the case of AR head-up display, and there is a limitation in the part where the questionnaire was conducted only with the image of each head-up display graphic style type. Therefore, this study is meaningful as an early stage study in defining what elements should be strategically utilized for the future vehicular AR head-up display development.

References

1. Head Up Display Market for Vehicles. INNOPOLIS Foundation, October 2017
2. IT   Chosun.   https://it.chosun.com/site/data/html_dir/2019/09/15/2019091500533.html. Accessed 15 Sept 2019
3. Autoview.co.kr. https://www.autoview.co.kr/content/article.asp?num_code=71092. Accessed 10 July 2020
4. Hwang, Y., Park, B., Kim, K.: The effects of toward AR-HUD system usage on drivers' driving behaviors. Institute Electron. Inf. Eng. **2015**(6), 1818–1820 (2015)
5. Cho, Y.: Augmented reality (AR) head-up display (HUD) design study for prevention of car accident based on graphical design, sensitivity and conveyance of meaning. Korean Soc. Des. Sci. **28**(3), 103–117 (2015)
6. Nadja, S., Katharina, W., Frederik, N., Alexandra, N., Bettina, L., Thomas, V.: An augmented reality display for conditionally automated driving. In: Automotive User Interfaces and Interactive Vehicular Applications. AutomotiveUI 2018, Toronto, Canada, pp. 137–141. ACM (2018)
7. Park, K., Ban, K., Jung, E., Im, Y.: A study on classification of the preferred areas of head-up display (HUD) while driving. Ergon. Soc. Korea **37**(6), 651–666 (2018)
8. Yoon, S., Yang, J., Jang, J., Lee, J., Choi, M., Park, W.: High quality in-vehicle hologram HUD AR content creation with driver view point tracking. Korean Inst. Electr. Eng. **2019**(10), 354–355 (2019)

IoT-Enabled Mobile Device
for Electrogastrography Signal Processing

Hakimjon Zaynidinov[1] , Sarvar Makhmudjanov[1] , Farkhad Rajabov[1] ,
and Dhananjay Singh[2(✉)]

[1] Tashkent University of Information Technologies named after Muhammad al Khwarizmi,
Tashkent, Uzbekistan
[2] ReSENSE Lab, Department of Electronics Engineering, Hankuk University of Foreign
Studies, Seoul, South Korea
dan.usn@ieee.org

Abstract. Electrogastrography (EGG) is a powerful instrument that represents
and provides high time precision for direct heart activities. After processing the
chest skin for the contact process by network electrodes, the conventional method
of collecting heart signals involves the presence of electrodes directly connected
to the patient's chest using gelatin. One of the most critical activities is tracking
the clinical signs of cardiac patients. A real-time EGG signal acquisition and processing device is developed to observe patients in various IoT-based environments
is discussed in this article. The EGG signals that are detected are passed to special
software. For further protection, it is ensured that the transmission of the EGG
signal from the patient without distortion to the doctors should take place. The
ambulatory and wearable device for the transmission of real-time EGG signals is
discussed in this article in order to monitor patients in various environments. For
further analysis, the captured EGG signal is transferred to software and signals
are transferred to the doctors for real-time evaluation of the patient.

Keywords: Internet of Things · Gastroenterological signal · Healthcare ·
Filtering · B-spline · Web technologies · Bluetooth · Sensors

1 Introduction

In recent times, research in the field of bioelectrical signal processing has garnered
increasing interest. EEG (electroencephalography), EGG (electrogastrography), infrared
signals with spectral analysis are some of the types of biomedical signals for real-world
applications. Medical consulting has been revolutionized with the advent of Internet
technology, removing the need for patients to visit hospitals in person. And in the case
of in-person visits, it is possible to use an online booking system to reserve a visit [1–5].
Using the mobile computer system on the smart phone of the patient or his caretaker,
the patient will display his health status. Following the transition from hospital to home,
early detection and diagnosis of potentially lethal physiological conditions such as heart
attack require constant monitoring of patient wellbeing. Health monitoring systems are

© Springer Nature Switzerland AG 2021
M. Singh et al. (Eds.): IHCI 2020, LNCS 12616, pp. 346–356, 2021.
https://doi.org/10.1007/978-3-030-68452-5_36

being proposed as a low cost solution in response to these types of needs [7, 8]. Such a device is made up of physiological knowledge that is stored, processed and shared locally, such as mobile phones and personal computers. These systems should fulfill strict criteria for safety, protection, reliability, and long-term real-time operation.

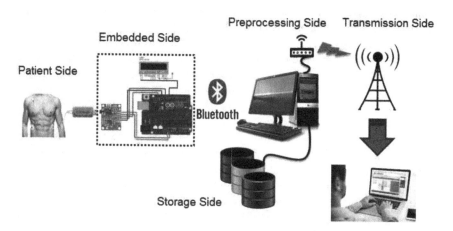

Fig. 1. Overall ECG signal surveillance system infrastructure

Pressure measurement tools, scales, and other portable devices are equipped with wireless transmitters that allow you to move data to a computer immediately and keep health records [17, 19]. Due to the conventional multiple electrodes used in EGG signal acquisition, continuous heart patient monitoring is impractical. The EGG signal is then switched from wired mode to wireless mode to be compact, light, and can be conveniently wearable without experienced assistance and pass these signals to physicians for diagnostic purposes in order to collect an ECG signal daily for heart disorders easily diagnosed. Special electrodes are used in Human-Computer Interaction (HCI)s based on the conventional methods of all EGG data, where these electrodes are difficult and require considerable time to be applied. The gel-free electrodes are designed to be linked to a spring pin and a smooth conductive polymer for greater patient convenience. Dry electrodes have been used in order to reliably calculate heart signals, but motion artifacts affect non-contact capacitive electrodes. The cardiac monitoring interfaces provide the cardiologist with a broad understanding of the role of heart disease as well as preventing or minimizing the disease. In recent years new specialized, computerized methods and means of processing and research for different biomedical signals have been intensively developed in various countries. New insightful data analysis approaches are being developed and implemented widely into clinical practice. In medicine, computers have long been used and many new methods of diagnosis are based on IoT. Examination procedures without machines are impossible, such as ultrasound or computer tomography. It is now difficult to find a medical area in which computers are not being used more and more actively. But in medicine, the use of computers is not limited solely to diagnostics. Increasingly, they are used in the treatment of different diseases, from the creation of an effective treatment plan to the management of different medical equipment during procedures [9–12]. The overall wearable EGG signal infrastructure surveillance system can be described in the Fig. 1. All reported EGG data is denoised by artifact filtering, amplifying the signal for evaluation purposes. Finally, a user-friendly graphical user interface (GUI) and a mobile computer with an Internet environment that offers the ability to quickly communicate with patients.

In this work, special server authentication was used with the original software application in order to ensure that the EGG signal was easily transmitted, carefully preserved without any distortion during transmission, and to ensure the compatibility of these signals during processing. The methods and means of electrocardiography (EGG) and electroencephalography (EEG) on the basis of computer technology have undergone the greatest growth at the present stage of the development of medical informatics systems [14, 15]. At the same time, major progress has been made in developing computer-based diagnostic systems for such medical fields as gastroenterology in many countries around the world in recent years. Non-invasive approaches give way to invasive methods of inspecting abdominal cavity organs, i.e. those associated with the insertion of sensors into the organism (for example, balloonography, gastroscopy, pH-metry), which causes patients to feel pain. The latter are primarily focused on the achievements of electrogastrography (EGG), i.e. the methods of documenting and subsequent study of the bio potential dynamics removed from the surface of the body and supplying the gastrointestinal organs with information on muscle activity (GIT). Further changes to the mathematics and software of the EGG information systems are required to enhance the characteristics of the equipment concerned [16–19].

2   Related Works

Research in adopting wearable IoT in healthcare has been extensive, many of the applications have already been commercialized and available in the market. Existing works often focus on assisting people who experience difficulties in maintaining independent living, for example, elderly or people with certain chronic diseases [14], such as Guo Ying, Wang Zhenzhen, Ke Li, Du Qiang, and Li Jinghui suggested a system for ECG signal detection and analysis based on digitized auscultation. For instance, Dr. Haitham Abbas Khalaf et al. [15] present a platform to enable student and end-users to learn the ECG basics principle, enable analysis the signal processing by using LabVIEW to validate and compiling data analysis, whilst Romero et al. [21] describe a system that diagnose and monitor Parkinson's disease. Nonetheless, it is problem to develop a diagnose application that can use gastroenterological signal and their filtration to different channels. Even nowadays some local hospitals does not have mobile devices which helps to take signals from human body using wireless technologies.

3   System Design and Methodology

The proposed system would calculate and screen a patient's basic physiological details, remembering the ultimate aim of accurately depicting the status of her/his wealth and health. The suggested structure will give the patient prosperity status on the web in the same way. The social protection master will offer critical psychological guidance by using that knowledge. Sensors, the data verification unit, the microcontroller, the client application and the server application are basically included in the device.

The EGG signal acquisition device block diagram can be represented in the following Fig. 2. There are three primary monitoring electrodes, which can be applied for use in both static and telemetry cardiac systems. The electrodes may be used for both fixed and ambulatory systems. It is possible to connect these electrodes to specific positions on the chest of the patient. The Arduino board acts as a micro-controller to access the computer's analog signal. In the following subsections, descriptions of device components will be defined. The approach of the suggested scheme can be seen based on Fig. 2.

EGG-related approaches are used for treating both medicinal and surgical diseases of the gastrointestinal organs. Their value is that it is possible to measure the markers of Human organs. This is particularly important for the prevention of diseases in the population, as well as for the timely treatment of detected GIT disorders.

Processes in gastrointestinal tract tissues, as opposed to other organs and non-stationary, are distinguished by the most low-frequency character fluctuations. For frequency-time analysis of gastro-and enterograms, a local stationary mathematical model of a gastro-signal based on a discrete short-term Fourier transform (FT) or spline method is commonly used which enables, in particular, the study of changes in spectral power density and other time signal characteristics within certain limits [3, 4, 6, 13].

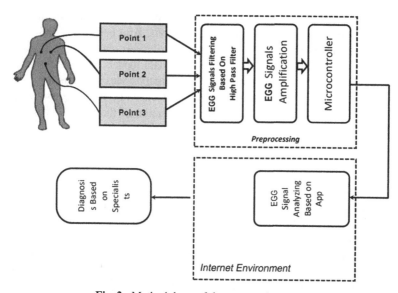

Fig. 2. Methodology of the proposed system

The possibility of more accurate processing of EGG signals arises with the use of wavelet analysis, in which off-sets and large-scale transformations of such oscillating functions localized in time and frequency form the basis of the signal space. The results of research in the direction of spectral processing of the most low-frequency biomedical signals (range 0.01–0.25 Hz) and disclosure of their local characteristics in both time and frequency regions are considered in this paper.

Table 1 shows the frequency ranges of gastroenterological signals for different (four organs) GIT group normal (i.e. healthy person).

Table 1. Distribution of organ parameters GIT for group normal

№	File name	Organ GIT	Frequency range
1	Common. Txt	Common GIT	0.03–0.22 Hz
2	Chn1.txt	Stomach	0.03–0.07 Hz
3	Chn2.txt	Ilenium	0.08–0.12 Hz
4	Chn3.txt	Skinny intestine	0.13–0.17 Hz
5	Chn4.txt	Duodenum	0.18–0.22 Hz

The functioning of the digestive tract in the work is illustrated by two types of graphs gastroenterograms, which differ from each other in a number of features:

1. Enterograms of a group of "norm" patients and a sick patient in which no pattern is basically present;
2. Gastrograms that include linear patterns in sick patients. They are distinguished by the inharmonious, irregular nature of oscillations. And before any information is given to the patient, the preliminary processing of gastro-enterograms is carried out automatically. This is due to the fact that the data loaded into the file processing software is represented by digital decimal or binary codes and thus it is important to convert the data to volts, bipolarize the signal, check for the maximum amplitude and so on for convenient and accurate visualization of the signal and subsequent analysis.

4   IoT Device and Experimental Setup

The Internet of Things (IoT) enables the reliability of medical devices to be enhanced by tracking the health of patients in real time, where special sensors collect patient data. This technology detects and conveys data about the health of the patient, thus minimizing manual interaction and therefore human error. Patient parameters are transmitted through a portal on the internet via medical devices, then processed and analyzed in a special computer program. In introducing the Internet of Things for healthcare applications, a serious challenge is to track all patients in various places. In the medical sector, IoT therefore offers a solution for reliable low-cost patient tracking and reduces the problems between patient outcome and disease management. In this research, particular attention was paid to the development of a method that uses a specially developed wireless mobile device to track signals from the patient's gastrointestinal tract in the Fig. 3.

Technologies for wireless networking will help enhance health care quality. It is commonly used in the quality of patient treatment, administration of medical care, enhancement of health care and operation, in medical follow-up procedures during surgery, as well as in contact with a doctor in emergencies.

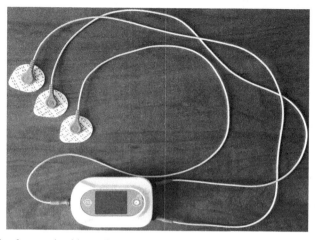

Fig. 3. A device for entering biomedical signals into a computer based on Bluetooth technology.

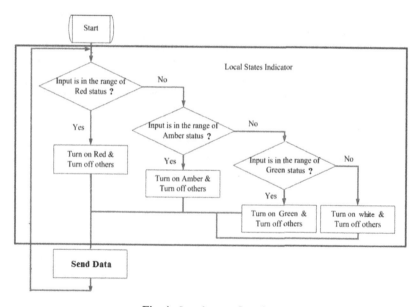

Fig. 4. Local states flowchart

Wireless networking systems are of several varieties. Blue-tooth Technology was one of them. The algorithm for connecting to a device through a Bluetooth module is illustrated in Fig. 4. The data is transmitted in serial form to the computer, which means that the information is transmitted to the computer program in real time. For data reception, the device uses the COM port. The program must configure the settings provided in Table 2 to start receiving data.

Table 2. Serial port configuration

Baud rate	9600
Flow control	None
Data Bits	8
Parity	None
Stop Bit	1

The computer sends the data via the COM port from the serial port to the control unit. The hardware signal range can be expected in such instances. Human body signals are given in Table 2.

5 Filtration of Gastroenterological Signals

In connection with the switching to fully digital gastro- and enterogram processing methods, the probability of software implementation of digital filtering over the entire channel collection must be considered in almost real time. It is possible to filter the wanted signal without directly going into the frequency domain. Multiplying the signal spectrum by the function of a rectangular window by the convolution theorem is equivalent to convolution in the time domain of the signal and the filter operator-the function obtained as a result of the window's inverse Fourier transform. In the filter window, the sum of the filter coefficients determines the average gain of signal values and remains constant over the entire data collection, and the sum of the filter coefficients is normalized.

The non-recursive band-pass filter performs the operation convolution of the samples to the discrete signal $\{x_j\}$ and the digital filter operator $\{h_j\}$:

$$y_i = \sum_{j=0}^{m-1} h_j x_{i-j}. \tag{1}$$

If you briefly characterize the smoothing windows used, you can see that the smoothing procedure using the Goodman-Enoxon-Otnes (GEO) window according to the 7-point algorithm:

$$\bar{x}_i = \sum_{l=-3}^{3} a_l x_{i+1}, \tag{2}$$

gives the following one-sided coefficients: a0 = 1, a1 = 0.1817, a2 = –0.1707, a3 = 0.1476. These equations used to filter gastroenterological signal into four channels and its result given in Fig. 5.

It is one of the most effective terms of the effectiveness suppressing the side components of the frequency response. 5-point and 3-point Hamming and Hanning windows are also used in digital filtering, but they weaken additional components to a lesser extent. Digital filtering gastroenterological signal over the entire set of channels was carried out

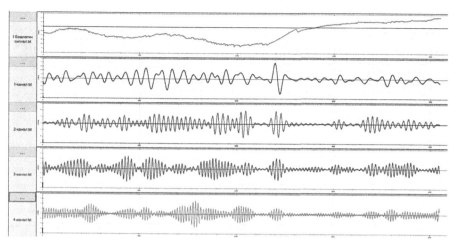

Fig. 5. Source signal, gastrogram and enterograms of healthy patients A B graphical form.

by a filtering program based on (1). The program is designed to filter and visualize the results in the form of gastrograms and enterograms. In Fig. 4. shows the initial signal, gastrogram and enterograms of healthy patient A in graphical form.

6   Experimental Results and Final Discussions

Here we are using the proposed system is shown in Fig. 1 board as our platform. It has sensors for receiving data from the human body, an analog to digital converter, and transmitting data to the Bluetooth module application. The application must be mounted on a mobile computer that collects and sends data from the user to the server application. Two applications are available, the server application and the client application. The software for the side client was built in the programming language Python. The sensors attached to the patient's body are connected to a computer. The software obtains and sends the data from the sensors to the client application. Python has a listener serial port library, which takes data and transforms data based on its algorithm using this library program. Finally, with the purpose that human resources practitioners have it will share information on the Internet. The side server program is devoted to gastroenterological signals being obtained and procured. It uses the method of B-spline approximation to interpret biomedical data from patients. The framework was developed using the PHP server-side programming language for web technologies have shown in the Figs. 6 and 7.

By asking them about different symptoms they may have the smart prediction module predicts the illness that the patient is suffering from and the alternatives are based on the previous symptom. The presented EGG signal which be nonsensical or reminiscent of a cardiac signal via the special software. This signal appears to be traveling at an irregular rate on the device, after which it may become stationary for a given period of time. For the purpose of obtaining a simple and understandable cardiac signal, the interference between the embedded device and the special program is changed.

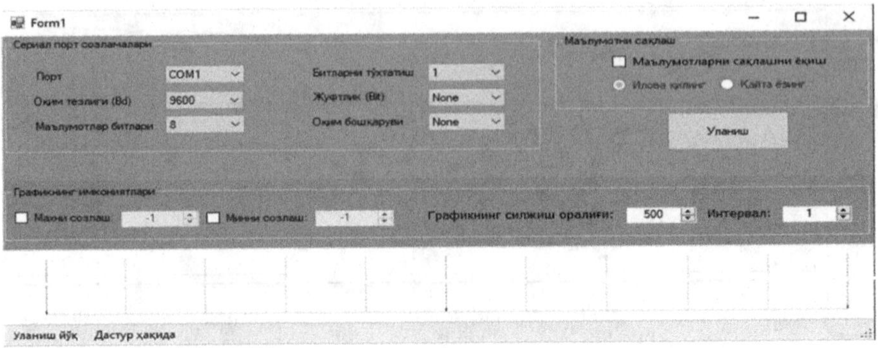

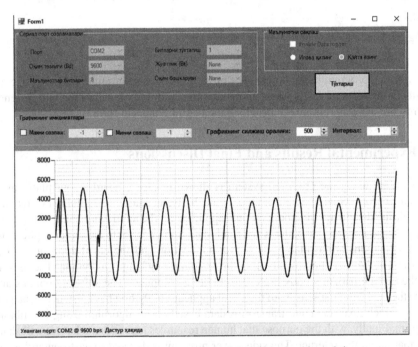

Fig. 6. Application interface.

Fig. 7. Graph of gastroenterological signal obtained in real time.

The Arduino plate, created by National Instruments and compatible with the application, can be replaced to allow overlapping and data acquisition. In this article, the developed pre-prototype is a primitive one to record and show the EGG signal of utilization. The gastrograph at the top and the frequency spectrum at the bottom are shown in Fig. 8. If there is more than one of the spectral critical points, then this patient has gastric disease. The prototype can be built with more functions for potential uses so that it is more convenient for the patient.

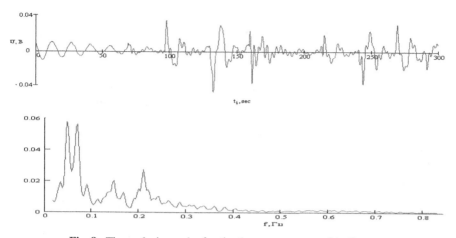

Fig. 8. The analysis graph of patient's gastrogramm and its Spector

7 Conclusions

The paper proposes and implements an algorithm based on the digital band pass filtering method and an algorithm for dynamic channel separation of analog gastrointestinal signals that are particularly pronounced in patients. The purpose of the recording and analysis of the EGG signal using a special program is to acquire and analyze the parameters of the EGG signal. In this article, a dedicated prototype of real-time EGG signal recording and analysis was developed, which is simple to wear and lightweight based on easily embedded systems. For urgent examination and diagnosis, gastroenterological data can be wirelessly transmitted to a specialist. This work focuses on the acquisition and analyzing of EGG signals, and determines gastricity and database maintenance using special applications. In the study and diagnosis of various real-time manifestations, this technology can assist physicians (gastroenterologists) and the healthcare system. In this study it can be concluded that the gastroenterological signal can be filtered across self-contained channels and the disease can be identified across its spectrum. It can be used for non-invasive therapy.

References

1. Х.Н. Зайнидинов, С.У. Махмуджанов, Р.Д. Аллабергенов, Д.С. Яхшибаев Интернет вещей (учебное пособие). Т., Aloqach», 2019, 206 стр
2. Banafa, A.: What is next for IoT and IIoT. Enterprise Mobility Summit. Int. J. Comput. Sci. Inf. Technol. (IJCSIT) **10**(2) (2018)
3. Singh, D., Singh, M., Hakimjon, Z.: B-Spline approximation for polynomial splines. In: Singh, D., Singh, M., Hakimjon, Z. (eds.) Signal Processing Applications Using Multidimensional Polynomial Splines, pp. 13–19. Springer, Singapore (2019). https://doi.org/10.1007/978-981-13-2239-6_2
4. Singh, M., Zaynidinov, H., Zaynutdinova, M., Singh, D.: Bi-cubic spline based temperature measurement in the thermal field for navigation and time system. J. Appl. Sci. Eng. **22**(3), 579–586 (2019). https://doi.org/10.6180/jase.201909_22(3).0019

5. Singh, D., Singh, M., Hakimjon, Z.: Evaluation methods of spline. In: Singh, D., Singh, M., Hakimjon, Z. (eds.) Signal Processing Applications Using Multidimensional Polynomial Splines, pp. 35–46. Springer, Singapore (2019). https://doi.org/10.1007/978-981-13-2239-6_5

6. Singh, D., Singh, M., Hakimjon, Z.: Parabolic splines based one-dimensional polynomial. In: Singh, D., Singh, M., Hakimjon, Z. (eds.) Signal Processing Applications Using Multidimensional Polynomial Splines, pp. 1–11. Springer, Singapore (2019). https://doi.org/10.1007/978-981-13-2239-6_1

7. Dimitrov, D.V.: Medical internet of things and big data in healthcare. Healthc. Inform. Res. 22(3), 156–163 (2016)

8. Zanella, A., Bui, N., Castellani, A., Vangelista, L., Zorzi, M.: Internet of things for smart cities. IEEE Internet Things J 1(1), 22–32 (2014)

9. Kim, Y., Lee, G., Park, S., Kim, B., Park, J.-O., Cho, J.-H.: Pressure monitoring system in gastro-intestinal tract. In: Proceedings - IEEE International Conference on Robotics and Automation 2005 (2005). https://doi.org/10.1109/ROBOT.2005.1570298

10. Beer, G., Marussig, B., Duenser, C.: Basis functions, B-splines. In: Beer, G., Marussig, B., Duenser, C. (eds.) The Isogeometric Boundary Element Method, pp. 35–71. Springer, Cham (2020). https://doi.org/10.1007/978-3-030-23339-6_3

11. Unser, M., Aldroubi, A., Eden, M.: B-spline signal-processing. 2. efficient design and applications. IEEE Trans. Signal Process. 41, 834–848 (1993). https://doi.org/10.1109/78.193221

12. Hao, X., Lu, L., Gu, W.K., Zhou, Y.C.: A parallel computing algorithm for geometric interpolation using uniform B-splines within GPU. Inf. Technol. J. 15, 61–69 (2016)

13. Lara-Ramirez, J.E., Garcia-Capulin, C.H., Estudillo-Ayala, M.J., Avina-Cervantes, J.G., Sanchez-Yanez, R.E., Rostro-Gonzalez, H.: Parallel hierarchical genetic algorithm for scattered data fitting through B-Splines. Appl. Sci. 9, 2336 (2019)

14. Ying, G., et al.: Heart sound and ECG signal analysis and detection system based on LabVIEW. In: 2018 Chinese Control and Decision Conference (CCDC), pp. 5569–5572. IEEE, June 2018

15. Khalaf, H.A., et al.: Wearable ambulatory technique for biomedical signals processing based on embedded systems and IoT. IJAST 29(04), 360–371 (2020)

16. Banka, S., Madan, I., Saranya, S.S.: Smart healthcare monitoring using IoT. Int. J. Appl. Eng. Res. 13(15), 11984–11989 (2018). ISSN 0973-4562

17. Almotiri, S.H., Khan, M.A., Alghamdi, M.A.: Mobile health (m-health) system in the context of IoT. In: 2016 IEEE 4th International Conference on Future Internet of Things and Cloud Workshops (FiCloudW), pp. 39–42, August 2016

18. Gupta, M.S.D., Patchava, V., Menezes, V.: Healthcare based on IoT using raspberry pi. In: 2015 International Conference on Green Computing and Internet of Things (ICGCIoT), pp. 796–799, October 2015

19. Xu, B., Xu, L.D., Cai, H., Xie, C., Hu, J., Bu, F.: Ubiquitous data accessing method in IoT-based information system for emergency medical services. IEEE Trans. Ind. Inform. 10(2), 1578–1586 (2014). ISSN 1551-3203

20. Parkman, H.P., Hasler, W.L., Fisher, R.S.: American gastroenterological association medical position statement: diagnosis and treatment of gastroparesis. Gastroenterology 127(5), 1589–1591 (2004)

21. Romero, L.E., Chatterjee, P., Armentano, R.L.: An IoT approach for integration of computational intelligence and wearable sensors for Parkinson's disease diagnosis and monitoring. Health Technol. 6(3), 167–172 (2016). https://doi.org/10.1007/s12553-016-0148-0

A Study on the Usability Test Method of Collaborative Robot Based on ECG Measurement

Sangwoo Cho[1] and Jong-Ha Lee[1,2]([✉])

[1] The Center for Advanced Technical Usability and Technologies, Keimyung University,
Daegu, South Korea
segeberg@kmu.ac.kr
[2] Department of Biomedical Engineering, School of Medicine, Keimyung University,
Daegu, South Korea

Abstract. Collaborative robots can work with operator, it used situations that handing things over or moving heavy objects next to the workers. It is most important to ensure the safety of users as a condition for using collaborative robots. Therefore, the use of ECG sensors, which can measure the body's autonomic nervous system reactions, in usability evaluations is a way to derive objective results from users. In this study, we tried to develop an objective usability evaluation index for collaborative robot products by conducting a usability evaluation of collaborative robot products based on heart rate measurement. We recruited 12 undergraduates with engineering majors (F = 6, M = 6, 23.75 ± 2.60 yrs). When the subjects expressed their intention to learn enough about how to use the collaborative robot, the subjects provided a 10-minute break to adjust the learning effect. Subjects wore ECG sensor (Bioharness, ZephyrTM Co.) to measure heart rate during break time. Subjects performed usability test of collaborative robot after the break (10 min) and continuously measured heart rate during the usability test. In comparisons between break time and the usability test of the collaborative robot, more heart rate was increased in usability test of collaborative robot than break time (p = 0.01). A significant positive correlation between the change value of heart rate and the time of usability test was shown (r = 601, p = 0.024). We found that heart rate can express the user's autonomic nervous system response to product use in usability evaluation, and check the convenience of using collaborative robot.

Keywords: Usability test · Collaborative robot · Heart rate

1 Introduction

1.1 A Subsection Sample

After the Industrial Revolution, the manufacturing industry grew by leaps and bounds. Along with the development of the manufacturing industry, many manufacturing robots have been developed. However, the recent entry into an aging society has reduced the

© Springer Nature Switzerland AG 2021
M. Singh et al. (Eds.): IHCI 2020, LNCS 12616, pp. 357–362, 2021.
https://doi.org/10.1007/978-3-030-68452-5_37

workforce in manufacturing. In order to solve the above problem, the development of collaborative robots has been actively carried out recently. Collaborative robots can work with operator, it used situations that handing things over or moving heavy objects next to the workers. Therefore, collaborative robot can secure workers' autonomy and safety in terms of working together [1]. It is most important to ensure the safety of users as a condition for using collaborative robots. Therefore, it should be taken into account that it is possible to efficiently assist workers in their work from the stage of development of collaborative robots.

As a way to evaluate the work efficiency of workers using cooperative robot products from the stage of product development, it is most important to actively utilize usability evaluation. Product usability evaluation methods include formation evaluation and overall evaluation. In addition, it is a way to evaluate the convenience of the actual user of the product [2]. Traditional usability evaluations used interview-based subjective scoring methods to validate product-use interface [3, 4]. In recent studies, usability assessment studies were conducted on the programming of cooperative robots with virtual reality technology [5].

The autonomic nervous system of the user's body reaction acts unconsciously within the body, and the reaction of the body organs is harmonized with the reaction of the sympathetic nerve and the parasympathetic nerve to maintain the function of the body [6–8]. Therefore, the use of ECG sensors, which can measure the body's autonomic nervous system reactions, in usability evaluations is a way to derive objective results from users.

The purpose of this study was to verify that ECG signals could be used as objective data in assessing the effectiveness of usability test. We tried to develop an objective usability evaluation index for collaborative robot products by conducting a usability evaluation of collaborative robot products based on heart rate measurement.

2 Methods

2.1 Subjects

We recruited 12 undergraduates with engineering majors (F = 6, M = 6, 23.75 ± 2.60 yrs) as shown in Table 1. All subjects that consented to participate in this study were informed about the experimental protocol. All subjects have experience in C/C++ language programming education.

Table 1. Demographics of subjects.

	Gender (M/F)	Age (years)	Major
Subject 1	Male	27	Electronic engineering
Subject 2	Male	27	Biomedical engineering
Subject 3	Female	25	Biomedical engineering

(continued)

Table 1. (*continued*)

	Gender (M/F)	Age (years)	Major
Subject 4	Male	27	Mechanical engineering
Subject 5	Female	24	Biomedical engineering
Subject 6	Male	25	Mechanical engineering
Subject 7	Female	20	Mechanical engineering
Subject 8	Female	22	Computer engineering
Subject 9	Female	21	Mechanical engineering
Subject 10	Male	24	Computer engineering
Subject 11	Male	23	Computer engineering
Subject 12	Female	20	Computer engineering
Total	6/6	23.75 ± 2.60	–

2.2 Procedure

After fully explaining the purpose and method of the experiment to the subjects, we received consent for participation and proceeded with the experiment. Subjects were trained an average of 40 min to explain the use of collaborative robot. The collaborative robot user manual consisted of power connection, robot arm motion implementation program interface use, and safety device (emergency stop button, anti-collision function). When the subjects expressed their intention to learn enough about how to use the collaborative robot, the subjects provided a 10-min break to adjust the learning effect. Subjects wore ECG sensor (Bioharness, ZephyrTM Co.) to measure heart rate during break time. Subjects performed usability test of collaborative robot after the break (10 min) and continuously measured heart rate during the usability test (Fig. 1).

The performance of the usability assessment of the collaborative robot consisted of five steps. First step is to turn on the collaborative robot and connect the app on the tablet PC. The second step is to connect the network between the tablet PC and the collaborative robot. The third step is to perform coding to make the movement of the collaborative robot. The fourth step is to use a safety device while checking the movement of a collaborative robot made of coding. The fifth stage of the evaluation is to shut down the power of the collaborative robot.

2.3 Data Analysis

The data measured in the usability test were time of usability test and heart rate. The paired t-test has been used to compare the heart rate value between break time and usability test time using the SPSSWIN 20.0 software package.

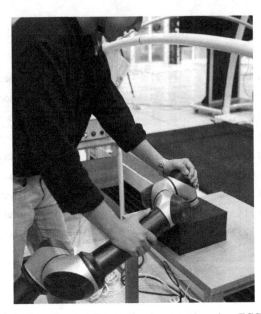

Fig. 1. Collaborative robot usability test environment based on ECG measurement

3   Results

3.1   Time of Usability Test

The subjects performed the usability test within a range of at least 536 s to a maximum of 945 s.

3.2   Heart Rate and Change Value of Heart Rate

The heart rate measured during the usability test of the collaborative robot was 86.16 ± 10.46 times, and the heart rate measured during the break time was 82.02 ± 11.34 times. In comparisons between break time and the usability test of the collaborative robot, more heart rate was increased in usability test of collaborative robot than break time ($p = 0.01$) (Fig. 2).

The change value of heart rate is the difference between the heart rate of usability test of collaborative robot and the heart rate of break time. The change value of heart rate was 4.63 ± 3.70 times. As shown in Fig. 3, a significant positive correlation between the change value of heart rate and the time of usability test was shown ($r = 601$, $p = 0.024$).

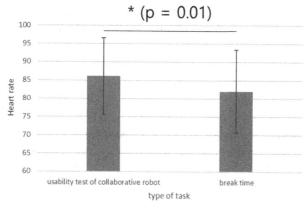

Fig. 2. A comparison of measured heart rate in break time and usability test

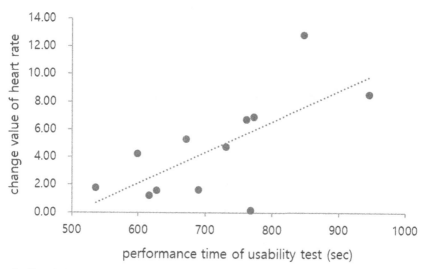

Fig. 3. Correlation between change value of heart rate and performance time of usability test (r = 601, p = 0.024)

4 Discussion

In this study, heart rate was measured during programming to implement motion of collaborative robot. Subjects performed all tasks within 16 min of the usability assessment. In particular, subjects tended to increase their heart rate during usability evaluations. The increased heart rate suggests that the task was performed with stress during the task.

A person's heart rate change was confirmed to have a strong association with cognitive loads in solving tasks [9]. In the result of change value of heart rate, subject who performed usability evaluations for a long time tended to change their heart rate. The results show that as the heart rate increases, stress occurs and the task performance time is extended. Therefore, we found that the change value of heart rate is related to the programming time. Heart rate data confirmed that it could be used to provide objective usability indicators.

5 Conclusion

In this study, usability test of collaborative robot used ECG sensor to verify objective usability assessment indicators. Heart rate increased about use of collaborative robot. In addition, change value of heart rate has correlate with time of usability test. We found that heart rate can express the user's autonomic nervous system response to product use in usability evaluation, and check the convenience of using collaborative robot. However, this study has limit to the lack of samples in the subject. Therefore, additional collaborative robot usability evaluations will have to be carried out.

Acknowledgments. This research is supported by "The Foundation Assist Project of Future Advanced User Convenience Service" through the Ministry of Trade, Industry and Energy (MOTIE) (R0004840, 2020) and Basic Science Research Program through the National Research Foundation of Korea(NRF) funded by the Ministry of Education (NRF-2017R1D1A1B04031182/NRF-2020R1I1A1A01072101) and a grant of the Korea Health Technology R&D Project through the Korea Health Industry Development Institute (KHIDI), funded by the Ministry of Health & Welfare, Republic of Korea (grant number: HI17C2594).

References

1. Choi, J.H.: Introduction of Cooperative Robots in Smart Media Sector and Technology Trends. Institute of Information & Communications Technology Planning & Evaluation (2019)
2. Dumas, J.S., Redish, J.: A Practical Guide to Usability Testing, Revised Edition, intellect (1994)
3. Ahmad, N., Boota, M.W., Masoom, A.H.: Smart phone application evaluation with usability testing approach. J. Softw. Eng. Appl. **7**, 1045–1054 (2004)
4. Lee, S.: Usability testing for developing effective interactive multimedia software: concepts, dimensions, and procedures. J. Educ. Technol. Soc. **2**(2), 1–12 (1999)
5. Materna, Z., Kapinus, M., Beran, V, Smrz, P, Zemcik, P: Interactive spatial augmented reality in collaborative robot programming: user experience evaluation. In: 2018 27th IEEE International Symposium on Robot and Human Interactive Communication (RO-MAN), pp. 80–87 (2018)
6. Hwang, M.C., Jang, G.Y., Kim, S.Y.: Research on emotion evaluation using autonomic response. Sci. Emot. Sens **7**(3), 51–56 (2004)
7. Kim, C.J., Hwang, M.C., Kim, J.H., Woo, J.C., Kim, Y.W., Kim, J.H.: A study on evaluation of human arousal level using PPG analysis. J. Ergon. Soc. Korea **29**(1), 1113–1120 (2010)
8. Lee, J.H., Kim, K.H.: A study of biosignal analysis system for sensibility evaluation. Korean Soc. Comput. Inf. **15**(12), 19–26 (2010)
9. Solhjoo, S., et al.: Heart rate and heart rate variability correlate with clinical reasoning performance and self-reported measures of cognitive load. Sci. Rep. **9**, 1–9 (2019)

The Human Factor Assessment of Consumer Air Purifier Panel Using Eye Tracking Device

Shin-Gyun Kim[1] and Jong-Ha Lee[1,2(✉)]

[1] The Center for Advanced Technical Usability and Technologies, Keimyung University,
Daegu, South Korea
segeberg@kmu.ac.kr
[2] Department of Biomedical Engineering, School of Medicine, Keimyung University,
Daegu, South Korea

Abstract. Among the various types of air purifiers available in the market, some products are more complex and difficult to use than others. This study aims to suggest a method to derive products, which are convenient to use, through the evaluation of an air purifier control panel using eye-tracking. An experiment was conducted with 28 housewives, in which various tasks using an air purifier were presented to them. The usage errors were measured and the tasks with low success rates were analyzed. As a result, the preferred icon and button names could be identified in Steps 5 and 6, where the success rates were low. Therefore, the convenience of using the control panel was accurately evaluated. Based on the obtained results, the study proposes the use of eye tracking as an evaluation method to derive qualitative data to increase the usability of control panels on home appliances such as air purifiers.

Keywords: Air purifier · Usability · Eye tracking

1 Introduction

In Korea, the air purifier market has been growing at the rate of more than three times since 2013 owing to the problem of fine dust [1]. In addition, in an analysis of the air purifier market, it was found that air purifiers are as an essential home appliance for Korean households, with more than 1 million units sold per year [2]. As products with various functions and designs are produced owing to increasing demands, products satisfying users' technical needs are increasing; however, users' understanding of products is decreasing and product operability is becoming increasingly complex, while the users' satisfaction for these products are gradually decreasing [3]. To increase user satisfaction, it is necessary to study the interface functionality, which is closely related to the presentation of visual information. Visual activity in humans greatly influences the visible environment, emotions, and their day to day life. Moreover, visual perception is the most crucial among the five senses as it determines the aesthetic quality of an environment; this is because 87% of human perceptions of an environment depends on the visual information presented to and received by them [4]. Among the various methods used in

© Springer Nature Switzerland AG 2021
M. Singh et al. (Eds.): IHCI 2020, LNCS 12616, pp. 363–369, 2021.
https://doi.org/10.1007/978-3-030-68452-5_38

the study of product interfaces, the most typical evaluation method using human visual cognitive characteristics is that of eye tracking. Eye tracking refers to the technology that allows for the analysis of visual behavior for a given visual stimulus by monitoring gaze tracking through pupil center recognition and cornea reflection [5]. When a person stares at a specific object, he recognizes and pays attention to it; this can be interpreted as a cognitive behavior of the person toward the object [6]. A review of existing studies on product design interfaces using eye tracking revealed the following. Eun-sun Seo (2016) investigated the use of mobile eye trackers to promote eye tracking research and demonstrated the usability of eye tracking in design studies such as user behavior analysis and user interface analysis [7]. Ho-Hyun Jang (2011) analyzed the graphical user interface (GUI) using an eye tracker while considering the design and device characteristics of various navigations and proposed 8 methodologies for developing a user interface design model. [8]. In addition, Jeong-yong Kim et al. (2017) obtained the appropriate design elements for an air purifier through a preference survey to derive its external design and attempted to design a physical interface through an icon eye-focusing experiment using eye tracking technology [9]. Similarly, several studies have been conducted to perform icon evaluation for air purifiers or navigation using eye trackers. However, only few studies have directly evaluated the UI usability of air purifiers available in the market. To this end, the present study evaluates the usability-related shortcomings of air purifiers and attempts to improve the usability of air purifiers through the evaluation of the UI of the control panels of the air purifiers currently available in the market.

2 Experiment Method

2.1 Participants for Experiment

The participants for this study comprised 28 housewives. They signed an agreement form after being informed about the scope of the study and meeting the selection criteria. None of the participants showed any visual impairments or cognitive malfunctioning. Their demographic information is presented in Table 1.

Table 1. Demographics of the participants.

Classification	Participants (Mean ± SD)
Age(year)	38.11 ± 5.76
Height(cm)	161.18 ± 5.26
Weight(kg)	57.82 ± 9.16

2.2 Operation Steps of Air Purifier

The air purifier used in this experiment has an intake port raised at its upper part for the intake of intensive harmful gases that are released during cooking food in a kitchen or table. The air purifier was operated in 7 steps, i.e., Step 1: Power on, Step 2: Air volume control (air cleaning start), Step 3: Pop-up operation (intake rise), Step 4: Swing control (intake port rotation), Step 5: Neutralize the intake port (return to its original position), Step 6: Pop-down operation (lower the intake and start air cleaning), and Step 7: Power off.

2.3 Configuration of UI Buttons of Air Purifier Control Panel

To understand the effects of the UI design of a control panel on the usability of the purifier, the control panel for each step was set in 4 button icon-word combinations, and the users were asked to select their preferred button types. The icons and words were suggested by the developers of the purifier and were amongst the most commonly used ones. Those with the highest user preference were selected.

2.4 Eye Tracking

Tobii X2 eye trackers, which provides a sampling rate of 30 Hz and an effective distance of approximately 40–90 cm, was used for eye tracking. The IMOTIONS (IMOTIONS INC.) analysis program was used to analyze the scan path and an AOI sequence and heat map were used for analyzing the measured data.

2.5 Experiment Procedure

The participants were asked to sit on chairs, each approximately 60 cm away from the display device and view the test video and experiment object on the screen. The laboratory was kept silent and clean to help the participants concentrate on the monitor. The participants performed 7 panel tasks, which involved selecting the correct (most appropriate) button corresponding to a prepared air purifier usage video from the 4 combinations of UI buttons (A, B, C, D). During the experiment, the visual responses of the participants were continuously measured and automatically stored using the IMOTIONS software.

2.6 Method of Analysis

In the present study, it was assumed that the UI of the control panel has a low success rate in operation tasks, and the preferred UI characteristics of the control panel were analyzed for an operation task with a success rate of 70% or less to determine the preferred control panel shape. To understand a preferred control panel UI, the gaze time for each area was compared through heat map analysis and comparison, and AOI sequence area setting (Fig. 1).

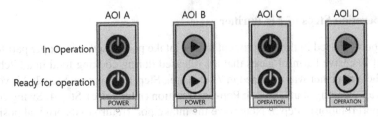

Fig. 1. Example of AOI area setting.

3 Results

3.1 Accuracy of Air Purifier Operation

The results of the task accuracy analysis show that Steps 1, 5, and 7 have accuracies less than 70% (Fig. 2). Additionally, several errors occurred during button selection when neutralizing the intake port and lowering the intake, which are Step 5 (accuracy: 39.29%) and Step 6 (accuracy: 60.71%), respectively. However, Step 1 (power on) and Step 7 (power off) had success rates of 60.71% and 89.92%, respectively. Considering that both tasks required using the same button, their average success rate was 75%, suggesting that the accuracy of recognizing the power button increased with experiences.

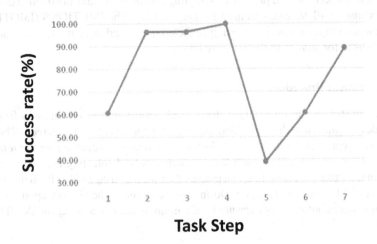

Fig. 2. Accuracies of each step.

3.2 Selection Time of Air Purifier Control Panel UI

3.2.1 Step 5 Task

From the results for Step 5, many participants selected the panel design in the order, C → B→A → D. When looking at the details, more users selected the button image in C than the existing image on the far left (A). This is because of the concise image that symbolizes the neutralization of the intake port. In addition, the heat map analysis revealed that most of the participants gazed at the Korean characters ("Neutral") displayed at the bottom of the button (Tables 2 and 3).

Table 2. Selection rate for different panel designs.

	Selection Rate (%)			
	A	B	C	D
Step 5	14.29	21.43	60.71	3.57
Step 6	7.14	32.14	17.86	42.86

Table 3. Gaze time per AOI area for different panel designs.

	Gaze time per AOI area (sec)			
	A	B	C	D
Step 5	2.3368	3.7814	6.3661	1.7429
Step 6	1.968929	4.330714	5.170714	5.547857

3.2.2 Step 6 Task

From the experiment result for Step 6, several participants selected buttons in the order, $D \rightarrow B \rightarrow C \rightarrow A$. D was the most selected button at 42.86%, and A was the least selected one at 7.14%. Compared to the existing image on the far left (A), participants preferred the image in D, which is a concise image representing the downward direction of the intake port, and the heat map analysis revealed that the participants gazed at the Korean characters ("Pop-down") at the bottom of the button.

4 Discussion

From the control panel UI measurement through eye tracking, the operation accuracy of the air purifier was mostly 70% or more. Two out of seven tasks had an accuracy less than 70%, and in particular, it was difficult to intuitively recognize the control panel when the intake port had to be lowered after it was raised and used. In addition, users' visual analysis heat map and AOI area gaze time were analyzed to select a preferred button for the task that obtained an accuracy of less than 70% in the prior operation task. In the task of neutralizing an intake port, the heat map analysis showed that placing the text 'Neutral' below the button attracted users' attention more than when the text 'Neutral' was placed in the image inside the button. This shows that the users obtained functional information regarding a button from the words below the button rather than the image on the button. In addition, in the task of lowering the intake port, the heat map analysis showed that more attention was on the text 'Pop-down' than the existing text 'Air volume.' This means that users did not immediately recognize the word 'Air volume' as having a function to lower the intake port when obtaining the information regarding the button function.

Jang-Seok Kim (2016) conducted an image survey for 30 male and female students to analyze emotion evaluation for household air purifier designs to build an indoor air purifier that fits well [10]. Jin-woo Lee (2008) conducted a questionnaire survey for 248 men and women over 20 years old on the design and development of a customized modular type air purifier to derive a design that applied mass customization to satisfy various consumer preferences in air purifier design [11]. In addition, Kyu-hong Min et al. (2012) proposed a qualitative design evaluation method focusing on consumers' consciousness and desire through a reaction survey on the usability of air purifier in the paper "Proposal of a qualitative method in design evaluation techniques" [12]. Similarly, all previous studies on air purifier design proposed designs based on qualitative questionnaire data, whereas in the present study, an efficient design concept for a control panel using eye tracking was proposed.

5 Conclusion

This study was conducted to evaluate the preference of the UI design for the control panel of air purifiers that are currently available in market. An eye tracking experiment with 28 housewives was conducted to understand the effect of the control panel shape and the word image on the usability of the air purifier. The results showed that the icon shape and the word significantly impact the effectiveness of a button when using an air purifier. For example, when the same word is used, a clear word displayed under the button is preferred over when the word is written inside the icon; in addition, the shape of the icon must be clearly displayed. It was concluded that icons and characters must be clearly displayed to improve readability, efficiency, and effectiveness of a control panel. If a product has various functions, its complex functions and buttons may make users feel uncomfortable. Therefore, products should be developed such that their roles are fulfilled, and all the functions are used efficiently. The evaluation of an air purifier panel through eye tracking therefore provided a highly quantitative result.

Acknowledgments. This research is supported by "The Foundation Assist Project of Future Advanced User Convenience Service" through the Ministry of Trade, Industry and Energy (MOTIE) (R0004840, 2020) and Basic Science Research Program through the National Research Foundation of Korea(NRF) funded by the Ministry of Education (NRF-2017R1D1A1B04031182).

References

1. Kim, J.B., Kim, J.H., Kim, K.H., Bae, G.N., Yoo, J.W.: Performance evaluation of air purifier by indoor air condition in the residential apartment. J. Odor Indoor Environ. **18**(2), 91–101 (2019)
2. KTB Investment & Securities: Companies favorable particulate matters. Issue & Pitch (2019)

3. Ha, Y., Choi, G.W., Kim, H.S., Ahn, J.H.: A study on developing user-centered interface of consumer electronics based on log analysis. J. HCI Soc. Korea 179–184 (2007)
4. Suh, J.H., Park, J.Y., Kim, J.O.: A comparative analysis of visual preference and cognitive processing characteristics using an eye-tracking method - with a focus on the 50 sites for rural development project. J. Digit. Des. 15(3), 335–343 (2015)
5. Maughan, L., Gutnikov, S., Stevens, R.: Like more, look more. Look more, like more: the evidence from eye-tracking. J. Brand Manage. 14(4), 335–342 (2007)
6. Kim, G.H., Boo, S.H., Kim, J.H.: Effects of depth perception cues in visual attention to advertising using eye tracker. Korean J. Advert. Pub. Relations 9(2), 277–310 (2007)
7. Seo, E.S.: mobile eye tracker and for use of the same for revitalizing studies on eye tracking. Int. J. Contents 16(12), 10–18 (2016)
8. Jang, H.H.: A study on interface design model from the eye-tracing of visual consciousness-focus on interface of navigation device. J. Digit. Des. 11(3), 333–343 (2011)
9. Kim, J.Y., Kim, D.J., Kim, H., Lee, J.C., Lee, Y.J.: Design of filter-less air purifier by using UI/UX metrics. Ergon. Soc. Korea 192–196 (2017)
10. Kim, J.S.: Emotion evaluation analysis on the design of air purifier for home. J. Korea Des. Forum 52, 157–168 (2016)
11. Lee, J.W.: A study on the development of customizing the modularized air cleaner of user. Master thesis. Graduate school of Hanyang University (2008)
12. Min, K.H., Chung, D.S.: Suggestion of qualitative method on the design evaluation method -with the cases of air cleaner and softener-. Korean Soc. Basic Des. Art 13(1), 143–152 (2012)

AI-Based Voice Assistants Technology Comparison in Term of Conversational and Response Time

Yusuph J. Koni[1], Mohammed Abdulhakim Al-Absi[1],
Seyitmammet Alchekov Saparmammedovich[1], and Hoon Jae Lee[2(✉)]

[1] Department of Computer Engineering, Dongseo University, 47 Jurye-ro, Sasang-gu,
Busan 47011, Republic of Korea
yusuphkoni@gmail.com, Mohammed.a.absi@gmail.com,
mslchekov@gmail.com
[2] Division of Information and Communication Engineering, Dongseo University, 47 Jurye-ro,
Sasang-gu, Busan 47011, Republic of Korea
hjlee@dongseo.ac.kr

Abstract. With the rapid development of artificial intelligence (AI) technology, the field of mobile phone technology is also developing in the direction of using artificial intelligence. Natural user interfaces are becoming popular. One of the most common natural users interfaces nowadays are voice activated interfaces, particularly smart personal assistants such as Google Assistant, Alexa, Bixby, and Siri. This paper presents the results of an evaluation of these four smart personal assistants in two hypothesis: The first hypothesis is presents the results of an evaluation of these digital assistants in how the application is very good in interacting based on conversation, means in how the app can be able to continue answer questions in depending on the previous answer, and the second hypothesis is to measure how long it take for an app to respond on those conversasion, doesn't matter if the app was able to keep on conversation or not. To determine this conversation and time taken on those conversation, four apps are install into four Samsung smartphone and defferent continuous question was asked after completed an answer previous question at the same time to all voice assistance. The overall result shows that, Google assistant was able to keep answer question by depending the previous answer without mis any question compare to the other digital assistants respectively but is very slowly app in responding to those conversation. Siri mis to keep on conversation on few question compare to Bixby and Alexa but is first app to respond to those conversation. Bixby was able to keep on conversation only in mathematical question but in other question was very poor but in case of time taken to respond is the thirsd app after Alexa. Alexa keeps on conversation in some question more than Bixby and in term of time taken to respond Alexa is second after Siri.

Keywords: Voice assistants · Google assistant · Alexa · Siri · Bixby · Digital assistants

© Springer Nature Switzerland AG 2021
M. Singh et al. (Eds.): IHCI 2020, LNCS 12616, pp. 370–379, 2021.
https://doi.org/10.1007/978-3-030-68452-5_39

1 Introduction

With the recent boost in Artificial Intelligence (AI) and Speech Recognition (SR) technologies, the Voice Assistant (VA), also known as the intelligent personal Assistant (IA), has become increasingly popular as a human-computer interaction mechanism. Voice assistants, such as Amazon Alexa, Samsung Bixby, Google Assistant, and Apple Siri, are widely adopted in smart devices and smartphones [1–4, 6] Fig. 1. Instead of interacting with the smartphones by touching the screen, users can send the voice commands to activate the voice assistant on smartphones, and ask him to perform the tasks, such as sending text messages, browsing the Internet, playing music and videos, and so on [5].

While voice assistants bring convenience to our daily lives, it also offers a backdoor for hackers to hack into our smartphones. Voice assistant on smartphones are usually granted high privileges to access various apps and system services, thus it can be utilized as a weapon to launch attacks to our smartphones, such as to unlock smartphones without fingerprints or PINs, forge emails, control smart home devices, and even transfer money [6]. For example, after activating the Google Assistant with the keywords "OK Google", you can send money to your contacts through Google Pay with the command such as "send Bob $50 for the dinner last night.

AI has suddenly become popular, and major technology companies have deployed one after another, and the current consumerized form of AI is undoubtedly the "voice assistant".

Virtual assistants are there, around all of us, waiting to get noticed, tried out and becoming a part of our lives and because talking is a fundamental part of being human, and because we interact with voice assistance and we try to imagine these assistances should design to talk and live like us and of course, there are voice assistants such as Amazon Alexa and Microsoft Cortana, but our focus today is to see which mobile phone-based voice assistant is better in terms of answering question by depending the previous answer or in other word conversation instead to base only on command and also we focus on their responding time on those conversation. While this will give us a good way to interact with our devices. And of course, there is very little conversation to our voice assistant than commands and yes, some voice assistant is much better to interact compare to other voice assistant.

2 Available Virtual Assistants

As I have already mentioned, there are a few assistants that most people have already heard of. Those are Apple Siri, Google Assistant, Amazon Alexa, Bixby and Microsoft Cortana.The market of virtual assistants is enormous, and we can see small companies or even groups of people trying to compete with the giants. Many AI assistant implementations are expensive, and we have to pay for them. There are a lot of other companies working on their virtual assistants like Samsung Electronics, Blackberry Limited even Facebook has their virtual assistant called M.

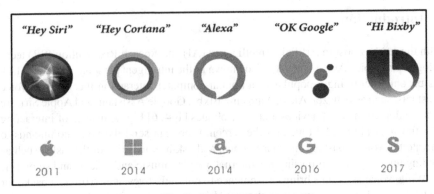

Fig. 1. Digital assistants

2.1 Amazon Alexa

Amazon Alexa was first introduced in November 2014 alongside Amazon Echo. At that time it was primarily a smart speaker with an option to control the user's music via voice. Alexa was the virtual agent that was powering the speaker and since then has evolved into the center of a smart home. Alexa was inspired by talking computer systems that appeared in Star Trek. Alexa's name is also a reminder of the library of Alexandria [7].

Alexa's transformation from a pure music-playing gadget into the center of a smart home wasn't as fast as people might think and even its developers did not think about this future use of Alexa. As the smart home devices came to market before Alexa and Echo, there was no simple way to use them, unless the user had a controller or a smartphone. The manufacturers of these smart home devices started looking for something that would unite and simplify the use of their products and thought that Alexa could be the solution to their problems since it was there in people's living room right next to their devices, so they started calling the developers of Alexa with offers to cooperate. Therefore Alexa slowly took its shape as we know it today [8].

It runs with a natural-language processing system, which is controlled just by talking to it. Alexa, like many other assistants, is always listening to what the user says, waiting for her time to wake up, and action the user's commands or answer his questions. By default, to wake Alexa up, the user has to call it by name "Alexa" and then he can follow with a question or command. This "Alexa" word is called a 'wake-up word'. The user can also change it to "Computer," "Echo" or "Amazon."

The listening system, if it is part of the Amazon Echo, is powered by many independent, sensitive microphones, which give Alexa the ability to hear the user across the room loud and clear. When the user wakes Alexa up with the wake-up word, it records anything the user says and sends it to Amazon's cloud computers which analyze the recording. Depending on the user's request the appropriate response is made. From playing music the user requested (if the user has a subscription for music streaming platform compatible with Alexa) through supplying any current weather conditions the user asks for, to operating the user's smart products such as thermostats, lights, computers or consoles, pretty much anything connected to Alexa [9].

2.2 Google Assistant

Google Assistant was first introduced as part of the Google Home smart speaker at Google's developer conference in May 2016. At first, it was just part of a chat application, and it was far from perfect. Proof of special treatment in the development of Google Assistant is that Google hired ex-Pixar animator Emma Coats to create and give the Assistant a little more personality. Google wanted to make the assistant more friendly and it has been done by giving it "personality", and a few funny answers. Google Assistant should behave like it had a life and people could relate to it [10]. Before the first appearance on Android smartphones running Android 6.0 Marshmallow or Android 7.0 Nougat in February 2017, it was exclusive to Pixel and Pixel NL smartphones. After that Google Assistant slowly found its way to Android Wear 2.0, Android TVs, laptops, cars and tablets. Google Assistant was also released to iOS operating system devices as a stand-alone app in 2017 [11].

Initially, Google Assistant was operated by voice. However, user interaction with Google Assistant can be done in two ways since 2017. The other way to command Google Assistant is by typing. That approach might be useful in certain situations. For example in a loud environment when the assistant would not understand the voice command correctly. On the other hand, this is a helpful feature in a quiet environment, when the user does not want to be heard by people around him. Furthermore, this feature completely changes the behavior of the assistant replying. The reply is not spoken as usual, it is strictly visual [12]. On May 08 2018 Google announced an extension to Google Assistant, called Google Duplex. Google Duplex is an AI system for conducting natural conversations. Google Duplex technology itself is not able to make general conversation without being intensely trained. The technology allows people to have as natural conversation with the machine as possible, without people having to adapt to a computer. Google Duplex technology was designed to perform tasks, such as scheduling individual appointments. For example, it can schedule.

2.3 Samsung Bixby

Bixby is the Samsung intelligence assistant. First it made its debut on the samsung Galaxy S8 and S8+. It's deeply integrated into the phone, meaning that Bixby is able to carry out a lot of the tasks you do on your phone but is designed to work across a range of Samsung products and is incorporated in numerous other devices like Samsang's family hub fridge and TVs.

Bixby is available on Galaxy S20, S20+, S20 Ultra, Note10, Note10+, S10e, S10, S10+, Fold, Note9, S9, S9+, Note9, S8, and S8+. Bixby service availability may vary depending on country; Bixby recognizes selected languages and certain accents/dialects; user interface may change and differ by device; availability of Bixby features and content providers may vary depending on country/carrier/language/device model/OS version; Samsung Account log-in and data network connection required.; Bixby controls selected apps, other apps to be supported.

2.4 Apple Siri

Siri is a virtual assistant created by Apple Inc. And is included on all their devices with watchOS, macOS, tvOS and of course iOS operating systems. But Siri was not initially created by Apple. In the beginning, it was an application for the iPhone developed by the 24-member startup team. Later this application was acquired by Apple for about $200 million [13]. The startup wanted something easy to remember, short to type, comfortable to pronounce, and a not-too-common human name. As a surprise, Apple kept the original name [14].

In 2009, before Steve Jobs contacted the Siri startup, they signed a deal to make Siri a default app for all Android phones, but when Apple bought Siri they insisted on Siri being exclusive to Apple devices. After Apple acquired Siri, they updated Siri's language and expanded its linguistic range from one to multiple languages. Apple also integrated Siri into the iPhone, so it could handle dozens of Apple's tools such as task scheduling, replying to emails, checking the weather, etc. Siri, as a part of the iPhone, was first introduced at a 2010 tech conference. It was able to buy tickets, register for a separate service, place a call or reserve a table, summon a taxi, all without the user having to open another app. At that time Siri was able to connect with 42 different web services, such as Wolfram Alpha and Yelp and return a single answer that integrated the best information obtained from those sources [15].

Currently, Siri is a part of all iPhones, Apple Watches, Apple HomePods, and other Apple devices. Since Siri is a smart virtual assistant that people buy and place in the middle of their homes, Siri's compatibility, options, and skills spread even more. Siri nowadays can control lights, thermostats, switches, outlets, windows, air conditioning and so many other smart devices the user can purchase. There is a full list of compatible devices on Apple's website [16].

3 Experimental Results

For my experiment, I have installed 4 different apps in smart phones. As you see on the Fig. 2 and I ask about 21 continuously questions according categories at the same time to all apps. If the app give me the succesfully respond to my question, I give the point to that apps as you can see on the Fig. 2 (b) on the left up corner of the figure, Where by Siri is succesful respond to my question and I give point 01 and I give 00 to other those Voice assistance that doesn't respond to my questions as you see of the Fig. 2 (d) on the left up corner of the figure, Where by Amazon Alexa is not succesful respond to my questions.

My first hypothesis is to see which phone will be able to keep on conversation and my second hypothesis is to measure the responding time of every apps to the specific questions compare to other apps, For example as you can see in the Fig. 2, I have already ask the question at the same time to all apps and the result came to show that Fig. 2(d) which is Amazon Alexa, is first to respond but it couldn't be able to keep on conversation and Fig. 2(b) which is Siri app, is second to respond with successful answer and is able to keep on conversation while Fig. 2(a) and (c) there are still in re-retraining process.

The categories that I was based on asking 21 questions to my apps are:

- "Mathematical question."
- "Internet Queries Question."
- "Control Within Apps."
- "General Questions."

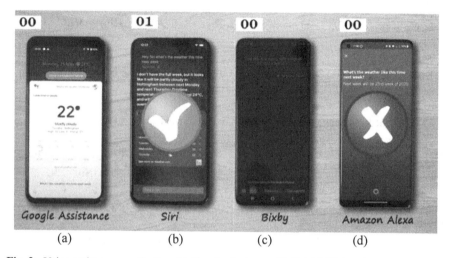

Fig. 2. Voice assistance application (a) Google Assistant (b) Siri (c) Bixby (d) Amazon Alexa.

Below tables are separated questions and the respond of voice assistance apps both succesfull and those which are not succesful according to different conversation. I separate this tables to show that some apps are good in keeping conversation on one categories and very poor to other categories, for example at Table 1, Bixby seems to be very good in keep conversation on Mathematical questions and is poor in keeping conversation on other categories, Also I separate this tables after some app failed to keep on conversation.

The word "(No)" in every table below represent the voice assistant app that was not be able to keep on conversation, all those number outside the brackets standing for position of app based on responding time in every question and the number inside the brackets stand for the time take for every app to respond to the specific question (Tables 2, 3, 4, 5 and 6).

Table 1. Mathematics conversation question and their responding time.

Question	Google assistance	Siri	Bixby	Alexa
"26*25" = 650	2(02.02 s)	1(01.09 s)	4(03.09 s)	3(02.23 s)
"26 + 25*26" = 676	3(02.70 s)	1(00.90 s)	2(02.43 s)	2(02.43 s)
"(26 + 25*26)/5" = 135.2"	3(02.11 s)	4(02.67 s)	2(01.55 s)	1(01.03 s) (No)
"135.2 * 999"	3(01.59 s)	1(00.79 s)	2(01.27 s)	2(01.27 s) (No)

Table 2. Internet Queries conversation question and their responding time.

Question	Google assistance	Siri	Bixby	Alexa
"Who is Robert Downey Jr."	3 (1.96 s)	1(1.37 s)	4(2.42 s)	2(1.47 s)
"How tall is he?"	1(1.55 s)	2(1.95 s)	3(2.32 s) (No)	1(1.55 s)
"Does he have any children?"	1(1.16 s)	2(1.33 s)	4(1.71 s) (No)	3(1.51 s)
"Take me to the Wikipedia page of Robert Downey Jr."	2(1.29 s)	1(1.07 s)	3(1.66 s) (No)	3(1.66 s) (No)
"Is he Spanish"	4(2.24 s)	1(1.09 s)(No)	2(1.55 s) (No)	3(1.86 s) (No)

Table 3. Internet Queries conversation question and their responding time.

Question	Google assistance	Siri	Bixby	Alexa
"Show me Jackie Chan Movies"	2(1.99 s)	2(1.99 s)	1(1.30 s) (No)	3(2.27 s)
"Tell me more about Rush hour"	4(2.36 s)	1(1.03 s)(No)	2(1.53 s) (No)	3(1.72 s)
"How much did the film gross"	2(1.54 s)	1(1.32 s)(No)	1(1.32 s) (No)	2(1.54 s)
"And which other actors are in it"	4(1.77 s)	1(1.12 s)	2(1.33 s) (No)	3(1.58 s)

Table 4. Control Within apps conversation question and their responding time.

Question	Google assistance	Siri	Bixby	Alexa
"Show me my photo from 2020?"	4(3.00 s)	1(1.85 s)	2(2.46 s)	3(2.69 s) (No)
"Which of them had cats in them?"	4(2.38 s)	1(0.82 s) (No)	2(1.09 s) (No)	3(1.35 s) (No)

The below tables show the average position of app based on responding time. The number outside the blackets represent the position of every app based on responding time and the number inside the blackest is the average of time taken for all apps to respond to the question doesn't matter if the app was not be able to keep answer questions depending on the previous answer. However, some apps take a short time to respond in the other hand some are taking a long time.

Table 5. General conversation question and their responding time.

Question	Google assistance	Siri	Bixby	Alexa
"What is my name"	**3**(1.46 s)	**3**(1.46 s)	**2**(1.26 s)	**1**(0.82 s)
"That is not how you pronounce it"	**3**(1.25 s)	**1**(0.77 s)	**2**(1.04 s) **(No)**	**2**(1.04 s) **(No)**

Table 6. General conversation question and their responding time.

Question	Google assistance	Siri	Bixby	Alexa
"Remind me to pick up a key in 10 min"	**4**(1.98 s)	**3**(1.79 s)	**2**(1.58 s)	**1**(1.25 s)
"Now delete that reminder"	**3**(1.39 s)	**1**(0.75 s)	**2**(1.08 s) **(No)**	**4**(1.63 s)

Table 7. Total position of app based in responding time.

Voice assistance application	Average responding time(sec)
Google assistance	**4**(1. 70 s)
Siri	**1**(1.20 s)
Bixby	**3**(1.55 s)
Amazon Alexa	**2**(1.47 s)

4 Conclusion

This paper described the results of an evaluation of four intelligent personal assistants, to identify the best assistant based on how a voice assistance can be able to keep conversation question and how faster the app can be able to respond. The study included the most popular personal assistants on the market: Siri, Bixby, Alexa, and Google Assistant. According to result on the table above.

Google Assistant: Is first app base on conversation which means it was able to keep on conversation without missing any question but according to time take to respond those conversation compare to other app is the last app as you can see on Table 7.

Siri: Unlike Google assistant, According to Table 7, Siri is the first app to respond even though in somehow it couldn't be able to keep on conversation compare to Google assistant and Its the second apps to use because somehow it was successful to keep on conversation compare to Samsung Bixby and Amazon Alexa apps.

Bixby: According to Table 7, Bixby is third app based on responding time but was very poor apps out of four because it was able to keep on conversation only mathematical question as you see on the Table 1 and it was not be able to keep on conversation on other questions.

Amazon Alexa: According to Table 7, Amazon Alexa is second app which take very short time after Siri but this app it couldn't be able to keep answer question depending on the previous answer compare to Google assistant and Sir.

According to my result there is both advantage and disadvantage in all four app for both user and attacker, because for user is very helpful to a lot of activity if the app is very good in answering question depending on previous answer and take short time to respond but also will be easy for attacker to attack because if the attacker can then launch a series of malicious operations through the same voice assistance like Siri or Google assistance, once it has been very good in continuity conversations compare to other voice assistance and because of that you can found that some attack are possible in one voice assistant and is not possible to other voice assistance mostly because of the time taken for apps to respond or mostly because one voice assistance can be best to interact with and can be able to keep conversation where by other apps can not be able to do that.

All in all, real human conversations are very difficult. Furthermore, research shows that making a machine human may be necessary and even unreliable. Instead, we may need to think about how and why we interact with these assistants and learn to accept their advantages as machines.

Acknowledgment. This work was supported by the Institute for Information and Communications Technology Promotion (IITP) grant funded by the Korea government (MSIT) (No.2018–0-00245). And it was also supported by the Basic Science Research Program through the National Research Foundation of Korea (NRF) funded by the Ministry of Education, Science, and Technology (grant number: NRF2016R1D1A1B01011908).

References

1. Alepis, E., Patsakis, C.: Monkey says, monkey does: security and privacy on voice assistants. IEEE Access **5**, 17841–17851 (2017)
2. Aron, J.: How innovative is Apple's new voice assistant, Siri? (2011)
3. Diao, W., Liu, X., Zhou, Z., Zhang, K.: Your voice assistant is mine: how to abuse speakers to steal information and control your phone. In: Proceedings of the 4th ACM Workshop on Security and Privacy in Smartphones & Mobile Devices (SPSM 2014), pp. 63–74 (2014)
4. Knote, R., Janson, A., Eigenbrod, L., Söllner, M.: The What and How of Smart Personal Assistants: Principles and Application Domains for IS Research (2018)
5. Kiseleva, J., et al.: Understanding user satisfaction with intelligent assistants. In: Proceedings of the 2016 ACM on Conference on Human Information Interaction and Retrieval (CHIIR 2016), pp. 121–130 (2016)
6. Google: What can your Google Assistant do. Googl (2018). https://assistant.google.com/explore?hl=en-AU

7. Amazon engineers had one good reason and one geeky reason for choosing the name Alexa. Insider Inc. (2016). https://www.businessinsider.com/whyamazon-called-it-alexa-2016-7. Accessed 09 Apr 2019

8. How Amazon's Echo went from a smart speaker to the center of your home. Insider Inc. (2017). https://www.businessinsider.com/amazon-echo-andalexa-history-from-speaker-to-smart-home-hub-2017-5. Accessed 28 Jan 2019

9. What Is Alexa? What Is the Amazon Echo, and Should You Get One? Wirecutter, Inc. (2018). https://thewirecutter.com/reviews/what-is-alexawhat-is-the-amazon-echo-and-should-you-get-one/. Accessed 28 Jan 2019

10. Google wants to make its next personal assistant more personable by giving it a childhood. Designtechnica Corporation (2016). https://www.digitaltrends.com/mobile/google-assistant-getting-childhood/. Accessed 09 Apr 2019

11. Google launches Google Assistant on the iPhone. Verizon Media (2017). https://techcrunch.com/2017/05/17/google-launches-google-assistant-on-the-iphone/. Accessed 09 Apr 2019

12. You can finally use the keyboard to ask Google Assistant questions. Vox Media, Inc. (2017). https://www.theverge.com/2017/5/17/15648472/google-assistant-keyboard-input-announced-io-2017. Accessed 09 Apr 2019. The National Vulnerability Database. https://nvd.nist.gov/. Accessed 14 Aug 2019

13. Apple acquires Siri, developer of personal assistant app for iPhone. Quiller Media, Inc. (2010). https://appleinsider.com/articles/10/04/28/apple_acquires_siri_developer_of_personal_assistant_app_for_iphone. Accessed 09 Apr 2019

14. How Did Siri Get Its Name?. Forbes Media LLC (2012). https://www.forbes.com/sites/quora/2012/12/21/how-did-siri-get-its-name/. Accessed 09 Apr 2019. https://nvlpubs.nist.gov/nistpubs/Legacy/SP/nistspecialpublication800-64r2.pdf. Accessed 24 Oct 2019

15. SIRI RISING: The Inside Story Of Siri's Origins — And Why She Could Overshadow The iPhone. Verizon Media (2017). https://www.huffpost.com/entry/siri-do-engine-apple-iphone_n_2499165. Accessed 09 Apr 2019

16. Home accessories. The list keeps getting smarter. Apple Inc. https://www.apple.com/ios/home/accessories/. Accessed 21 Apr 2019

HCI Based In-Cabin Monitoring System for Irregular Situations with Occupants Facial Anonymization

Ashutosh Mishra$^{(\boxtimes)}$ ⓘ, Jaekwang Cha ⓘ, and Shiho Kim ⓘ

Yonsei Institute of Convergence Technology, Yonsei University, Incheon, South Korea
{ashutoshmishra,chajae42,shiho}@yonsei.ac.kr

Abstract. Absence of human in-charge in fully autonomous vehicles (FAVs) impose multi-pronged in-cabin monitoring in real-time to monitor any irregular situation. Moreover, occupant's visual information (such as video, images) transmission from in-cabin of vehicle to the data center and control-room is required for various in-cabin monitoring tasks. However, this information transfer may cause a substantial threat to the occupants. Though applying some patch on face of the occupants protects personal information however, simultaneously it deteriorates the important facial information which is very crucial in monitoring various tasks such as emotion detection. Therefore, a human-computer-interaction (HCI) is required to manage this trade-off between facial information and facial anonymization without any important information loss. Also, this HCI based in-cabin monitoring system should be implicit to retain the users trust in the intelligent transportation system (ITS). In this paper, we have proposed generative adversarial network (GAN) based in-cabin monitoring approach for emotion detection in FAV. In this method, we proposed to generate an artificial (virtual) face having real-facial expressions to provide facial anonymity to the occupants while monitoring FAV cabin. Therefore, this method provides anonymous facial information which is essentially required in detection and monitoring tasks to avoid irregular situations. We have used both publicly available and our own in-cabin dataset to assess our proposed approach. Our experiments show satisfactory results.

Keywords: Facial anonymization · Generative adversarial network · In-cabin monitoring · Human-computer-interaction · Irregular situation

1 Introduction

Fully autonomous vehicles (FAVs) are the demand of increasing population, smart cities, over-crowded transportation system, and urbanization [1–3]. In emerging intelligent transportation system (ITS) the FAV provides extra-time, luxury, and stress-free transportation experiences to the occupants in their daily commute [1]. Though, there are many previous reported research works which involve the monitoring of in-cabin. However, different work has been reported with different motive. Marcondes et al., reviewed in-vehicle monitoring for violence detection inside the FAV [4]. Khan et al., reviewed

© Springer Nature Switzerland AG 2021
M. Singh et al. (Eds.): IHCI 2020, LNCS 12616, pp. 380–390, 2021.
https://doi.org/10.1007/978-3-030-68452-5_40

for driver monitoring in autonomous vehicles [5]. Bell et al., have evaluated the in-vehicle monitoring system (IVMS) to reduce the risky driving behaviors [6]. Szawarski et al., patented the idea of monitoring a vehicle cabin [7]. In their claim, they reported to receive the FAV in-cabin image data that includes a vehicle seat, occupant within a vehicle cabin, orientation of the vehicle seat and an orientation of the occupant using the camera and the computer based system. Further, Song has considered the safety and cleaning issues of the in-cabin [8]. Fridman et al., have recommended the monitoring task for monitoring driver behavior [9]. They performed the in-cabin monitoring to study the driver behavior and interaction with automated driving so that similar approach can be implemented in FAV driving. However, the absence of in-charge enforces the in-cabin monitoring task to provide safety and security to both vehicle and occupants. These two features are most essential and challenging task related to in-cabin monitoring to achieve satisfaction of the passengers in FAVs. Safety of a person or thing is to ensure security against some unintended accidents. On other hand, security is the protection of a person from some deliberate threats. FAVs are vulnerable to both these risks mostly, because it has no any in-charge inside the vehicle, and all the occupants are passenger only. Therefore, in-cabin monitoring in real-time is essentially required to provide protection against any safety or security related risk.

1.1 In-Cabin Monitoring in Fully Autonomous Vehicles

Fully autonomous vehicles require to be monitored from inside for various purposes. Vehicle safety, passengers' safety, and passengers' security are the main goals of in-cabin monitoring [31]. These three main goals are confining many sub-goals as well. Broadly, related to vehicle safety, the sub-goals are to monitor any anomaly inside cabin of FAV, vehicle-control related issues, malicious intentions towards FAV, and/ or unintentional impairment situations related to FAV, etcetera. In case of the occupants, safety, and security are the goals of in-cabin monitoring. It further consolidates many sub-goals broadly, related to the object and event/ behavior detection. Emotion detection is necessary and an important task while monitoring the in-cabin of FAV. State-of-the-art researches in automotive driving recommends the occupant monitoring for safety and security related issues (e.g., in-vehicle violence, etc.) [4]. Further, the occupant monitoring become challenging task in transport vehicles to detect any abnormality such as unauthorized passengers, chaos-like situation inside the cabin, child detection, etc. [4]. However, emotion detection is the major task to achieve behavior monitoring and providing early safety related countermeasures in the Fully autonomous vehicles [10]. In-cabin monitoring tasks require local as well as cloud-based monitoring of the cabin of connected FAVs. However, this real-time image monitoring is severely affected by the revealing of important personal information such as the human identity [11, 12].

1.2 Privacy in Public Domain

Privacy in public domain has been considered as an issue because of the continuous technological development. Privacy in FAV mainly become an important concern because breach in the privacy (especially related to our life practices) may be used by others with some malicious intention and may leads to some potential harm [13–16]. Therefore, secrecy in privacy in public domain provide security and protection. Also, it protects an individual or group against any anomaly/ irregular situation. The Intelligent Transportation Systems (ITSs) are desired to be safe against privacy related threats [32, 33]. It is because of the vital information exchanged between the automotive vehicles and the infrastructures (i.e., data centers and control-rooms). Therefore, the ITS are supposed to be developed by considering mandatory provisioning for the privacy related threats. Because, privacy is the critical issue which must be taken into account while monitoring in-cabin of FAVs. Anonymity is one common way to secure the privacy. Therefore, creation of patches on face (such as mosaic) is the common solution used to hide real-face. Recently, generative adversarial network (GAN) in deep models is very popular approach in creating such patches [17, 18]. However, facial expressions are highly required features in emotion detection [19]. Figure 1. Illustrate the in-cabin monitoring schema. Here, the video sequences are transmitted to the data center and the control-room for detailed in-cabin monitoring regarding various aspects including monitoring suspicious and/or malicious activities of the occupants.

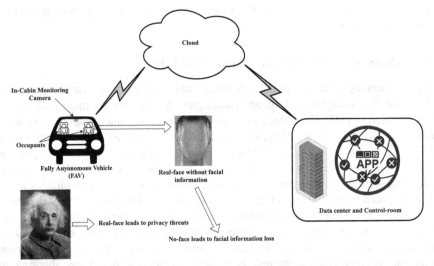

Fig. 1. Illustration of the in-cabin monitoring schema. The in-cabin monitoring (surveillance) camera captures the real-time video sequence and transmit it to the data center and control-room for further processing. The information of occupants without facial information is illustrated to show the privacy challenge related to in-cabin monitoring.

Here, main challenge is to protect the personal privacy while monitoring in-cabin of FAV. Therefore, in this paper, we have proposed the in-cabin monitoring of FAV for security purpose with privacy. In this work, we have maintained the real-facial expression with non-real faces to prevent security with privacy. We have utilized the concept of face swapping to real-world human face with suitable non-real-world human face to preserve the facial information. Because, facial information is the vital parameter in emotion detection. We have applied the state-of-the-art face swapping technique using recurrent neural network (RNN)–based GAN approach. In this method, we have superimposed a suitable source image on a target image to preserve its facial information of further emotion and thereby intent detection to provide security with privacy in FAV monitoring.

2 Related Work

Blanz et al., pioneered the face exchanging technique in 2004 [20]. They used the *'Morphable Model'* based approach to exchange the faces with changes in viewpoint and illumination. Thereafter, Bitouk et al., presented a complete system for automatic face replacement in 2008 [21]. Lin et al., shown 3D head model based face swapping [22]. Zhang et al., suggested to utilize machine learning for face swapping task [23]. Nirkin et al., have utilized fully convolutional network (FCN) for face segmentation and swapping [24]. Further, the research and development on the GAN based deep models facilitates the fake face generation. Region-separative generative adversarial network (RSGAN) has been introduced by Natsume et al., in 2018 [25]. Their system integrates the automatic generation and editing of the face images through face swapping. Further, Korshunov et al., presented a GAN based approach to detect 'DeepFake' videos [26]. They have used autoencoder-based Deepfake algorithm for face swapping in videos. Encoder-decoder network using GAN based FSNet has been represented by Natsume et al., to swap the human face [27]. Bailer et al., suggested face swapping for solving collateral privacy issues [28]. They have used GAN to de-identify original face from the image to secure privacy related threats. Nirkin et al., provided FSGAN for subject agnostic face swapping and reenactment between a pair of faces [29]. Naruniec et al., presented a fully automatic neural face swapping in images and videos [30]. Most of these approaches either utilize the face swapping or face reenactment. However, FSGAN has both capabilities. Therefore, in this work, we have utilized the concept of FSGAN to secure the privacy while monitoring emotion inside the cabin of FAV [12].

2.1 Face-Swapping Generative Adversarial Network (FSGAN)

FSGAN algorithm is introduced by Nirkin et al., for face swapping and reenactment [29]. They implemented the *'Face Swapping'* task for seamlessly replacing a real face with any non-real face such that after replacement (swapping) also it resembles a realistic face. In similar fashion, they introduced the *'Face reenactment'* task. It is a facial transfer or puppeteering such that a similar facial expression can be produced on the swapped face in the video or images. They utilized generative adversarial network (GAN) based approach to transfer a face from source image to target image. In their approach, they have considered a source image and a target image to create a new image based on the considered target image while retaining the same pose and expression as that of the target image. FSGAN is helpful in creating an agnostic face from pairs of faces with facial expressions (emotions) intact. They utilized RNN–based approach for face-reenactment. FSGAN based face- reenactment can be applied either to a single image or on a video sequence. Therefore, we have used this concept in our own in-cabin monitoring task to provide security with privacy.

3 Proposed Work

The emotion monitoring is a critical task which requires human facial expressions as the important features to deduce the behavioral aspects. However, use of real-human face as the monitoring input leads to the privacy threat in public domain. Therefore, there is an essential requirement of suitable countermeasure to maintain the privacy while monitoring in public domain. In this work, we have proposed a novel framework to maintain privacy secrecy while monitoring real-human emotions. We have used the human-computer-interaction based system to superimposition the artificial human-like face on the real-human face to protect both privacy of the real-human, and the facial information of the real-human. Concept of face swapping to maintain the privacy while keeping the original emotion on the face.

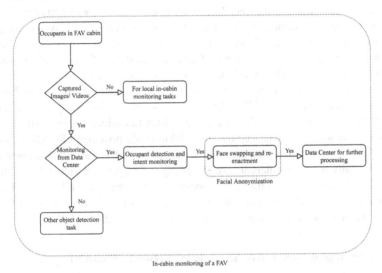

In-cabin monitoring of a FAV

Fig. 2. Flowchart of the proposed in-cabin monitoring approach of a fully autonomous vehicle (FAV). This approach provides facial anonymization to maintain facial privacy of the occupants in public domain while monitoring the in-cabin of an autonomous vehicle to avoid any irregular circumstances.

Figure 2 Provide the flowchart of the proposed approach and the framework has been presented in the Fig. 3. In this work, we have swapped the real-world human face with an artificial face which is not existing in the real-world to maintain privacy while monitoring in-cabin from control-room. Actually, the absence of in-charge inside the cabin of FAV put-forth the essential requirement of in-cabin monitoring to provide safety and security of the occupants and vehicle both. Further, in the connected domain of such FAVs all the occupants' information physical information has to be fed to the data center for rigorous monitoring to avoid any chance of irregularity. Though, there are certain monitoring task which are to be performed locally inside the FAV itself. However, these local monitoring task have no any facial privacy related threat. Further, the monitoring task which need to be performed at the data center require exact facial information as that of the passenger/occupant for accurate monitoring. Such as emotion, behavior, and/or intent monitoring of the occupants. Therefore, it requires the transmission of facial information from the vehicle to the data center. However, personal information breach is the challenge in this situation. Proposed framework helps in this situation to protect the personal information by using facial anonymization. The same is clearly depicted by the flowchart and framework of the proposed approach.

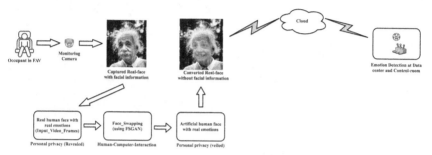

Fig. 3. Proposed framework to maintain privacy while monitoring in-cabin from control-room for in-cabin monitoring of fully autonomous vehicle.

We have considered the face swapping and reenactments by considering two factors gender-wise and as look-wise. This consideration enhances the exactitude into the newer generated face. In our experiment, we have considered male and female faces with eastern and western looks as the source images. Then applied on our own in-cabin dataset. Step-by-step description of the work is briefly explained below:

1. Capturing in-cabin visual data: Firstly, the in-cabin monitoring camera captures the visual information inside the cabin of FAV.
2. Facial anonymization: Face swapping and re-enactment is performed by applying GAN technique to retain the facial information after facial anonymization.
3. Data transfer to the data center: To proper monitor the emotion, behavior, and/or intent of the occupants inside FAV to avoid any irregular circumstances.
4. In-cabin monitoring: Finally, essential in-cabin monitoring is performed in FAV.

4 Results and Discussions

We have considered two females and two males as the source image each with eastern
and western looks respectively. For target images we have chosen different images with
various emotions from public domain as well as from our own dataset related to in-cabin
images. Figure 4 represents a western target image having 'sad emotion' on face and its
related artificial (virtual) generated faces.

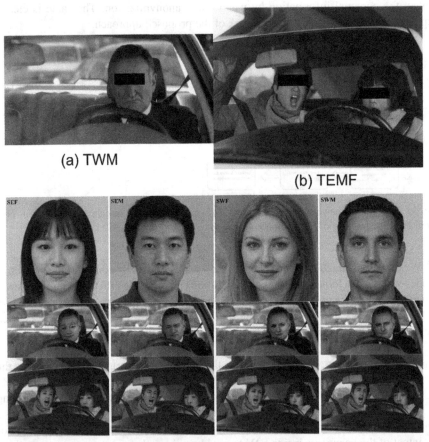

Fig. 4. Row#1: Original target image (a) containing western male face (TWM) with 'sad emotion,'
(b) eastern target image containing both male and female faces (TEMF) with 'shocking emotion,'
(face is masked for privacy). Row#2: Source images (from left) eastern female (SEF), eastern male
(SEM), western female (SWF), and western male (SWM), respectively. Row#3: Corresponding
face swapped images for a target image (TWM). Row#4: Corresponding face swapped images for
a target image (TEMF).

We have obtained the source images by an artificial intelligence (AI) based online platforms (Generated photos). It creates images from scratch by AI systems such that these images can be used as it is without any privacy or right related issues (e.g., copyrights, distribution rights, infringement claims, or royalties, etc.). Further, it is observed by Fig. 4 that sad emotion is perfectly resembled by the superimposition of SWM face on TWM image. Even SWF face on TWM image could not able to provide exact emotional representation. Furthermore, we have considered the TEMF having both male and female faces inside of the vehicle with same combination of source images. Here, we have considered the 'shocking emotion' on their face. Interestingly, the similar finding is observed in this experiment also. SEF source image has shown better result for eastern female target whereas SEM source image has performed better result on eastern male target.

Further, we have applied the proposed approach on our own in-cabin dataset by considering same source images (i.e., SEF, SEM, SWF, and SWM). The original target images have been shown in Fig. 5. Faces are masked for privacy purpose.

Fig. 5. Original eastern target image with 'shocking emotion,' from our own in-cabin dataset.

In Fig. 6 we have presented the results of the proposed approach on our own dataset. Again, the similar finding is observed here in this experiment. As our dataset belongs to the Asia subcontinent and all the occupants in the original target images are of eastern looks. Female faces on the target images are better represented with corresponding female face of eastern source image. Similarly, the male faces are found to carry appropriate facial information while applied with eastern male face source images.

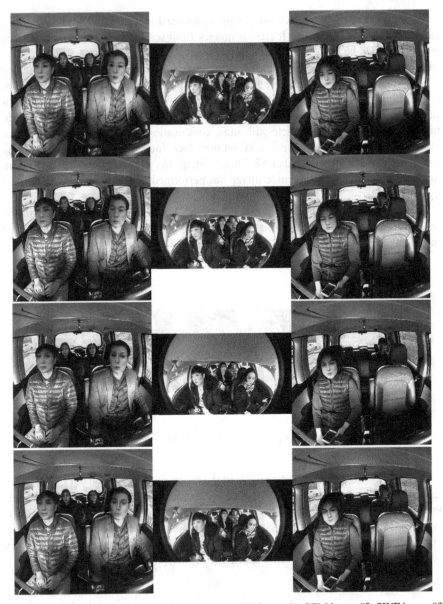

Fig. 6. Face swapped image corresponding to the: SEF in row#1, SEM in row#2, SWF in row#3, and SWM in row#4 for target images shown in Fig. 5, respectively.

5   Conclusions

In this work, we have proposed a human-computer-interaction based approach to preserve the facial information without any breach in privacy in public domain while monitoring in-cabin of the fully autonomous vehicle. In our experiment we have found that look-wise and gender-wise generation of artificial (virtual) face incorporate more appropriate facial information. Here, we have used GAN based approach to generate the non-real-word face of a real-world occupant such that the facial information remains intact. This eventually helps in further detailed and complex related to intent/behavior monitoring tasks from the data center and control-room. We have applied emotional faces in our experiment and found the proposed approach has provided satisfactory results. Furthermore, obtained results suggest that suitable deep models are required to be augmented in our proposed method to monitor the in-cabin of the FAV in detail.

Acknowledgment. This research was supported by Korea Research Fellowship program funded by the Ministry of Science and ICT through the National Research Foundation of Korea (NRF-2019H1D3A1A01071115).

References

1. Koopman, P., Wagner, M.: Autonomous vehicle safety: An interdisciplinary challenge. IEEE Intell. Transp. Syst. Mag. **9**(1), 90–96 (2017)
2. Yurtsever, E., Lambert, J., Carballo, A., Takeda, K.: A survey of autonomous driving: common practices and emerging technologies. IEEE Access **8**, 58443–58469 (2020)
3. Skrickij, V., Šabanovič, E., Žuraulis, V.: Autonomous road vehicles: recent issues and expectations. IET Intel. Transport Syst. **14**(6), 471–479 (2020)
4. Marcondes, F., Durães, D., Gonçalves, F., Fonseca, J., Machado, J., Novais, P.: In-vehicle violence detection in carpooling: a brief survey towards a general surveillance system. In: Dong, Y. (ed.) DCAI 2020. AISC, vol. 1237, pp. 211–220. Springer, Cham (2021). https://doi.org/10.1007/978-3-030-53036-5_23
5. Khan, M.Q., Lee, S.: A comprehensive survey of driving monitoring and assistance systems. Sensors **19**(11), 2574 (2019)
6. Bell, J.L., Taylor, M.A., Chen, G.X., Kirk, R.D., Leatherman, E.R.: Evaluation of an in-vehicle monitoring system (IVMS) to reduce risky driving behaviors in commercial drivers: comparison of in-cab warning lights and supervisory coaching with videos of driving behavior. J. Saf. Res. **60**, 125–136 (2017)
7. Szawarski, H., Le, J., Rao, M.K.: Monitoring a vehicle cabin. U.S. Patent 10,252,688, issued April 9 (2019)
8. Song, X.: Safety and clean vehicle monitoring system. U.S. Patent 10,196,070, issued February 5 (2019)
9. Fridman, L., et al.: MIT advanced vehicle technology study: large scale naturalistic driving study of driver behavior and interaction with automation. IEEE Access **7**, 102021–102038 (2019)
10. Bosch, E., et al.: Emotional GaRage: a workshop on in-car emotion recognition and regulation. In: Adjunct Proceedings of the 10th International Conference on Automotive User Interfaces and Interactive Vehicular Applications, pp. 44–49 (2018)
11. Qin, Z., Weng, J., Cui, Y., Ren, K.: Privacy-preserving image processing in the cloud. IEEE Cloud Comput. **5**(2), 48–57 (2018)
12. Xia, Z., Zhu, Y., Sun, X., Qin, Z., Ren, K.: Towards privacy-preserving content-based image retrieval in cloud computing. IEEE Trans. Cloud Comput. **6**(1), 276–286 (2015)
13. Taeihagh, A., Lim, H.S.M.: Governing autonomous vehicles: emerging responses for safety, liability, privacy, cybersecurity, and industry risks. Transp. Rev. **39**(1), 103–128 (2019)

14. Glancy, D.J.: Privacy in autonomous vehicles. Santa Clara L. Rev. **52**, 1171 (2012)
15. Collingwood, L.: Privacy implications and liability issues of autonomous vehicles. Inf. Commun. Technol. Law **26**(1), 32–45 (2017)
16. Lim, H.S.M., Taeihagh, A.: Autonomous vehicles for smart and sustainable cities: an in-depth exploration of privacy and cybersecurity implications. Energies **11**(5), 1062 (2017)
17. Goodfellow, I.J., Shlens, J., Szegedy, C.: Explaining and harnessing adversarial examples (2014)
18. Braunegg, A., et al.: APRICOT: A Dataset of Physical Adversarial Attacks on Object Detection. arXiv preprint arXiv:1912.08166 (2019)
19. Fox, E., Lester, V., Russo, R., Bowles, R.J., Pichler, A., Dutton, K.: Facial expressions of emotion: are angry faces detected more efficiently? Cogn. Emot. **14**(1), 61–92 (2000)
20. Blanz, V., Scherbaum, K., Vetter, T., Seidel, H.P.: Exchanging faces in images. In: Computer Graphics Forum, vol. 23, no. 3, pp. 669–676. Blackwell Publishing, Inc., Oxford and Boston (2004)
21. Bitouk, D., Kumar, N., Dhillon, S., Belhumeur, P., Nayar, S.K.: Face swapping: automatically replacing faces in photographs. In: ACM SIGGRAPH Papers, pp. 1–8 (2008)
22. Lin, Y., Wang, S., Lin, Q., Tang, F.: Face swapping under large pose variations: a 3D model based approach. In: IEEE International Conference on Multimedia and Expo, pp. 333–338. IEEE (2012)
23. Zhang, Y., Zheng, L., Thing, V.L.: Automated face swapping and its detection. In: IEEE 2nd International Conference on Signal and Image Processing (ICSIP), pp. 15–19. IEEE (2017)
24. Nirkin, Y., Masi, I., Tuan, A.T., Hassner, T., Medioni, G.: On face segmentation, face swapping, and face perception. In: 13th IEEE International Conference on Automatic Face & Gesture Recognition, pp. 98–105. IEEE (2018)
25. Natsume, R., Yatagawa, T., Morishima, S.: Rsgan: face swapping and editing using face and hair representation in latent spaces. arXiv preprint arXiv:1804.03447 (2018)
26. Korshunov, P., Marcel, S.: Deepfakes: a new threat to face recognition? Assessment and detection. arXiv preprint arXiv:1812.08685 (2018)
27. Natsume, R., Yatagawa, T., Morishima, S.: Fsnet: an identity-aware generative model for image-based face swapping. In: Jawahar, C.V., Li, H., Mori, G., Schindler, K. (eds.) ACCV 2018. LNCS, vol. 11366, pp. 117–132. Springer, Cham (2019). https://doi.org/10.1007/978-3-030-20876-9_8
28. Bailer, W.: Face swapping for solving collateral privacy issues in multimedia analytics. In: Kompatsiaris, I., Huet, B., Mezaris, V., Gurrin, C., Cheng, W.-H., Vrochidis, S. (eds.) MMM 2019. LNCS, vol. 11295, pp. 169–177. Springer, Cham (2019). https://doi.org/10.1007/978-3-030-05710-7_14
29. Nirkin, Y., Keller, Y., Hassner, T.: FSGAN: subject agnostic face swapping and reenactment. In: Proceedings of the IEEE International Conference on Computer Vision, pp. 7184–7193 (2019)
30. Naruniec, J., Helminger, L., Schroers, C., Weber, R.M.: High-resolution neural face swapping for visual effects. In: Computer Graphics Forum, vol. 3, no. 4, pp. 173–184 (2020)
31. Mishra, A., Kim, J., Kim, D., Cha, J., Kim, S.: An intelligent in-cabin monitoring system in fully autonomous vehicles. In: 17th International SoC Conference (ISOCC 2020), South Korea (2020)
32. Kim, S., Shrestha, R.: Automotive Cyber Security: Introduction, Challenges, and Standardization. Springer Singapore, Singapore (2020). https://doi.org/10.1007/978-981-15-8053-6
33. Shrestha, R., Kim, S.: Integration of IoT with blockchain and homomorphic encryption: challenging issues and opportunities. In: Advances in Computers, vol. 115, pp. 293–331. Elsevier (2019)

Achievement of Generic and Professional Competencies Through Virtual Environments

Zhoe Comas-Gonzalez[1], Ronald Zamora-Musa[2,3]([✉]), Orlando Rodelo Soto[1], Carlos Collazos-Morales[4], Carlos A. Sanchez[4], and Laura Hill-Pastor[5]

[1] Universidad de la Costa, Barranquilla, Colombia
`zcomas1@cuc.edu.co`
[2] Universidad Cooperativa de Colombia, Barrancabermeja, Colombia
`ronald.zamora@campusucc.edu.co`
[3] Universidad Pontificia Bolivariana, Medellín, Colombia
[4] Universidad Manuela Beltrán, Bogotá, Colombia
[5] Benemérita Universidad Autónoma de Puebla, Puebla de Zaragoza, Mexico

Abstract. This study is aimed to prove how Virtual Environments (VE) and Information and Communication Technologies (ICT) can be used as a tool to verify professional competencies. The incursion of virtual environments in education has shown that there is much potential in distance learning development. To find out how it influences the achievement of competencies, there was made an experimental study with a post-test design and control group. Students were divided into two groups; each of them was submitted to a different test. The results demonstrate that with the implementation of VE using ICT, the students who used the VE had a better performance than students who used the traditional evaluations. Confirmed with the 83% of the sample who achieved the highest levels (50% got strategical professional competencies, and 33% got autonomous professional competencies). Considering the study, the authors could notice that students do develop professional competencies along virtual environments, reflected not only in the level of competence achieved by the ones tested on the virtual environment but also in the average time they spend to do the test. Therefore, virtual environments have positives effects in the education field.

Keywords: Virtual education · Virtual environment · Pedagogical tools · Learning process · Self-learning · Competences evaluation

1 Introduction

Technology is an essential axis for the development of a country. There are thousands of projects in the education field related to the use of Information and Communication Technologies whose main purpose is to encourage the learning process's autonomy, covering a wide spread of disciplines, including engineering. It has generated changes in social and pedagogical schemes, including the way engineering classes are addressed [1]. On the other hand, the teaching process is the unity of teaching, whose main objective is the interaction; in e-learning, this interaction is bidirectional due to communication between

© Springer Nature Switzerland AG 2021
M. Singh et al. (Eds.): IHCI 2020, LNCS 12616, pp. 391–405, 2021.
https://doi.org/10.1007/978-3-030-68452-5_41

individuals, systems and other users, aimed to solve problems, complete learning task or challenges [2].

Using e-learning systems, teaching and learning has become common in the last years [3–7]. The Communications and Information Technologies will conduce to a new revolution that will involve more than just robotics, artificial intelligence [2] and data mining to improve implementation of virtual environment [8], taking us to a bigger change that the expected.

These virtual environments are supported by technological platforms oriented to distance learning, known as E-learning, which provide access to information without space/time boundaries [9]. These platforms can be used in any field besides bioengineering, like teaching domotics, which is a branch of automation [10].

According to what was related, this study is aimed to prove if Information and Communication Technologies influences positivity the development of competencies of electronics engineering students at Universidad de la Costa, using the virtual environment, built as part of the project. The paper is organized as follows: in Sect. 2, we presented several related works about Virtual Environments. In Sect. 3, we introduced the materials and methods to show the steps considered in the methodology. In Sect. 4, we presented the results and discussions, analyzing the results achieved in the groups defined for the experiment. This work concludes in Sect. 5, by the exposition of a series of conclusions and the acknowledgment for the contribution to this study to the individuals involved, and the identification of future work.

2 Related Work

There are several studies where researchers implemented novel immersive environments, but the main contribution is new engineering techniques applied to the development of virtual environments that improve the use of technological resources [11]. On the other hand, other studies show the effectiveness of using environments that make use of ICT in education [12–17], but these studies do not account for all the areas required in a competencies evaluation, for example, conceptual and attitudinal aspects.

According to [18] in virtual education in the last decade, immersive environments have emerged, which results from the effort to improve the effectiveness of learning processes in this type of education. Some of the pioneers [19–21] in integrating immersive environments into virtual education affirm that they promote effective learning. In the same way in [22, 23], they show that the learning outcomes in these environments are equivalent to the results in traditional education environments and concludes that research should focus on models that allow the evaluation of competencies or specific educational objectives.

Similarly, [24–26] show that these learning environments are beginning to apply extensive analysis regarding the interaction´s effectiveness and the continuous evaluation that the students develop. Current and future studies must consider the component associated with using models to acquire competencies [27, 28], enabling the complement between models such as the constructivist and the socio-formative to generate projects that facilitate the development of skills [29, 30].

The implementation of immersive environments based on Information and Communication Technologies (ICT) generates a positive impact when it is planned and designed according to the processes related to the user, that is, researchers must implement technological strategies, but they must also develop skills students [31–33]. The use of ICT in education without planning generates e-learning resources without the necessary elements to achieve the desired objectives [17, 25].

To summarize, we observed that current research had been developed referring to the immersive environments, which go from the novelty in its implementation, the effectiveness in the usability, favorable comparisons concerning environments of traditional education and proposal of possible competencies that can be developed by students. Therefore, the use of immersive environments should not be seen as a novel way of generating knowledge; its implementation should ensure the acquisition of skills in different knowledge areas how Engineering.

3   Materials and Methods

The sample, population and type of experiment were defined. The test was based on a quasi-experiment with a post-test design and control group [34]. It is applied research where technology is used and has a correlational scope because they associate concepts or variables; in this case, virtual environments and the achievement of generic and professional competencies.

In the quasi-experimental design, the subjects are not randomized; the groups are already formed because they are groups of students organized previously and independently of the experiment.

The control group was denominated "group A" and the experimental group as "group B". Students of "group A" (control group) made a paper test while students of "group B" (experimental group) used the virtual environment. The experimental condition is only applied to group B. Table 1 shows the quasi-experimental design.

Table 1. Quasi-experimental design.

Student groups	Experimental condition	Post-test measure
Group A (Control)	–	O1
Group B (Experimental)	X	O1

In the quasi-experimental design, "X" means stimulus or experimental condition. "O" means the measurement of the subjects of a group; if it appears before the experimental condition, it is a pre-test, and if it appears after the experimental condition, it is a post-test. Also, if the "O" symbols are aligned vertically, they mean they take place simultaneously. "-" means the absence of experimental condition and indicates that it is a control group.

As population, there were selected students of ninth semester of electronics engineering at Universidad de la Costa. As sample, there were selected the eleven students of ninth semester. According to these, five students were part of the "group A" and six students of "group B".

3.1 Selection of the Course

Bioengineering is a ninth semester class of the electronic engineering program at Universidad de la Costa (Unicosta). It is focused on the sense and measure of the human body's electrical behavior, understanding human anatomy and the medical equipment setup.

The literature review has shown a sort of virtual environment application in education and specific field issues. For example, there are projects related to telemedicine, telecare, tele-homecare and medical networks [35–37]. And some others, like biomedical equipment calibration, digital health, anatomy, medicine, biomedical robotics, among others, using virtual environments [36, 37].

Therefore, it is considered essential to carry out a study aimed at a specific situation such as developing skills in a subject of trend and practical theoretical as Bioengineering. This is focused on using biomedical equipment, for which students must know about the electrical signals of the human body; based on this, they are capable of making them maintenance and proposing technological improvements to prevent diseases previously diagnosed by doctors. The topics selected to be evaluated are the human eye's anatomy, the electrophysical behavior of the eye and the electrooculogram.

3.2 Instruments for the Evaluation

As part of this study, there were made pedagogical tools for tracing the development of student's competencies [38]. Rubrics and observation guides are useful for these cases. Rubrics allow to analyze student's achievements and abilities [39, 40]. Observation guides, by the other hand, allow to register and describe student's behavior at the moment they are doing an activity [39].

The rubric was made according to [40] methods, previously validated. Evaluation is based on proficiency levels: that guide the student in the achievement of skills, from the simplest to the most complex. The levels of proficiency allow verifying the professional skills enabling to improve the performance of the students. The categories established to test the professional skills achieved are: strategical, autonomous, resolutive, and receptive, and their description is the following:

- Strategical: ability to solve problems. Creative. Outstanding.
- Autonomous: autonomy person, understand problems.
- Resolutive: understand elemental process. Ability to solve easy problems.
- Receptive: low understanding of elemental problems. Operative person and mechanical acting.

The rubric of Table 2 allows evaluating the categories as mentioned earlier (strategical, autonomous, resolutive, receptive), by monitoring generic and professional competencies through virtual environments, from "strategical" the highest level of developed competence to "receptive" the level lowest developed competition.

Table 2. Bioengineering rubrics.

Category	Strategical	Autonomous	Resolutive	Receptive	Indicator
Understanding the information and problem solving	The student understands the information provided and has a strategy behavior in front of problems	The student understands the information provided and has an autonomous behavior in front of problems	The student understands elemental information and solve just easy situations	The student cannot understand the information and do not proposes solutions to problems	% of comprehension of the test in the virtual environment
Scientific knowledge (the anatomy of the human eye)	The student demonstrates high knowledge of the anatomy of the human eye	The student demonstrates certain knowledge of the anatomy of the human eye	The student demonstrates basic knowledge of the anatomy of the human eye	The student do not demonstrate knowledge of the anatomy of the human eye	% of knowledge of the anatomy of the human eye
Scientific knowledge (electro physical behavior)	The student demonstrates high knowledge of the electro physical behavior	The student demonstrates certain knowledge of the electro physical behavior	The student demonstrates basic knowledge of the electro physical behavior	The student does not demonstrate knowledge of the electro physical behavior	% of knowledge of the electro physical behavior
Scientific knowledge (The electro oculogram)	The student demonstrates high knowledge of the electro oculogram	The student demonstrates certain knowledge of the electro oculogram	The student demonstrates basic knowledge of the electro oculogram	The student do not demonstrate knowledge of the electro oculogram	% of knowledge of the electro oculogram
Time spent solving the test in the virtual environment	The student answers the test in less time	The student answers the test in the given time	The student answers the test in a longer time	The student does not answer the test	Average of time used to answer the test

3.3 Elaboration of Questions

As the methodology used for this study is quasi-experimental, there were elaborated two questionnaires. Each test has nine questions, divided into three sub-themes: the human eye's anatomy, the electro-physical behavior and the electro-oculography management. Each of them allows verifying the levels of generic and professional competencies.

From the most under "receptive" where it solves problems with a low and mechanical performance, only following instructions without knowing the reason for the actions carried out, to the highest level "strategic" where it solves problems achieving impact in the context.

For the virtual environment was selected green as the main color because it indicates that something is "uptrend" and also indicates "go" [41]. Both groups A (paper-based test) and B (virtual test) made the evaluation on the same day and under the same conditions, like a similar classroom, a supervisor and twenty minutes to do the test. The purpose of this experiment is to identify whether or not if an individual does develop competencies in virtual environments and how ICT impacts it.

3.4 The Virtual Environment

The virtual environment is built through the OpenSim platform, that researchers used for teaching interaction, group collaborations, distance learning, among others. Any individual can develop applications for a variety of fields [42, 43]. Its main features are that it works as open-source management software and allows multiple users, which means that users represented through avatars can interact and carry out collaborative activities. Because of these benefits, Opensim was selected as the technological tool for creating of the Bioengineering virtual environment.

The following specifications were considered for the design of the virtual environment: To show a welcome message in the virtual environment, To explain the purpose of the experiment, To show on a board slides presentations of the topics to be evaluated, To make boards with the test. Figure 1 and Fig. 2 display parts of the virtual environment developed for the Bioengineering class.

Fig. 1. Welcoming message.

Figure 1 illustrates a welcoming message in the virtual world. It gives directions of the experiment and the evaluation

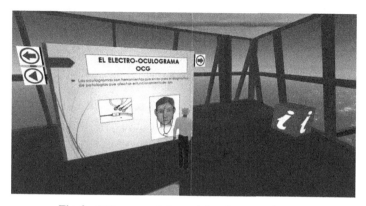

Fig. 2. Slides presentations of the electro oculogram.

Figure 2 illustrates an example of the slides presentation. Specifically, this one refers to a review of the EOG concepts.

4 Results and Discussion

Both groups A (paper-based test) and B (virtual test) made the evaluation the same day at the same hour and under the same conditions. Both of them had twenty minutes to solve the test and were supervised by the researchers of this study's researcher

4.1 General Results

Graphic 1 shows the general results of the test applied on group A and group B. Students of group B had a better performance than students of Group A. It indicates that virtual environments have a more significant impact on people than traditional evaluations, which is confirmed by 83% of the sample who achieved the highest levels (50% got strategical and 33% got autonomous). Comparing the strategic outcome between groups A and B, it mentions that "the student understands the information provided and has a strategy behavior in front of problems." It evidences a percentual difference of 30% benefiting group B.

Alike, comparing the autonomous competence that mentions "the student understands the information provided and has an autonomous behavior in front of problems" there is a difference of 13% benefiting group B. Last, when the authors compare the "resolutive" outcome, which mentions "the student understand elemental information and solve just easy situations" it evidences a 3% difference in favor of group A. Analyzing these results, it is demonstrated that the application of the independent variable aka virtual environment, generated that group B obtained better results in the highest categories of professional outcomes contrary to most students of group A who reached the lowest categories.

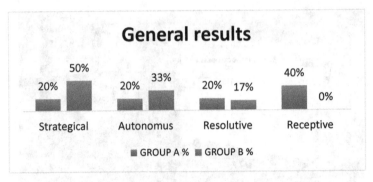

Graphic 1. General results of group A and group B.

4.2 Results for the Anatomy of the Human Eye

Graphic 2 shows the results obtained by group A and group B on the human eye evaluation's anatomy. It shows that students of group B had a better performance than Group A students, which is confirmed with the 100% of the sample who achieved the highest levels (67% got strategical and 33% got autonomous). It means that Group B students had specialized knowledge of the topic "anatomy of human eye" and on using virtual environments. Only 60% of Group A got strategical; the remaining 40% got resolutive, which also shows that some students have a lower knowledge of the topic and have to reinforce it.

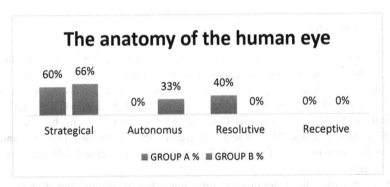

Graphic 2. Resuts for the anatomy of human eye.

Results of Group B also prove that Information and Communications Technology, ICT, impacts the know-know process.

4.3 Results for the Electro Physical Behaviour

Graphic 3 shows the results obtained by group A and group B on the human eye evaluation's electro physical behavior. It indicates that group B students had a better performance than students of Group A, which is confirmed with the 83% of the sample who achieved the highest levels (66% got strategical and 17% got autonomous); nevertheless, a 17% of them got resolutive. For this case, both groups demonstrate knowledge of the topic evaluated; however, the best performance is still in group B. Moreover, even these results show appropriate use of ICT, both groups have to reinforce their knowledge on the electrophysical behavior topic.

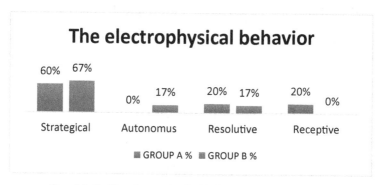

Graphic 3. The electro physical behavior of the human eye.

4.4 Results for the Use of the Electro Oculogram (EOG)

Graphic 4 shows the results obtained by group A and group B on the electrooculogram evaluation. It displays that students of group B had a better performance than students of Group A, which is confirmed with the 83% of the sample who achieved the highest levels (50% got strategical and 33% got autonomous) that indicates students of this group have certain knowledge on the topic EOG. The 100% of the sample of Group A achieved resolutive and receptive levels, which means lower management and expertise of EOG.

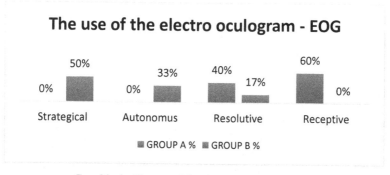

Graphic 4. The use of the electro oculogram EOG.

Once again, it is demonstrated that students who belong to the experimental group aka group B, can better comprehend the test, answer situations in virtual environments and have an outstanding performance.

4.5 Competencies Reached

Graphic 5 illustrates the competencies reached by group B. It indicates that the population of group B tested in the virtual environment are in the highest levels of performance. A 60% of the students are in the high level in the "ability to understand a virtual tool" (A1) and a 40% of them are in the high level in the same category; for the competence ability to identify and understand issues (A2), it was obtained the same results for the high and superior levels. However, in the competence ability to identify, formulate and solve engineering issues (A3), the results vary, and it is reflected in the 40% in the superior level and 60% in the highest one.

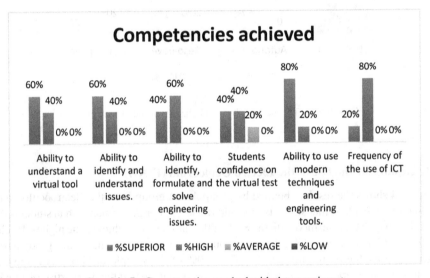

Graphic 5. Competencies reached with the experiment.

Respecting to (A1) and (A2) outcomes, most students of group B (60%) are in the superior level, which is related to the autonomous category; and the A3 got a 40% in the high level, which is associated with the strategic category.

4.6 Test Durability

Graphic 6 illustrates the time spent by each tested group to take the test. There was set twenty minutes for both cases and it shows that group B employed less time than the provided to do the evaluation, which was in the virtual environment. Even it was evaluated the same topics, group A spent more time than provided to do it; their results also demonstrated an average performance.

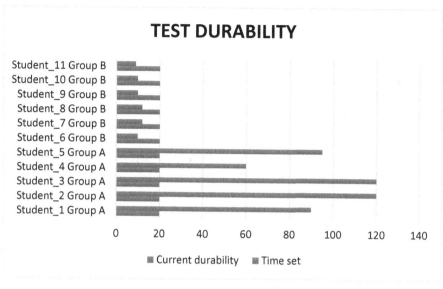

Graphic 6. Test durability in minutes.

According to the results obtained, virtual environments impact people as they promote concentration and professional skills development.

5 Conclusions

The built-in environment Opensim demonstrated that students could develop specific competencies with superior or equal results than those obtained with traditional methods, which ensures virtual environments and ICT impact not only in the learning process; it also does in the achievement of competencies. A virtual environment promotes to focus on results by encouraging students to understand situations, identifying clear objectives and providing earlier knowledge. Plus, the more immersive a person is, the less time they will take to react.

Virtual environments are necessary tools nowadays. They promote autonomous learning, a higher apprehension of concepts, development of professional and technical skills and a unique experience where students can deepen their understanding of certain notions. The more remarkable and eye-catching the virtual world is, the more immersive will be the person and the greatest achievement will be reached. Furthermore, as a technological tool they become a significant influence in many areas if they are correctly used.

To guarantee reliable results the course needs to be coherent with the formative plan, as in this study. It has to reflect evaluative strategies used for the learning process, like using ICT, and promoting responsibility, self-learning, self-control and self-regulation.

It was evidenced that the use of virtual environment produces a positive difference of 30% in the strategy outcome, which is the result of the 50% of the students of group

B (students that use virtual environments) reaching competence versus just the 20% of students of group A (students in the traditional education).

The authors highlight that the competences referred to the ability of understand the virtual environment and identify issues, associated to an autonomous level or "know-know", in which are the 60% of students of group B, the percentage goes down to 40% in the ability of identify, formulate and solve engineering situations associated to a strategic level or "know-do" that, even it is higher than traditional education. It can be improved in future works implementing virtual environments with models that ensure better "know-do" outcomes aka strategy levels.

Lastly, there was evidence that ICT influences and impacts positivity in the learning process, which can be measured through both pedagogical and technological tools. They promote the ability to understand facts and formulate and solve engineering issues, which is reflected in the development of professional skills achieved by the students of the experimental group of this study.

These results also evidenced that virtual environments can be implemented in any course to support the learning process and the evaluation of competencies and knowledge.

Acknowledgments. The authors would like to thank students of electronic engineering of Universidad de la Costa for their collaboration as a sample for this study. Also, to professors José Simancas, Elkin Ramirez, Gabriel Piñeres and Farid Meléndez that gave us technical tips, read the article and provided us language help at any time it was needed.

References

1. Comas-González, Z., Echeverri-Ocampo, I., Zamora-Musa, R., Velez, J., Sarmiento, R., Orellana, M.: Recent trends in virtual education and its strong connection with the immersive environments. Espacios **38**(15), 4–17 (2017)
2. Picatoste, J., Pérez-Ortiz, L., Ruesga-Benito, S.M.: A new educational pattern in response to new technologies and sustainable development. Enlightening ICT skills for youth employability in the European Union. Telemat. Informat. **35**(4), 1031–1038 (2018). https://doi.org/10.1016/j.tele.2017.09.014
3. Islam, A.K.M.N.: E-learning system use and its outcomes: moderating role of perceived compatibility. Telemat. Informat. **33**(1), 48–55 (2015). https://doi.org/10.1016/j.tele.2015.06.010
4. Yu, H., Zhang, Z.: Research on mobile learning system of colleges and universities. In: El Rhalibi, A., Pan, Z., Jin, H., Ding, D., Navarro-Newball, A.A., Wang, Y. (eds.) Edutainment 2018. LNCS, vol. 11462, pp. 308–312. Springer, Cham (2019). https://doi.org/10.1007/978-3-030-23712-7_42
5. Deguchi, S.: Case Studies of Developing and Using Learning Systems in a Department of Engineering. In: Zaphiris, P., Ioannou, A. (eds.) HCII 2020. LNCS, vol. 12205, pp. 34–48. Springer, Cham (2020). https://doi.org/10.1007/978-3-030-50513-4_3
6. Nuñez, M.E., Rodriguez-Paz, M.X.: A real-time remote courses model for the improvement of the overall learning experience. In: Zaphiris, P., Ioannou, A. (eds.) HCII 2020. LNCS, vol. 12205, pp. 132–143. Springer, Cham (2020). https://doi.org/10.1007/978-3-030-50513-4_10
7. Klímová, B., Pražák, P.: Mobile blended learning and evaluation of its effectiveness on students' learning achievement. In: Cheung, S.K.S., Lee, L.-K., Simonova, I., Kozel, T., Kwok, L.-F (eds.) ICBL 2019. LNCS, vol. 11546, pp. 216–224. Springer, Cham (2019). https://doi.org/10.1007/978-3-030-21562-0_18

8. Zamora-Musa, R., Velez, J.: Use of data mining to identify trends between variables to improve implementation of an immersive environment. J. Eng. Appl. Sci. **12**(22), 5944–5948 (2017)
9. Peng, J., Tan, W., Liu, G.: Virtual experiment in distance education: based on 3D virtual learning environment. In: 2015 International Conference of Educational Innovation through Technology (EITT) Wuhan, pp. 81–84 (2015). https://doi.org/10.1109/EITT.2015.24
10. Meléndez-Pertuz, F., et al.: Design and development of a didactic an innovative dashboard for home automation teaching using labview programming environment. ARPN J. Eng. Appl. Sci. **13**(2), 523–528 (2018). http://www.arpnjournals.org/jeas/research_papers/rp_2018/jeas_0118_6699.pdf
11. Chen, X., et al.: ImmerTai: immersive motion learning in VR environments. J. Vis. Commun. Image Represent. **58**, 416–427 (2019). https://doi.org/10.1016/j.jvcir.2018.11.039
12. Cabero-Almenara, J., Fernández-Batanero, J., Barroso-Osuna, J.: Adoption of augmented reality technology by university students. Heliyon **5**(5), e01597 (2019)
13. Talib, M., Einea, O., Nasir, Q., Mowakeh, M., Eltawil, M.: Enhancing computing studies in high schools: a systematic literature review & UAE case study. Heliyon **5**(2), e01235 (2019). https://doi.org/10.1016/j.heliyon.2019.e01235
14. Apuke, O., Iyendo, T.: University students' usage of the internet resources for research and learning: forms of access and perceptions of utility. Heliyon **4**(12), e01052 (2018). https://doi.org/10.1016/j.heliyon.2018.e01052
15. Hamari, J., Shernoff, D., Rowe, E., Coller, B., Asbell-Clarke, J., Edwards, T.: Challenging games help students learn: an empirical study on engagement, flow and immersion in game-based learning. Comput. Hum. Behav. **54**, 170–179 (2016)
16. Arantes, E., Stadler, A., Del Corso, J., Catapan, A.: Contribuições da educação profissional na modalidade a distância para a gestão e valorização da diversidade. Espacios **37**(22), E-1 (2016)
17. Zamora-Musa, R., Vélez, J., Paez-Logreira, H.: Evaluating learnability in a 3D heritage tour. Presence Teleoper. Vir. Environ. **26**(4), 366–377 (2018). https://doi.org/10.1162/pres_a_00305
18. Heradio, R., de la Torre, L., Galan, D., Cabrerizo, F., Herrera-Viedma, E., Dormido, S.: Virtual and remote labs in education: a bibliometric analysis. Comput. Educ. **98**, 14–38 (2016). https://doi.org/10.1016/j.compedu.2016.03.010
19. Garcia-Zubia, J., Irurzun, J., Orduna, P., Angulo, I., Hernandez, U., Ruiz, J. et al.: SecondLab: a remote laboratory under second life. Int. J. Online Eng. (IJOE) **6**(4) (2010). https://doi.org/10.3991/ijoe.v6i4.1312
20. Shen, J., Eder, L.B.: Intentions to use virtual worlds for education. J. Inf. Syst. Educ. **20**(2), 225 (2009)
21. Kemp, J., Livingstone, D., Bloomfield, P.: SLOODLE: connecting VLE tools with emergent teaching practice in second life. Br. J. Educ. Technol. **40**(3), 551–555 (2009). https://doi.org/10.1111/j.1467-8535.2009.00938
22. Brinson, J.: Learning outcome achievement in non-traditional (virtual and remote) versus traditional (hands-on) laboratories: A review of the empirical research. Comput. Educ. **87**, 218–237 (2015). https://doi.org/10.1016/j.compedu.2015.07.003
23. Cruz-Benito, J., Maderuelo, C., Garcia-Penalvo, F., Theron, R., Perez-Blanco, J., Zazo Gomez, H., Martin-Suarez, A.: Usalpharma: a software architecture to support learning in virtual worlds. IEEE Revista Iberoamericana De Tecnologias Del Aprendizaje **11**(3), 194–204 (2016). https://doi.org/10.1109/rita.2016.2589719
24. Bawa, P., Lee Watson, S., Watson, W.: Motivation is a game: massively multiplayer online games as agents of motivation in higher education. Comput. Educ. **123**, 174–194 (2018). https://doi.org/10.1016/j.compedu.2018.05.004

25. Zamora-Musa, R., Velez, J., Paez-Logreira, H., Coba, J., Cano-Cano, C., Martinez, O.: Implementación de un recurso educativo abierto a través del modelo del diseño universal para el aprendizaje teniendo en cuenta evaluación de competencias y las necesidades individuales de los estudiantes. Espacios **38**(5), 3 (2017)

26. Chen, J., Tutwiler, M., Metcalf, S., Kamarainen, A., Grotzer, T., Dede, C.: A multi-user virtual environment to support students' self-efficacy and interest in science: a latent growth model analysis. Learn. Instr. **41**, 11–22 (2016)

27. Guerrero-Roldán, A., Noguera, I.: A model for aligning assessment with competences and learning activities in online courses. Internet High. Educ. **38**, 36–46 (2018). https://doi.org/10.1016/j.iheduc.2018.04.005

28. Bhattacharjee, D., Paul, A., Kim, J., Karthigaikumar, P.: An immersive learning model using evolutionary learning. Comput. Electr. Eng. **65**, 236–249 (2018). https://doi.org/10.1016/j.compeleceng.2017.08.023

29. Cardona, S., Vélez, J., Tobón, S.: Towards a model for the development and assessment of competences through formative projects. In: XXXIX Latin American Computing Conference, vol. 17, no. 3, pp. 1–16 (2013)

30. Lucas, E., Benito, J., Gonzalo, O.: USALSIM: learning, professional practices and employability in a 3D virtual world. Int. J. Technol. Enhanced Learn. **5**(3/4), 307 (2013). https://doi.org/10.1504/ijtel.2013.059498

31. Mustami, M., Suryadin and Suardi Wekke, I.: Learning Model Combined with Mind Maps and Cooperative Strategies for Junior High School Student. Journal of Engineering and Applied Sciences, 12(7), pp. 1681 – 1686 (2017)

32. Freire, P., Dandolini, G., De Souza, J., Trierweiller, A., Da Silva, S., Sell, D., et al.: Universidade Corporativa em Rede: Considerações Iniciais para um Novo Modelo de Educação Corporativa. Espacios **37**(5), E-5 (2016)

33. Tawil, N., Zaharim, A., Shaari, I., Ismail, N., Embi, M.: The acceptance of e-learning in engineering mathematics in enhancing engineering education. J. Eng. Appl. Sci. **7**(3), 279–284 (2012)

34. Hernández, R., Fernández, C., Baptista, P.: Metodología de la investigación. J. Chem. Inform. Model. **53** (2014). https://doi.org/10.1017/CBO9781107415324.004

35. Tsay, L.S., Williamson, A., Im, S.: Framework to build an intelligent RFID system for use in the healthcare industry. In: Proceedings - 2012 Conference on Technologies and Applications of Artificial Intelligence, TAAI 2012, pp. 109–112 (2012). https://doi.org/10.1109/TAAI.2012.58

36. Kovács, P., Murray, N., Gregor, R., Sulema, Y., Rybárová, R.: Application of immersive technologies for education: state of the art. In: 2015 International Conference on Interactive Mobile Communication Technologies and Learning (IMCL) IEEE, pp. 283–288 (2015). https://doi.org/10.1109/IMCTL.2015.7359604

37. Estriegana, R., Medina-Merodio, J., Barchino, R.: Student acceptance of virtual laboratory and practical work: an extension of the technology acceptance model. Comput. Educ. **135**, 1–14 (2019)

38. Valverde, J., Ciudad, A.: El uso de e-rúbricas para la evaluación de competencias en estudiantes universitarios. Redu Revista de Docencia Universitaria **12**(1), 49–79 (2014)

39. Banerjee, S., Rao, N.J., Ramanathan, C.: Rubrics for assessment item difficulty in engineering courses. In: Proceedings - Frontiers in Education Conference, FIE, (2014)

40. Kim, G., Lui, S.M.: Impacts of multiple color nominal coding on usefulness of graph reading tasks. In: Proceeding - 5th International Conference on Computer Sciences and Convergence Information Technology, ICCIT 2010, pp. 457–463 (2010). https://doi.org/10.1109/ICCIT.2010.5711101

41. Zhao, H., Sun, B., Wu, H., Hu, X.: Study on building a 3D interactive virtual learning environment based on OpenSim platform. In: 2010 International Conference on Audio Language and Image Processing (ICALIP), pp. 1407–1411 (2010)

42. Sitaram, D., et al.: OpenSim: a simulator of openstack services. In: Proceedings - Asia Modelling Symposium 2014: 8th Asia International Conference on Mathematical Modelling and Computer Simulation, AMS Taipei: IEEE 2014, pp. 90–96 (2014). https://doi.org/10.1109/AMS.2014.28

43. OpenSimulator (2020). http://opensimulator.org

AARON: Assistive Augmented Reality Operations and Navigation System for NASA's Exploration Extravehicular Mobility Unit (xEMU)

Irvin Steve Cardenas, Caitlyn Lenhoff, Michelle Park, Tina Yuqiao Xu, Xiangxu Lin, Pradeep Kumar Paladugula, and Jong-Hoon Kim[✉]

The Advanced Telerobotics Research Lab,
Kent State University, Kent, OH 44240, USA
{icardena,jkim72}@kent.edu
http://www.atr.cs.kent.edu/

Abstract. The AARON system is a, reactive, integrated augmented reality (AR) system developed with considerations for collaborative activity. The underlying software architecture and design allows for bi-directional interactions and collaboration between human crewmembers, virtual agents that manage the system and EVA timeline, and robotic systems or smart tools that an astronaut may interact with during an exploration EVA (xEVA). We further present an AR user experience testbed developed to monitor and assess human interactions with the AR system. This testbed is composed of a custom telepresence control suit and a software system that throttles the communication through-put and latency between the AR headset and a control unit, or between simulated crewmembers. This paper outlines the technical design of the system, it's implementation, the visual design of the AR interface, and the interaction model afforded by the system.

Keywords: Mixed reality · Augmented reality · Human-robot interaction · Human-computer interaction · Telepresence · NASA · Artemis

1 Introduction

A prevalent assumption disproven by the field of informatics and human-computer interaction (HCI) has been that, providing individuals with large amounts of information will lead to more efficient task completion or effective system management. This assumption is also addressed in the field of human-robot interaction (HRI), which has identified that complex control interfaces and convoluted information displays can lead to high cognitive task load and fatigue experienced by robot operators. This is further considered by the

---

This work was presented at the 2020 NASA SUITS Challenge.

M. Singh et al. (Eds.): IHCI 2020, LNCS 12616, pp. 406–422, 2021.
https://doi.org/10.1007/978-3-030-68452-5_42

National Aeronautics and Space Administration (NASA) in [1], which addresses the future development of Decision Support Systems (DSS) for extravehicular activity (EVA). It makes the point that an approach where astronauts are solely reliant on information displayed on a helmet-mounted or arm-cuff display is presently infeasible, and would require further development of DSS. Furthermore, the experience and cognitive capacity provided by a group of experts observing an EVA cannot be simply replaced by the addition of information displays.

But, overall, we can consider that in a near-future integrated, intelligent, Decision Support Systems may leverage real-time EVA data to: (1) assist in the management of EVA timelines, (2) allow introspection into the state of an Exploration Extravehicular Mobility Unit (xEMU), and (3) facilitate the interaction between crewmembers, as well as facilitate the interaction with other agents involved in an EVA - e.g. smart tools, robots - rovers, virtual representations of the systems. All of this would take place not to simply increase the autonomy of the astronaut, but to improve coordination and collaboration between all parties involved in a lunar or Martian exploration EVA (xEVA) – where environmental conditions are unpredictable and network latency and bandwidth is constrained.

With this in mind, the system we present applies the concepts and research from the field of human-robot interaction. In particular, research related to teaming, collaborative activity, and telepresence. In the following sections we described the design and development of the AR interface system, as well as our tests. In accordance to Risk Gap 10 - this testbed includes the development of a telepresence suit and an underlying monitoring system that can assess the health and performance of a user interacting with our AR system.

Design Context: The stark differences in International Space Station (ISS) EVA and lunar xEVA is discussed in [5]. ISS EVA operations involve a fleet of ground support personnel using custom console displays, and performing manual tasks - e.g. taking handwritten notes to monitor suit/vehicle systems and to passively adjust EVA timeline elements. On the other hand, lunar xEVA is more physically demanding, more hazardous, and less structured than the well-rehearsed ISS EVAs. Further details on the differences and needs can be found in [1]. Most critically, the reactive approach of, ground-personnel, providing real-time solutions to issues or hazards (e.g. hardware configuration, incorrect procedure execution, life support system diagnosis) will not be feasible in the conditions of lunar xEVA, i.e. limited bandwidth and latency between ground support and inflight crewmembers. This requires more consideration on the list of factors that contribute to the risk of injury and compromised performance during EVA operations, covered in NASA's HRP EVA evidence report [4].

2   Literature Review

The work presented on this paper can be divide into the fields of HCI, HRI, and Telepresence. Although these fields are interrelated we make our best effort to delineate the work and provide further understanding into how these fields merge to support the work presented.

Human-Computer Interaction: Beyond using AR for entertainment, one of the key focuses over the years has been to reduce the cognitive load or training required to interact with a complex physical system. For example, [9] applied AR technology to the maintenance of power plants, where AR text dialogues and indicators indicating failures or required maintenance. Other earlier work such as [14] used object detection to achieve a markerless AR system for maintenance and repair of a automotive vehicles. Both used 3D object overlays and external cameras to track real-world object. This is contrary to present compact AR devices that integrate a range of sensors, e.g. depth-sensing cameras, IR rangefinders and IMUs. For example, the Magic Leap uses a six-layer waveguide display, a depth sensor, an infrared projector, and a camera to create depth-map.

One of the most attractive use cases for AR is remote or in-person collaboration. For example, [18] presents a software system where remote participants can interact with a virtual object in VR and AR, while maintaining voice communication. More interestingly, work such as [22], present an algorithm to synchronize the three-dimensional scene information shared by all participants. This algorithm is applied to an AR application running on a the Hololens 1. A thorough survey on AR and VR interactions is presented in [13].

Human-Robot Interaction: The literature also highlights efforts to make robot actions explainable and easy to control. For example, [19] uses an AR system that tracks a user's hand position - allowing the user to set start and goal positions that are represented as virtual markers. These markers define start and goal positions that a manipulator uses to execute a motion plan. Similar work is presented in [15]. But, instead the authors implement a gesture detection system using the Hololens 1. This is used to place target waypoints instead of tracking QR marker on the user's hand with cameras. A combination of head orientation and speech is also used to issue commands to the robot. Further work is focused on using AR to visualize trajectories, goal status and intents of robots [20], other work focuses on gestures [6] – enhancing the kinds of collaborative interactions that could take place between the human and robot.

Telepresence: Our work in particular focuses on immersive telepresence robotics - the development of technologies that allow a user to see, hear, and see as if he/she is present in the remote environment. This work has led to the development of the Telesuit [3], a telepresence robot control suit that immerses the operator in the remote environment of a humanoid robot. The suit allows the operator the control the humanoid robot via natural motions, Other work has focused on using camera-based motion capture systems, external IMU sensing device, and even sEMG to keep track of human joint positions for a robot to replicate [7,17]. Other work focuses on less physically immersive telepresence, e.g. embodied video conference calls. For example, [12] uses a Microsoft Kinect sensor to collect cloud point data of a remote user, and reconstruct the remote user's body and project it in the local physical environment using commodity projector – allowing the remote and local users to virtually interact without using any wearable devices.

3   Our Approach

Within the risk categories listed in the HRP EVA evidence report, the "EVA Factors", "Physical State" and "Mental State" categories contain factors that can be actively monitored and whose impact may be minimized through proactive countermeasures. We propose using an approach similar to our work on immersive telepresence control modalities [2,3] in which we implement a monitoring system that is situational aware and which allows interfaces to be reactive to mission tasks and reactive to type the interaction between the agents involved in the mission (e.g. ground support team, inflight crew, or virtual assistant). For example, it might be desirable that certain widgets of the interfaces are enlarged or remain focused depending on the criticality of information. Similarly, a summary of information (e.g. suit vitals) might be preferred during an on-going mission over fine-detail information, to reduce any cognitive overload.

Most of the interactions are mediated by a voice assistant. The use of this voice assistant in turn allows the AARON system to gain situational awareness through back-and-forth dialogue with the astronaut and access to contextual and situational information. This also allows for different components of the system to gain information to self-improve an interaction or the performance of a task. For example, if the vision recognition module implemented in the AARON system incorrectly identifies an object, the astronaut can verbally correct the system by speaking the correct classifier or pointing to an object that represents the system's initial classification. In essence, the more the astronaut interacts with the AARON system through the voice assistant, the more the system improves. This is a similar concept we apply in our work on the underlying control architecture of [2], which allows: (1) an AI to operate the robot and for a human to correct the AI in real-time, and (2) for the human to operate the robot and for the AI to correct/aid the human when the AI observes that operational performance has degraded or when the human's bio-signals show anomalies. More formally, this approach relates to shared autonomy and learning from experience, e.g. [16].

This type of high-level **interaction model** which focuses on coactivity/interdependence and collaboration, in turn requires that: (1) the interfaces adequately react to the state of the activity - being data rich when requested or when necessary to provide **transparency/observability into the mission and the system**, and that the (2) the interfaces provide a means to **communicate and collaborate with other agents** (i.e. human or synthetic), as well as (3) share/receive enough information to maintain communication legible, all while respecting the constraints of the system and mission.

In summary, contrary to the manual, passive and reactive approach taken in ISS EVAs, lunar EVAs must leverage a degree of automation, actively process data, and apply proactive risk management. The interaction model and interfaces we propose play a key role in achieving this.

4 System Design Description

4.1 Design Concept

Basic Assumptions for Design and Implementation Details: paginationWe implement an assistant named EEVAA (Enhanced ExtraVehicular Activity Assistant). The implementation is an in-house chatbox that uses a custom voice generated through an extension of the Tacotron model/algorithm [21]. We do this to generate a personalized voice. Bandwidth constraints and data transfer overhead (e.g. video) are also consider in the underlying data architecture We use a publish and subscribe communication middleware - DDS, and a real-time processing system - Kafka. A similar underlying architecture has been implemented for low-latency messaging and analytics processing in a number of our telepresence robots, e.g. [2].

3D Environments, Apps and UX Elements: Two 3D Environments in which the astronaut will interact were developed. The first one is the EVA Hub, which allows the astronaut to access all apps and customize the apps used during an EVA. The second one is the EVA Environment, which is the main environment used during exploration and task-execution. Each environment affords the astronaut with a set of apps/widgets. Each app has a different view according to the environment and the critically of information needed to be displayed. For example, Fig. 1 shows a detailed view of the vitals app during non-critical communication with the virtual assistant, while Fig. 2a the app widget located in the upper left, resembles a "toast menu" similar to that on mobile devices.

Table 1. Table with UI/view state transition

From state	input/Action/Intent	To state
Empty View	Gaze Upper Left	Show Vitals, Apps/Tools Widget
Minimized Vitals View	Voice: "Show vitals" Gesture: "Swipe to top left"	Extended Vitals View
Empty View	Crew Member Message	Communication Widget Shows

Interaction and Experience: Our interaction model uses hand gestures and hands-free interaction (i.e. natural language and gaze).

- Gaze, Region of Interest, Attention Directors: *(Gaze)* Eye gaze detection, overall, can be imprecise. In our case we use gaze detection for gross movement such as gazing to the upper left region of the display in order to bring up the vitals app widget, or for eye-gaze-based auto scroll where the user can read through instructions which automatically start scrolling once the user gets to the bottom of the text box. *(Region of Interest)* With coactive design and collaboration in mind, region of interest markers can be placed by any party involved in a task. For example, the Enhanced ExtraVehicular Assistant described in Sect. 4.1 can highlight a tool required to fix the tire of a rover. Figure 2c, depicts the assistant recognizing the rover's tire and using a circular marker to highlight it. Similarly, a remote crewmember can monitor the astronaut's view and highlight a region of interest through the AARON Monitoring Unit. In a collaborative scenario, where the astronaut is communicating or receiving instruction from another party, the astronaut can also point to a region of interest through gaze and commit to the identified region through a voice command or gesture. Table 1 summarizes a few interface flows that involve gaze and detecting region of interest. *(Attention Detectors)* Other elements, like attention detectors (e.g. arrows, warning signs) will be implemented for guidance and navigational purposes.

- Natural Language Input and Gestures: Both, natural language and gestures will be used. A custom voice assistant will be implemented using Tacotron [21] to generate the assistant's voice.
- Content Type: We focus on only using static and procedural content types while avoiding or minimizing the display of animated, 3D, and dynamic content. The only exception to this norm is in the EVA Hub 3D environment and apps/tools that require a 3-dimensional perspective, such as the visualization of elevation data from a moon mountain in Fig. 2a.
- Personalized Experience and Training - The system has a set of core components (i.e. apps and environments), but *allows each user to have a personalized experience by encouraging exploration* of the apps, and interactions modes (i.e. gestures, voice) while in the EVA Hub 3D environment. We once again draw a corollary to our work in immersive telepresence control modalities such as the Telesuit [3], where the Telesuit system learns a user's body map, degree of freedom and postures during virtual training; as well as allows the user to customize how each interface widget is displayed in the VR head-mounted display. Similarly, the EVA Hub 3D environment can serve as an environment that allow astronauts to accustom to viewing content through the "helment-mounted display" free of any critical tasks or specific gestures and also allow the user to progress through the stages of the different apps we provide (e.g. xEMU vitals app, visualization tools, etc.).

Coactivity and Collaboration: Future technology roadmaps envision systems that don't simply do things for us or that are plain reactive. Rather they envision systems that are collaborative or cooperative. These systems can work together with people and can enhance the human experience and productivity [8].

Applying this vision/paradigm to the development of end-to-end systems that will support lunar EVAs can be a key to enhancing the experience of EVAs and which may drive productivity of astronauts during EVA tasks. In the context of our work with telepresence robotic systems, we apply this paradigm to reduce the cognitive overload experienced by an operator after extended us of the system, and to allow the operator to assist or teach robot tasks such as autonomous navigation. A state machine depicting the latter is shown in [2].

4.2   Design Requirements

Suit Vitals Requirement: Suit vitals and other critical information is shown in the form of static multi-view astronaut profile widgets. A high-fidelity design of the vitals widget is shown in Fig. 1, information regarding suit vitals was obtained from [11]. A condensed view of the widget can be seen in Fig. 2a. In Fig. 1, the widget also includes a visualization of the astronaut speaking, this is in case of communication failures. The widget also shows feedback of the virtual assistant speaking or presenting critical information Fig. 2a.

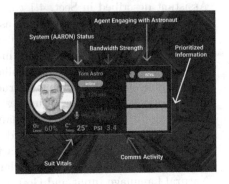

Fig. 1. High-fidelity detail-view vitals app widget

Science Sampling Support Requirement: In the EVA Environment, the astronaut can leverage different apps and utilize widgets accordingly - customization of the widgets can take place in the EVA Hub 3D Environment. For example, in Fig. 2a we show a high-fidelity wireframe of a possible view during a space walk, or a science sampling mission. In this Figure there's another astronaut in sight that is speaking, this is denoted by the widget on the top right that displays the other astronaut's information, the communication information and transcription of what is being said. This is a temporary widget that leaves the view as soon as the other astronaut stops speaking to the astronaut whose view we are depicting. The bottom right corner shows a static navigational widget that remains there during EVAs - similar to other widgets, the astronaut has the option to customize the view. The top left area, shows the minized profile/vitals widgets as well as an analytics widget, and widget used to summon apps or virtual tools. The center part of the Figure shows an app that allows the astronaut to visualize the elevation information from the moon mountain depicted in the back. This is a non-static tag-along 3D object which the astronaut can interact with. We implement virtual apps/tools such as a data visualizer. Figure 2b, on the bottom left, shows a mission widget that displays a set of instructions. In this Figure, the communication widget is no longer visible.

Rover Repair Support Requirement: A mix between a procedural and static-type content is used to display rover repair instructions. A high-fidelity wireframe is shown in Fig. 2c. It shows a set of content cards on the right - containing a video panel and short expandable instructions, which can be extended with gaze, voice input or gestures. Additionally, the virtual assistant's attempt at recognizing the tire, as discussed in Sect. 3. Given our coactive - collaborative approach, the virtual assistant would verbalize it's attempt, if the assistant makes a mistake that astronaut has the opportunity to correct it.

(a) Exploration environment with moon mountain & communication visualization

(b) High-fidelity wireframe of instruction widgets for geological sampling task

(c) Instruction display, rover telemetry and auto-targetting by AARON

(d) Color palette for illuminated & dark environments

Fig. 2. Examples of science sampling support requirement

Color, Light and Materials: Another critical aspect to the design is reactiveness to the environments. The UI is implemented with the color palette in Fig. 2d. Chosen to function in well-illuminated environments and dark environments.

5 Concept of Operations (CONOPS)

The AARON system supports NASA's need to provide astronauts with augmented/mixed reality interfaces that assist astronauts with their responsibilities during lunar EVAs, and that enhance the interaction with critical and noncritical informatic systems; all, in an effort to minimize the cognitive workload experienced during an EVA and to proactively (rather than reactively) manage an EVA mission. The AARON system accomplishes this through: its (1) UI/UX design, (2) the implementation of an AI-enabled EVA assistant, and (3) the underlying coactive software design, and the use of analytics.

UI/UX Design: We implement a data-driven, situational aware, system that drives the user interfaces (UI). This in turn leads to a richer user experience (UX) during EVAs by making the display of information reactive to events and anomalies, and by allowing a virtual assistant to provide situational to task-specific feedback. For example, each app widget is non-persistent and has a master-detail view to avoid visual clutter. Mission critical widgets (e.g. vitals, navigational tool) are implemented as static or "tag-along" objects to avoid disorientation during continuous movement. Given technological constraints (e.g. bandwidth limitation, battery, on-board computer) the UI elements provide feedback to the astronaut during the transfer of data. Most importantly, given the constraints of the environment the system proactively monitors the physical state of the astronaut and aims to infer on the mental state of the astronaut and the performance of the task, this is inspired by our work on control interfaces for a humanoid search and rescue robot in [2,3].

EVA Virtual Assistant: The use of a EVA-specific virtual assistant that is mission-aware (by leveraging the system's analytics) can make it more intuitive for the astronaut to navigate through the informatics system, simply by speaking verbal commands and listening to the assistants verbal output.

Coactive Design: Applying the concepts of coactive design and collaborative control (Sect. 4.1) to this lunar EVA-specific virtual assistant, can allow an astronaut to progressively offload tasks as the virtual assistant learns from the astronaut's feedback; Overall, this can also progressively improve the interaction between the astronaut and the virtual assistant by shaping the interaction to support interdependence - similar to the interaction that takes place between teammates (both depend on each other), rather than an interaction with explicit uni-directional dependence between a human and automaton (the human depends on the automaton, the automaton does not).

Analytics: Proactively monitoring the environmental state and an astronaut's physiological state creates opportunities to make a system that progressively learns from experience and that can optimize the interaction with UI elements based on the context or situation in which they are used.

CONOPS Overview: Figure 3 shows an overview of the concept of operations of the AARON System. Bi-directional communication between the astronaut and the assistant can continuously take place. Otherwise, the astronaut can also opt to use gestures to interact with the apps and navigation menus of the system. Data from both interactions is collected by the Data Distribution Service. Data from the suit monitoring system and the environment monitoring systems is also collected by the

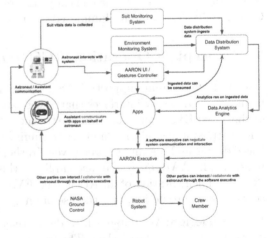

Fig. 3. CONOPS diagram

Data Distribution System. The data collected by the system can in turn be processed by the Data Analytics Engine. Both the raw data from the Data

Distribution System and the Data Analytics Engine can be consumed by the apps and widgets. An underlying executive software module, the AARON Executive, monitors the interactions that take place and can proactively tune properties of the apps or widgets. For example, given our underlying data architecture implementation and middleware - during low bandwidth scenarios the AARON Executive can tune the quality-of-service of a high-bandwidth data request such as video file or live-stream. Other parties (e.g. crew members or synthetic agents) can also interact/collaborate directly with a given astronaut or manipulate an app being used by an astronaut by first bridging through the executive.

6   Human-in-the-Loop (HITL) Test Measurement

The HITL test measures the efficiency of coactivity/interdependence and collaboration between human crewmembers and the interaction with a virtual assistant, as afforded by our interfaces and the underlying system architecture. Two scenarios with three protocols are tested. We further assess the different methods for achieving situational awareness, through seven different mechanisms.

Subject Pool, Demographics and Metrics: The subject pools includes 50 students from the following departments which the Advanced Telerobotics Research lab collaborates with at Kent State University: (1) Aeronautics Department - student pilots, (2) Computer Science - undergraduate and graduate students, (3) College of Public Health - mental health and nursing students. The ages of the students will range from 18–30, and efforts will be made to include an equal number of female and male students.

Task completion time and bio-indicators correlating to stress, exhaustion or discomfort are measured - as later discussed. Additional metrics include the rate of communication between the virtual assistants, and the rate of communication with a secondary crew member. Other artifacts such as gaze is monitored, as well as verbal cues that indicate stress or discomfort.

Test Scenario and Protocol: Rover Repair: This test scenario virtually simulates the need to repair a rover's tire. The approach in which the instructions are presented is as followed. *(Approach: Legacy/Traditional)* A list of instructions is displayed in front of the astronaut, similar to that in Fig. 2c. *(Approach: Human-to-Automaton Collaboration)* The virtual assistant is involved in disseminating the instructions and can also assist in identifying tools or parts. Extending this approach to a "Wizard of Oz" interaction is currently under consideration. *(Approach: Human-to-Human Collaboration)* A remote crew member is allowed to assist the astronaut in identifying parts or tools verbally or by assigning markers to the main astronaut's view. The other crew member may also assist in reading through instructions.

Test Scenario and Protocol: Robot/Rover Collaborative Work: This scenario presents the astronaut with opportunity to collaborative control the rover (shared autonomy), a control flow similar to that in [2]. It is assumed that the rover has enough autonomy to navigate on its own but can benefit from real-time control feedback. This scenario is inspired by our work in [2]. *(Approach: Legacy/Direct, Programmed, Robot Instructions)* The astronaut directly programs the robot. *(Approach: Visual/Virtual Robot Collaborative)* The astronaut can provide instructions to the rover through visual waypoints markers that guide the robot's navigation to a specific location. *(Approach: Coactive/Collaborative Remote Telepresence)* Similar to the latter, but a remote crewmember can perform waypoint placement and give feedback to the rover.

Test Scenario and Protocol (Assessing Physical State and Environmental State): This scenario aims to test whether sharing with the astronaut any identified anomalies in physiological state or environmental state has an impact on the astronaut's performance or physiological state.

- Numerical Warning Markers: Anomalies in an astronaut's physiological state are reported through numerical markers (bio-signals - heart-rate, GSR).
- Graphical Warning Level Markers: Anomalies in an astronaut's physiological state, bio-signals, are discretized to three severity levels low, mid and high.
- Synthetic Verbal Numerical Warning: EEVAA provides a verbal warning of a bio-signal anomaly level. The numerical warning markers are shown.
- Synthetic Verbal Discrete Warning: EEVAA provides a verbal warning of a bio-signal anomaly level. The graphical warning markers are shown.
- Virtual, Embodied, Avatar Numerical Warning: The virtual assistant is presented as a 3D avatar that provides a verbal warning of a given bio-signal anomaly. The numerical warning markers are shown.
- Virtual, Embodied, Avatar Discrete Warning: The virtual assistant is presented as a 3D avatar that provides a verbal warning of a given bio-signal anomaly level. The graphical warning markers are shown.
- Human Verbal Warning: A human crew member provides a warning of an anomaly in a bio-signal. The dialogue is unconstrained.

Assessments of Physiology in Hybrid Reality Environment: Preliminary HITL testing is done through a virtual reality simulation, similar to our work in [3], that assess average performance time to complete a task such as (fixing a rover). The physical and mental state of the subject during the performance of the task is evaluated during the HITL test. Physiological data is collected through a wearable, non-EEG, signal-based monitoring device such as the Empatica medical device. Similar to our work we can analyze such data in real-time or post-process to detect anomalies and correlate those to the user's performance. For example, a Galvanic Skin Response (GSR) sensor measures the electrical conductance of the skin, which can then be used to detect stress. A summary of the data analyzed is listed in Table 2.

Table 2. Bio-signal monitored during HITL with VR environment

Type of Bio-signal	Type of Sensor	Description of Measured Data
Electrocardiogram (ECG)	Skin/Chest electrodes	Electrical activity of the heart
Respiration rate	Piezoelectric /piezoresistive sensor	Number of movements indicative of inspiration and expiration per unit time (breathing rate)
Heart rate	Pulse Oximeter /skin electrodes	Frequency of the cardiac cycle
Perspiration (sweating) or skin conductivity	Galvanic Skin Response	Electrical conductance of the skin is associated with the activity of the sweat glands
Body Movements	Accelerometer	Measurement of acceleration forces in the 3D space

7 Implementation Overview

The AR interfaces were developed using the Unity game engine and the Magic Leap SDK. The Telesuit software was developed in Python and uses a custom motion tracking library. The robot software is developed using the Robot Operating System (ROS) framework and libraries. The underlying infrastructure of our system uses RTI's DDS implementation. DDS allows the system to agnostically communicate without additional networking libraries. Below we provide an overview of some of the interfaces as they are displayed in the augmented space.

Quick Looks (Vitals Display): Figure 4(a) shows the vitals panel. It's displayed on-demand, expanded, or collapsed through voice commands. Rather than displaying all data received from the vitals telemetry, the user has the ability to select what data to display. Communication status is shown on the top. When the vitals display is collapsed, the system can still display alerts. Anomaly alerts are spoken by EEVAA. The risk level of the alert is also highlighted by using two colors: red for high-risk alerts, orange for mid-risk. Blue colors icons are used for transient alerts. Presently, the risk levels are solely classified on the variance of the data. E.g., a sharp sudden drop in the fan tachometer data would be considered a high-risk alert. Whereas, a fluctuating fan RPM would be considered an mid-risk alert. The data is analyzed by our data processing engine, and the alert is sent to the AR interface system. Figure 4(b) shows a red alert, additional on the far right EEVAA's voice indicator is displayed.

Feedback: The AR interface system provides interaction feedback through both visual and auditory indicators. For example, when EEVAA is speaking an oscillating audio wave is displayed. This wave oscillates as long as EEVAA is speaking. A similar wave, is also presented when the user is performing a voice call. Both indicators to provide the user with feedback on the performance of microphone and speakers. This is similar to the user experience presented by commercial voice/video communication applications. In situations where the opposing party

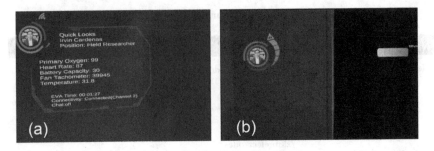

Fig. 4. (a) Quick looks: AR vitals display, (b) Minimal alert

claims the user cannot be heard, the visual feedback provides assurance to the user that their microphone is functional and it's an issue on the opposing party. The visual indicator is shown in Fig. 5(a). Auditory feedback is given after every command acknowledged by EEVAA, and after every action of with the tools - e.g. note taker, camera, image recognition.

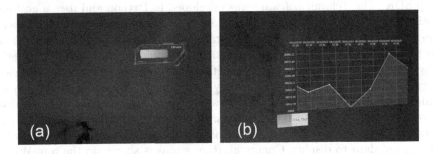

Fig. 5. (a) EEVAA Voice Indicator, (b) Data visualization tool

Visualization Tools: We implement a visualization tool that can plot real-time and after-the-fact telemetry data. The user can request EEVAA to visualize a specific telemetry parameter, or plot historical data. In our implementation, the real-time telemetry data is stored for later use. Similar to other displays, the user can request EEVAA to lock the display in 3D space, allowing the user to perform other tasks while maintaining the data visualization within the field of view. A visualization of the fan tachometer reading is displayed in Fig. 5(b).

Robot Control via AR Markers and Auditory Feedback: We implemented the use case of smart tools/the ability for a user to control a robot. Waypoint markers are set using the hand controller which raycasts to a target location. Placement of each waypoint triggers a sound within the pentatonic scale according to the distance the waypoint was placed, providing spatial feedback. The robot has a predefined map of the environment. This map along with

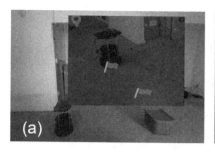

Fig. 6. (a) Robot control using AR markers, (b) Robot state display

the starting waypoint, which is placed where the robot is located in reference to the augmented space, allows the robot to attempt to navigation between each waypoint. Figure 6(a) shows a set of waypoint markers. The center image is the augmented space overlayed over the real-world space. The robot utilizes ROS and the standard Turtlebot packages for autonomous navigation. Figure 6(b) shows the robot navigation state panel. Similar to others, it can be anchored in 3D-space and collapsed.

NASA XR Telesuit: paginationThis work extends [3]. The suit collects motion and biometric data which could be transmitted and displayed on an AR device. Similar to the referenced work, the NASA XR Telesuit is leveraged to better understand the user's physical state during the interactions that take place with the AR interfaces. During simulated xEVAs this would give us better insight into repetitive postures or gestures that may cause the user discomfort during long-term interactions. Polymer-based strain sensors placed across the suits allowed to detect flexion and inflexion. Other sensors ECG and GSR are included (Fig. 7).

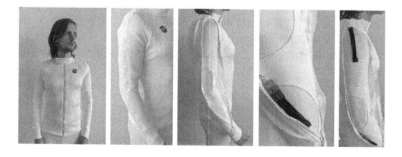

Fig. 7. NASA XR (AR/MR/VR) telesuit

8   Design and System Constraints

The FOV is constrained to Magic Leap 1's specifications, which has a horizontal, vertical, and diagonal FOV of 40, 30, 50 degrees respectively - with a Near Clipping Plane set at 37 cm from the device. Our system leverages simulated data generated by the NASA SUIT Telemetry Stream Server [10]. Although, the server can simulate failures, the data produced does not simulate real anomaly patterns that might take place - e.g. loss of data, latency, non-linear variance. To overcome this we collected the telemetry data generated during a simulated EVA, and added statistical variance to the data. Further work is needed to truly simulate different scenarios that might take place during an xEVA.

Limitations also exist at the networking level. To simulate bandwidth constraints and delays, we leverage RTI's Data Distribution Service. Our architecture implements a data relay service, this data relayer is implemented since native binaries of RTI core software are not available for ARM computers. Lastly, our current interface system does not implement in-situ multi-party AR collaboration. The current system only considers collaboration between the astronaut performing the xEVA and inflight/ground crew and robots/smart tools.

9   Conclusion and Future Work

In conclusion, we presented the design and development of AARON: Assistive Augmented Reality Operations and Navigation System, an AR-based system developed for the 2020 NASA SUITS Challenge. The system was presented at the SUITS Challenge and evaluated by a panel of NASA engineers and astronauts – as part of the top 10 teams selected to visit Johnson Space Center. However, the presentation took place virtually due to the COVID-19 pandemic. The system considers collaborative activity, and leverages research related to the field of HRI and Telepresence. Additionally, an AR testbed was developed to monitor and assess human interaction with AR system. It is composed of a telepresence control suit (Telesuit) and a software system that throttle the communication throughput and latency across all subsystems. Future work further applies immersive telepresence research towards engineering AR interfaces for NASA's xEMU.

References

1. Abercromby, A., et al.: Integrated extravehicular activity human research and testing plan, July 2019
2. Cardenas, I.S., Kim, J.H.: Design of a semi-humanoid telepresence robot for plant disaster response and prevention. In: 2019 IEEE/RSJ International Conference on Intelligent Robots and Systems (IROS), November 2019

3. Cardenas, I.S., et al.: Telesuit: an immersive user-centric telepresence control suit. In: 2019 ACM/IEEE International Conference on Human-Robot Interaction HRI (2019)

4. Chappell, S.P., et al.: Risk of injury and compromised performance due to EVA operations (2015)

5. Coan, D.: Exploration EVA System ConOps. EVA-EXP-0042, December 2017

6. Gromov, B., Guzzi, J., Gambardella, L.M., Giusti, A.: Intuitive 3D control of a quadrotor in user proximity with pointing gestures. In: 2020 IEEE International Conference on Robotics and Automation (ICRA) (2020)

7. Huang, D., Yang, C., Ju, Z., Dai, S.-L.: Disturbance observer enhanced variable gain controller for robot teleoperation with motion capture using wearable armbands. Autonom. Robots **44**(7), 1217–1231 (2020). https://doi.org/10.1007/s10514-020-09928-7

8. Johnson, M., et al.: Coactive design: designing support for interdependence in joint activity. J. Hum. Robot Interact. **3**(1), 43–69 (2014)

9. Klinker, G., Creighton, O., Dutoit, A.H., Kobylinski, R., Vilsmeier, C., Brugge, B.: Augmented maintenance of powerplants: a prototyping case study of a mobile AR system. In: IEEE & ACM International Symposium on Augmented Reality (2001)

10. NASA: SUITS telemetry stream server (2020). https://github.com/SUITS-teams/BackEnd

11. Patrick, N., Kosmo, J., Locke, J., Trevino, L., Trevino, R.: Wings in orbit: extravehicular activity operations and advancements, April 2011

12. Pejsa, T., Kantor, J., Benko, H., Ofek, E., Wilson, A.: Room2room: enabling life-size telepresence in a projected augmented reality environment. In: ACM Conference on Computer-Supported Cooperative Work & Social Computing (2016)

13. Pidel, C., Ackermann, P.: Collaboration in virtual and augmented reality: a systematic overview. In: De Paolis, L.T., Bourdot, P. (eds.) AVR 2020. LNCS, vol. 12242, pp. 141–156. Springer, Cham (2020). https://doi.org/10.1007/978-3-030-58465-8_10

14. Platonov, J., Heibel, H., Meier, P., Grollmann, B.: A mobile markerless AR system for maintenance and repair. In: 2006 IEEE/ACM International Symposium on Mixed and Augmented Reality (2006)

15. Quintero, C.P., Li, S., Pan, M.K., Chan, W.P., Machiel Van der Loos, H.F., Croft, E.: Robot programming through augmented trajectories in augmented reality. In: IEEE/RSJ International Conference on Intelligent Robots and Systems (IROS) (2018)

16. Reddy, S., Dragan, A.D., Levine, S.: Shared autonomy via deep reinforcement learning (2018)

17. Rolley-Parnell, E., et al.: Bi-manual articulated robot teleoperation using an external RGB-D range sensor. In: 2018 15th International Conference on Control, Automation, Robotics and Vision (ICARCV) (2018)

18. Shang, J., Wang, H., Liu, X., Yu, Y., Guo, Q.: VR+AR industrial collaboration platform. In: 2018 International Conference on Virtual Reality and Visualization (ICVRV) (2018)

19. Thanigaivel, N.K., Ong, S.K., Nee, A.: AR assisted robot programming system for industrial applications. Robot. Comput. Integr. Manuf. **61**, 101820 (2020)

20. Walker, M., Hedayati, H., Lee, J., Szafir, D.: Communicating robot motion intent with augmented reality. In: Proceedings of the 2018 ACM/IEEE International Conference on Human-Robot Interaction (2018)
21. Wang, Y., et al.: Tacotron: towards end-to-end speech synthesis (2017)
22. Xu, C., Wang, Y., Quan, W., Yang, H.: Multi-person collaborative interaction algorithm and application based on HoloLens. In: Jain, V., Patnaik, S., Popentiu Vladicescu, F., Sethi, I.K. (eds.) Recent Trends in Intelligent Computing, Communication and Devices. AISC, vol. 1006, pp. 303–315. Springer, Singapore (2020). https://doi.org/10.1007/978-981-13-9406-5_37

Socio-Cognitive Interaction Between Human and Computer/Robot for HCI 3.0

Sinae Lee, Dugan Um[(⊠)], and Jangwoon Park

Texas A&M-CC, Corpus Christi, TX 78413, USA
dugan.um@tamucc.edu

Abstract. With an increasing number of AI applications in everyday life, the discussion on how to improve the usability of AI or robot devices has been taking place. This is particularly true for Socially Assistive Robots (SAR). The presented study in this paper is an attempt to identify speech characteristics such as sound and lexical features among different personality groups, specifically per personality dimension of Myers-Briggs Type Indicator (MBTI), and to design an Artificial Neural Network reflecting the correlations found between speech characteristics and personality traits. The current study primarily reports the relationship between various speech characteristics (both mechanical and lexical) and personality dimensions identified in MBTI. Based on significant findings, an ANN (Artificial Neural Network) has been designed in an effort to predict personality traits only from speech processing. The current model yielded 75% accuracy in its predictive ability, warranting further attention to the applicability of speech data in developing and improving various domains of human-computer interactions.

Keywords: HMI · HCI · Personality analysis · Artificial intelligence · Natural language processing

1 Introduction

With an increasing number of AI applications in everyday life, the discussion on how to improve the usability of AI or robot devices has been taking place. This is particularly true of Socially Assistive Robots (SAR), which are developed to work as caregivers alongside doctors, nurses, and physical therapists (Kang et al. 2005). Studies have shown that humans respond more positively to robots that accommodate them in some way. For example, Andrist et al. (2015) observed that the amount of time spent during a therapeutic task significantly increased upon matching the SAR's gaze behavior to the user's personality. Additionally, users interacting with a gaze-matching SAR completed their tasks more frequently than those interacting with a non-gaze-matching SAR. Similarly, Goetz et al. (2003) found that a robot's demeanor (friendly vs. serious) in an entertaining task significantly influences users' perceptions of a robot (e.g., friendly, playful, and witty) and their willingness to comply with the robot's instruction. Nass et al. (2005) showed that pairing the voice of the car (e.g., energetic voice) with the driver's emotion (e.g., happy) had a positive effect on the perceived attention to the road. Thus, for

© Springer Nature Switzerland AG 2021
M. Singh et al. (Eds.): IHCI 2020, LNCS 12616, pp. 423–431, 2021.
https://doi.org/10.1007/978-3-030-68452-5_43

future robots, especially in HCI 3.0, the function of firstly detecting the personality of a user, and secondly adjusting its behaviors in order to accommodate the user's intention might be of use. The current study is an attempt to identify speech characteristics such as sound and lexical features among different personality groups, specifically per personality dimension of Myers-Briggs Type Indicator (MBTI), and to design an Artificial Neural Network reflecting the correlations found between speech characteristics and personality traits.

2 Proposed Methods

2.1 Data Collection

The data was collected in a semi-controlled naturalistic interview setting, at times that work best for each participant and a moderator. Prior to each interview, the purpose was explained to the participants by a moderator. For each session, the moderator facilitated a naturalistic conversation with a participant for 40 min on average, covering various subjects in a Q&A format. Each question was followed by an answer varying in length, after which the moderator moved onto the next question item. The conversations between a moderator and participants were audio-recorded. The sampling rate of the recorded audio was 44,000 data points per second. 17 female and 11 males participated in the interview. Following the interview, participants were requested to fill out the MBTI test survey. Each interview recording was transcribed for further data processing and analysis. Therefore, the input dataset for the personality analysis includes MBTI test results, voice recordings, and its transcriptions (See Fig. 1).

2.2 Extracted Features

From the voice recording and corresponding transcription, response time (the duration between the end of a question and the beginning of an answer), intensity (in dB), pitch (in Hz), speech rate (word per minute), and lexical attributes were extracted using Praat (Boersma and Weenink 2020) scripting. The output dataset was then compared with the MBTI test results for personality matching between input dataset and participants' personality. Figure 1 illustrates the data collection and processing procedure.

In order to measure the time delay, or the response time, between the moderator's question and the participant's answer, a Praat script was written to process the interview datasets. The script allowed us to measure the time delay between the question and answer using Praat's annotation file that contains transcription of the recording segmented into words. The main feature for comparison in response time is the response time average. Therefore, each response time between a question and a corresponding answer was measured, which yielded an average value of response time per participant.

Pitch data was extracted per person by averaging the fundamental frequency (F0) at the midpoint of each stressed vowel throughout the interview. Per participant, minimum F0, maximum F0, mean F0, and median F0 were logged. Similarly, intensity (or amplitude) was measured by extracting dB at each stressed vowel at the midpoint, and per participant, minimum, maximum, mean, and median dB were logged.

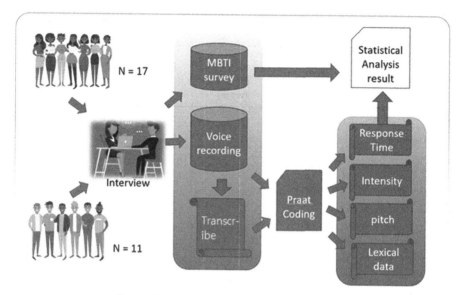

Fig. 1. Data collection and processing procedure

Additionally, speech rate as measured by the number or words per minute (wpm) was extracted per participant. Each interview was segmented into intonational phrases, each of which has information on its start time and end time. Within each intonational phrase, the number of words were counted and divided by the duration (end time - start time). From this, wpm for each intonation phrase was extracted throughout the interview, after which the average wpm per participant was calculated.

The words used by each participant were analyzed in terms of frequency. We considered several categories of words, such as fillers, hedges, approximators, etc., and the participant frequency of each category was generated for comparison.

3   Findings

Table 1 shows statistical testing results on each mechanical characteristic of the speech data. Two-sample t-test shows that the difference between male and female was significant in terms of pitch (unit: Hz) ($p < 0.001$), which is expected given the physiological sex differences. For the females, the average pitch was 187.9 Hz (SD = 20.4), and for the males, the average pitch was 136.3 Hz (SD = 11.4). In addition, the average pitch of the females, who were classified as intuiting individuals on the sensing-intuiting dimension of MBTI, was about 23 Hz higher than sensing individuals ($p = 0.018$).

There was no significant difference between men and women in the words per minute, and there was no significant difference in the remaining four personality categories. In terms of reaction time (unit: ms), there was no gender difference, but the extroverts answered about 200 ms faster on average than the introverts ($p = 0.090$). The participants that were classified as feeling individuals responded about 600 ms faster on average than the participants who were classified as thinking individuals ($p = 0.039$). Lastly, in terms

Table 1. Statistical analysis results (*p*-values) on the mechanical characteristics of the speech data

	Pitch (Hz)	Words per minute (wpm)	Reaction time (ms)	Intensity (dB)
Female (17)/Male (11)	< 0.001*	0.683	0.566	0.796
Extraversion (14) /Introversion (14)	F: 0.854 M: 0.404	0.151	0.090*	0.692
Sensing (12)/Intuition (16)	F: 0.018* M: 0.718	0.439	0.390	0.486
Thinking (4)/Feeling (24)	F: n/a** M: 0.521	0.843	0.039*	0.843
Judging (16)/Perceiving (12)	F: 0.812 M: 0.407	0.104	0.586	0.012*

*p-value < 0.1 or < 0.05; **Because there was only one female in the Thinking/Feeling category, a statistical analysis was not available. Numbers in the parenthesis in the first column are the number of participants who are associated with the category.

of intensity (unit: dB), there was no significant difference between males and females. However, the participants who were classified as judging individuals were found to be about 4.5 dB louder on average than the participants who were classified as perceiving (p = 0.012). In conclusion, we can presume the three mechanical characteristics of speech (pitch, reaction time, and intensity) could be used to distinguish a person's personality.

Lexical analysis shows that the intuiting group as opposed to the sensing group uses hedge words or approximating words at a higher frequency (p = 0.001). The use of the down grader (just) is used with different frequency between the intuiting group and the sensing group as well, but only among male participants. Specifically, intuiting men are using 'just' more frequently than sensing men.

4 ANN Modeling

ANNs are composed of units of structure called neurons. These artificial neurons are connected by synapses with weighted values. As the computational power has been multiplied with the accumulation of big data, ANN becomes a powerful tool as an optimization problem solver. Among the core enablers is the backpropagation mechanism that helps train an ANN useful to solve specific non-linear MIMO (Multi-Input, Multi-output) problems.

When the result is unexpected, an error is calculated and the values at each neuron and synapse are propagated backwards through the ANN for the next train cycle. This process takes big data for an ANN to be trained and later used as a problem solver. In the proposed Artificial Neural Network (ANN), there is an input layer which is composed of several personality signature data, a hidden layer, and an output layer estimating the MBTI of a person. From the perspective of the problem scope of the cell classification, we start with an assumption that one hidden layer suffices to solve the problem. After several iterations, we finalized the ANN structure with 13 input nodes, 20 hidden nodes and 4 output nodes for MBTI estimation as shown in Fig. 2.

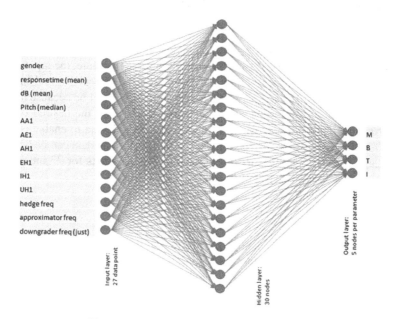

Fig. 2. Artificial Neural Network (ANN) design

5 Experiments

5.1 ANN Training

Among many ANN training methods, the supervised training is used. In the training phase, the correct class for each record is kept and used for the training process for the supervised training. During the training, the true MBTI values obtained from the survey are compared with the output value with 0.5 of initial weighting on each propagation. The error terms between target and actual output are then used to adjust the weights in the network. A key feature of neural networks is an iterative learning process in which data cases are presented to the network one at a time, and the weights associated with the input values are adjusted at each epoch. After all cases are presented, the process often starts over again. During this learning phase, the network learns by adjusting the weights so as to be able to predict the correct class label of input samples. Advantages of neural networks include their high tolerance to noisy data, as well as their ability to classify patterns on which they have not been trained. We train the ANN 100 times at each epoch and compare the total root mean square error with the desirable goal. The goal value of the root mean square error is set to below 0.5 with 4 sigmodal outputs.

5.2 Input Layer

We collected time domain and various frequency domain traits as well as lexical traits in conjunction with the MBTI survey of each participant during the interview. Among them are gender, response time (mean), Power in dB (mean), Pitch (median), Lax vowel pitch mean (AA1, AE1, AH1, EH1, IH1, and UH1), and Lexical data such as hedge frequency, approximator frequency, and downgrade frequency which are statistically the most meaningful to the MBTI survey results (Table 1). Therefore, the input layer of the ANN is composed of 13 vocal and lexical data of a person's vocal data and annotated data (Table 2, 3). Note that some networks never learn. This could be because the input data do not contain enough trend nor discrepancy from which the desired output is derived. Networks also don't converge if there is not enough data to enable complete learning. Ideally, there should be enough data so that part of the data can be held back as a validation set. Original solution data contains 27 data points for 27 interviewees among which 11 interviewees are male and the rest are female.

5.3 Output Layer

In order to take advantage of the sigmoid function at the output layer, we tested two possible configurations of the output layer. One was the 2 nodes of 00, 01, 10, and 11 for 4 different values of MBTI.

Table 2. Input data structure I

Sex	Gender
Time domain traits	Response time (mean)
Frequency domain traits	dB (mean)
	Pitch (median)
	AA1
Lax Vowel	AE1
Pitch (mean)	AH1
	EH1
	IH1
	UH1

Table 3. Input data structure II

Lexical terms	Examples
Hedge frequency	like, well, uh, um, just, kinda, sorta
Approximator frequency	like, kinda, sorta, kind of, sort of
Downgrader frequency	just

The other one was the 4 nodes with digital value from 0001 to 1000 linking each decimal point to MBTI output. As a result, the latter choice turned out being more accurate in classification. Therefore, we set the output layer of the ANN with 4 neurons as listed in Table 4.

Table 4. Output data structure

Output	MBTI traits
0001	M
0010	B
0100	T
1000	I

5.4 Data Analysis

In order to train the ANN, we collected 28 sets of data by interview with participants. In an effort to make the ANN converge and, at the same time, yield the highest accuracy, various combinations of hyper-parameters (learning rate, momentum, epoch, etc.) were tried out.

The root mean square error for the convergence evaluation is shown in the equation below.

$$RMS = \sqrt{\left(M^E - M^T\right)^2 + \left(B^E - B^T\right)^2 + \left(T^E - T^T\right)^2 + \left(I^E - I^T\right)^2} \qquad (1)$$

where M^E, B^E, T^E, and I^E are prediction values of MBIT, while M^T, B^T, T^T, and I^T are ground truths of MBIT from the survey. Over the period of experiments, we discovered that 0.4 for the learning rate and 0.4 for the backpropagation moment work best and yield optimum performance for both convergence and training time. In order to obtain the best result, we processed the raw data by following the data processing sequence as shown in Fig. 3. All of the time domain data and frequency data are normalized for the best ANN training efficiency and the training result is shown in Fig. 4. The ANN started reacting fast for the first 400 epochs, then it slowed down eventually yielding 0.5 of RMS error after 1,000 epochs. Assuming the maximum error is 2 in RMS value, the trained ANN demonstrated 75% MBTI classification accuracy.

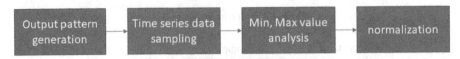

Fig. 3. Data processing sequence for ANN training

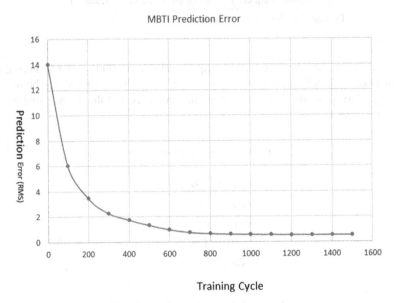

Fig. 4. RMS trend per training epoch

6 Conclusion

The current study reports the relationship between various speech characteristics (both mechanical and lexical) and personality dimensions identified in MBTI test. Based on significant findings, AI modeling has been designed in an effort to predict the personality traits from processing speech input data. The modeling yielded 75 percent accuracy in its predictive ability, warranting further attention to the applicability of speech data in developing and improving various domains of human-computer interactions.

References

Andrist, S., Mutlu, B., Tapus, A.: Look like me: matching robot personality via gaze to increase motivation. In: Proceedings of the 33rd Annual ACM Conference on Human Factors in Computing Systems, Seoul, Korea, 18–23 April 2015, pp. 3603–3612 (2015)

Boersma, P., Weenink, D.: Praat: doing phonetics by computer [Computer program]. Version 6.1.26 (2020). Accessed 5 Oct 2020. http://www.praat.org/

Goetz, J., Kiesler, S., Powers, A.: Matching robot appearance and behavior to tasks to improve human-robot cooperation. In: Proceedings of the IEEE International Workshop on Robot and Human Interactive Communication, Millbrae, CA, USA, 2 November 2003, pp. 55–60 (2003)

Kang, K.I., Freedman, S., Mataric, M.J., Cunningham, M.J., Lopez, B.: A hands-off physical therapy assistance robot for cardiac patients. In: Proceedings of the 9th IEEE International Conference on Rehabilitation Robotics, Chicago, IL, USA, 28 June–1 July 2005, pp. 337–340 (2005)

Nass, C., Jonsson, I.M., Harris, H., Reaves, B., Endo, J., Brave, S., Takayama, L.: Improving automotive safety by pairing driver emotion and car voice emotion. In: Proceedings of the CHI 2005 Extended Abstracts on Human Factors in Computing Systems, Portland, OR, USA, 2–7 April 2005, pp. 1973–1976 (2005)

Park, J., Lee, S., Brotherton, K., Um, D., Park, J.: Identification of speech characteristics to distinguish human personality of introversive and extroversive male groups. Int. J. Environ. Res. Public Health **17**, 21–25 (2020)

Author Index

.

Printed in the United States
By Bookmasters